POSITIVE SEMIGROUPS OF OPERATORS, AND APPLICATIONS

POSITIVE SEMIGROUPS OF OPERATORS, AND APPLICATIONS

Edited by

OLA BRATTELI

Mathematics Institute, University of Trondheim, Norway

and

PALLE E. T. JØRGENSEN

Dept. of Mathematics/El, University of Pennsylvania, Philadelphia, U.S.A.

Reprinted from

Acta Applicandae Mathematicae, Vol. 2, Nos. 3/4

D. Reidel Publishing Company

Dordrecht / Boston

ISBN 90–277–1839–3

Printed in The Netherlands

ACTA APPLICANDAE MATHEMATICAE

Volume 2 Nos. 3 and 4 Sept./Dec. 1984

POSITIVE SEMIGROUPS OF OPERATORS, AND APPLICATIONS

Edited by Ola Bratteli and Palle E. T. Jørgensen

LIST OF FORTHCOMING PAPERS

Françoise Brossier: Mathematical Modelisation of Equatorial Waves

John H. Cushman: Multiphase Transport Based on Compact Distributions

Philippe Destuynder: A Classification of Thin Shell Theories

Jerome A. Goldstein: Bound States and Scattered States for Contraction Semigroups

Minoru Kanehisa and Charles DeLisi: Prediction of Protein and Nucleic Acid Structure: Problems and Prospects

V. V. Lychagin: Singularities of Multivalued Solutions of Nonlinear Differential Equations, and Nonlinear Phenomena

Catherine A. Macken and Alan S. Perelson: Some Stochastic Models in Immunology

Servet Martínez A.: Non-Equilibrium Entropy on Stationary Markov Processes

Moshe Zakai: The Malliavin Calculus

Acta Applicandae Mathematicae is published 4 times per annum: March, June, September, and December.
Subscription prices, per volume: Institutions $76.00, Individuals $36.00.
Application to mail at second-class postage rates is pending at New York, N.Y. ISSN 0167-8019
U.S. Mailing Agent: Expediters of the Printed Word Ltd., 515 Madison Avenue (Suite 917), New York, NY 10022.
Published by D. Reidel Publishing Company, Voorstraat 479-483, P.O. Box 17, 3300 AA Dordrecht, Holland, and 190 Old Derby Street, Hingham, MA 02043, U.S.A.
Postmaster: please send all address corrections to: c/o Expediters of the Printed Word Ltd., 515 Madison Avenue (Suite 917), New York, NY 10022, U.S.A.

Acta Applicandae Mathematicae **2**, 213–219. 0167–8019/84/0023–0213$01.05

Positive Semigroups of Operators, and Applications: Editors' Introduction

OLA BRATTELI
Mathematics Institute, University of Trondheim, 7034 Trondheim, Norway
and

PALLE E. T. JØRGENSEN
Department of Mathematics/E1, University of Pennsylvania, Philadelphia, PA 19104, U.S.A.

(Received: 28 February, 1984)

AMS (MOS) subject classifications (1980). Primary: 47B44, 47D05, 47D07, 47D45, 47H06, 46H02, 20M20, 35K22; Secondary: 06F20, 26A33, 28D10, 31A35, 31C12, 34C35, 34H05, 35J10, 35P25, 41A35, 46A40, 47B55, 47C15, 53C20, 58G11, 58G20, 58G32, 60H99, 60J35, 60J40, 60J45, 60J60, 60J65, 81E05, 82A15, 82A70

Key words. One-parameter semigroups of operators, positivity, dissipative and dispersive operators, diffusion processes, Markov fields, Poisson and Weierstrass transforms, Dirichlet problem, Schrödinger equation, Riemannian manifold, Ricci curvature, Laplace operator, Fokker–Planck equation.

1. Introduction

In this collection of papers the editors have attempted to present a state of the art review of the field of positive semigroups of operators with special emphasis on new (and recent) research developments and applications.

The special issue begins with an article by C. J. K. Batty and D. W. Robinson on positive one-parameter semigroups of operators on ordered Banach spaces. The first half of this paper gives an updated treatment of ordered Banach spaces with many illustrating examples and applications. Generation of positive semigroups, strict positivity, irreducibility, and spectral properties (in parallel with the Perron–Frobenius theory) are items from the second (main) part of the paper.

A wide interpretation of the positive semigroup concept would include all of Markov processes. This, of course, is not our intention. Instead, we have made an effort to focus on those newer mathematical theories which apply to classical and quantum statistical mechanics. We regard the applications to operator algebras (i.e., the setting of quantum models) as being particularly promising.

The Batty–Robinson paper is followed by five slightly more specialized articles covering the following list of topics: asymptotic behavior, transport theory, quantum

★ Work supported in part by NSF.

dynamical semigroups, open systems and stochastics dilations, interacting particles, and spin-systems.

We have encouraged an exposition of a discursive kind accessible to nonsuper-specialists.

The issue concludes with two long book reviews.

2. Historical Notes

The first major theorem on semigroups of operators was obtained in the 1940s by Hille and Yosida (independently). We refer to the books [2, 3, 5, 11, 17, 19, 21, 29, 31, 34, 42, 51, 54 vol. 1] for background material. With vigorous activity in research and applications up to the present day, the analytic theory of semigroups has, by now, become a standard tool in probability, potential theory, harmonic analysis, partial differential equations, functional analysis, ergodic theory, mathematical physics and, more recently, biology and genetics.

Perhaps, it is even possible to notice different research trends in the field for each of the post-war decades: (1) Diffusion processes, potential theory, and ergodic theory in the Fifties with W. Feller, J. L. Doob, J. Deny, N. Dunford, E. Hille, S. Kakutani, N. Kolmogorov, and K. Yosida; (2) parabolic partial differential equations and scattering theory in the Sixties with A. Friedman, E. Nelson, E. Dynkin, G. Hunt, P. Lax, J. Moser, R. Phillips, and T. Kato; (3) mathematical physics, optimal control theory, and population processes in the Seventies with I. E. Segal, E. Nelson (the Markov fields), J. Glimm, A. Jaffe; J. L. Lions, A. V. Balakrishnan; and T. G. Kurtz.

Mention should also be made of the connections to approximation theory (P. L. Butzer, H. Berens, T. G. Kurtz), to classical infinite particle models (F. Spitzer, T. M. Ligget, R. A. Holley, D. W. Stroock, L. Helms) and quantum models (D. Ruelle, D. W. Robinson, H. Araki), to stochastic differential equations (K. Itô, H. P. McKean, T. Hida), to nonlinear partial differential equations (I. E. Segal, F. Browder, M. Crandall, A. Pazy, and T. Kato), to geometry of complete Riemannian manifolds (S.-T. Yau [45, 46]), and to constructive quantum field theory ([53, 55]).

3. The Selection of Papers

We also mention work on hypercontractive semigroups, the infinite-dimensional Orenstein–Uhlenbeck semigroups, and Schrödinger semigroups, by E. Nelson, I. E. Segal, B. Simon, R. Høegh-Krohn, E. B. Davies, P. R. Chernoff, K. L. Chung, R. Carmona, L. Gross, and others. While many of these specializations have been exposed in book form, this is not so for the developments in the Eighties centering around positive semigroups. This journal issue attempts to present a balanced exposition of these results and applications. But even this rather recent research literature is large and scattered, and it goes without mention that the topics which have been included represent a definite choice on the part of the editors and authors. We apologize for any omissions. We feel, however, that the present selection of papers is consistent and logically connected. But there is certainly room for a different and disjoint selection of papers on related topics.

4. Three Examples

In 1935, it was noticed by Hille that both the Poisson transform

$$(P_y f)(x) = \frac{y}{\pi} \int_{-\infty}^{\infty} \frac{f(u + x)\,du}{u^2 + y^2} \tag{1}$$

and the Weierstrass transform

$$(W_t f)(x) = (\pi t)^{-1/2} \int_{-\infty}^{\infty} e^{-u^2/t} f(u + x)\,du \tag{2}$$

satisfy the semigroup law

$$S_{t_1 + t_2}(f) = S_{t_1}(S_{t_2} f), \quad t_1, t_2 \geqslant 0. \tag{3}$$

In itself, a relatively simple fact which is now (and perhaps was also then) taken for granted: In the same paper [18] which appears to have initiated the study of semigroups, Hille also noted that (3) undoubtedly had been well-known earlier to mathematical physicists since $(P_y f)(x)$ provides a solution to Dirichlet's problem for the upper half-plane corresponding to the boundary function $f(x)$ on the x-axis, whereas $W_t f$ is a solution to the heat equation in a single dimension with initial temperature configuration $f(x)$. The semigroup law (3) was noted as early as 1892 by P. Appell (*J. de Mathématique* (4) **8**, pp. 187–216). Hille also credited similar observations to E. Cesàro (1902), and to G. Doetsch (1926).

The infinitesimal generator, A say, of a semigroup S_t is defined as follows: It is assumed that S_t acts in a linear space (which could consist of integrable, or square integrable functions, bounded measurable, or continuous functions, just to mention a few examples). A vector f (alias a function) is said to belong to the domain of the generator A, and $A(f) = h$ if the limit $t^{-1}(S_t f - f) \to h$ exists, as $t \to 0$.

Except for a few very special cases, A is only partially defined, i.e., the domain of A is not the whole space. Typically, it will not be a bounded operator relative to the norm on the space under consideration. In the model examples (1) and (2) above, we have the formulas,

$$A(f) = \tfrac{1}{4}\left(\frac{d}{dx}\right)^2 f = \tfrac{1}{4} f'' \quad \text{in case (2)}$$

and

$$A(f) = \tilde{f}' \quad \text{in case (1)},$$

where $\tilde{g}(\cdot)$ denotes the M. Riesz conjugate function to g.

For the one-dimensional diffusion equation satisfied by the probability densities $f(x)$, the infinitesimal generator A is formally equal to a second-order differential operator [12],

$$(Df)(x) = \frac{d}{dx}\left(\frac{d}{dx} a(x) f - b(x) f\right) + c(x) f(x) \tag{4}$$

where the functions a, a', b and c are defined in an open interval $-\infty \leqslant r_1 < x < r_2 \leqslant \infty$, and $a > 0$. The function c satisfies a certain technical condition [12]. (Actually, Feller reduces the analysis to the case $c = 0$.)

Since the semigroup S_t is known to be determined uniquely by the generator A, the analytic determination of S_t (or equivalently the solution of the Cauchy problem) therefore amounts to a specification of the domain for A. In his pioneering work [12], Feller determined precisely the restrictions of the operator D in (4) which generate semigroups of positive contraction (relative to the L^1-norm) operators. Moreover, this classification of the family of generators in terms of the domains (or the so-called boundaries) had a direct interpretation in terms of the physical diffusion problem: the regular boundary, the exit boundary, the entrance boundary, and the natural boundary [12], Section 23. It made it possible for Feller to state the appropriate boundary conditions for certain peculiar diffusion processes which had for years resisted all attempts at a mathematical formulation.

5. Partial Differential Equations and Geometry

Here we wish to call attention to two separate early developments which appear to have provided a particularly rich source of inspiration for later work, and which focused attention on the above-mentioned domain question.

In 1954, Lax and Milgram [32] proved that if an alliptic operator $P(x, \partial/\partial x)$ on an open set Ω in $\mathbf{R}^n$ is specified with Dirichlet boundary conditions on $\partial\Omega$, then the resulting operator with this domain generates a semigroup on $L^2(\Omega)$.

The second result, due to Nelson in 1964 [39], provides a semigroup solution to the Cauchy problem for the Schrödinger equation (in any dimension) with a large class of singular potentials, including the attractive $1/r^2$ potential. In the construction of S_t, the infinitesimal generator A occurs as an extension of the Schrödinger operator $-(1/2m)\Delta + V(x)$, or equivalently, a restriction of the corresponding maximal operator, and the (time-irreversible) solution agrees, for Nelson's semigroup, with classical mechanics in the correspondence limit.

Generalizing Itô [24], Yau constructed in [46], the diffusion semigroup S_t which is generated by a second-order elliptic operator L without a zero-order term on a specified complete Riemannian manifold M. If $L = \Delta$ is the Laplacian of M, and if M has Ricci curvature bounded from below, Yau's main theorem implies that the generator A of S_t is equal to the graph completion of Δ on the space of smooth functions with compact supports on M (i.e., the minimal operator). Moreover, S_t, acting on bounded measurable functions, maps the space $C_0(M)$ of continuous functions on M vanishing at infinity into itself, $S_t(C_0(M)) \subset C_0(M)$, $t \geqslant 0$. As an application, Yau notes that, for a bounded function f on M, the following two conditions are equivalent:

(i) Δf is bounded in the distribution sense.

and

(ii) $t^{-1}((S_t f)(x) - f(x))$ converges boundedly as $t \to 0$.

This means that semigroup theory may be applied directly to the study of the equation $\Delta f = h$ on M. In [45] Yau proves that, for $h \geqslant 0$, there are no nonconstant, nonnegative solutions f in L^p for $1 < p < \infty$. From this, Yau gets the geometric fact that complete noncompact Riemannian manifolds with nonnegative Ricci curvature must have infinite volume, a result which was announced earlier by Calabi [4].

6. Concluding Remarks

In several of the above results, positivity of the semigroup plays an important role. This was also true, although only implicitly, for the early work of Hille and Yosida on the Fokker–Planck equation, i.e., Equation (4) with $c = 0$. But it was Phillips [41], and Lumer and Phillips [37] who first called attention to the importance of dissipative and dispersive properties of the generator in the context of linear operators in a Banach space.

The generation theorems in the Batty–Robinson paper appear to be the most definitive ones, so far, for this class of operators. The fundamental role played by the infinitesimal operator, also for the understanding of order properties, in the commutative as well as the noncommutative setting, are highlighted in a number of examples and applications in the different papers, and it is hoped that this publication will be of interest to researchers in a broad spectrum of the mathematical sub-divisions.

Acknowledgements

We are grateful to Professor Rhonda Hughes for reading the manuscripts and making helpful suggestions.

References

An effort has been made to limit the size of this opening reference list, since the more specialized references will be given in the individual papers to follow.

1. Balakrishnan, A. V.: *Stochastic Differential Systems: Filtering and Control. A Function Space Approach*, Springer-Verlag, Berlin, Heidelberg, New York, 1973.
2. Balakrishnan, A. V.: *Applied Functional Analysis*, Application of Math. vol. 3, Springer-Verlag, Berlin, Heidelberg, New York, 1976.
3. Butzer, P. L. and Berens, H.: *Semi-Groups of Operators and Approximation*, Die Grundlehren bd. 145, Springer-Verlag, Berlin, Heidelberg, New York, 1967.
4. Calabi, E.: 'On Manifolds with Non-negative Ricci Curvature II', *Notices Amer. Math. Soc.* **22** (1975), A205.
5. Davies, E. B.: *One-parameter Semigroups*, Academic Press, London New York, 1980.
6. Deny, J.: Familles fondamentales Noyaux associes', *Ann. Inst. Fourier* **3** (1951), 73–101.
7. Deny, J.: 'Noyaux de convolution de Hunt et noyaux associes a une famille fondamentale', *Ann. Inst. Fourier* **12** (1962), 643–667.
8. Doob, J. L.: *Stochastic Processes*, Wiley, New York, 1953.
9. Doob, J. L.: 'A Probability Approach to the Heat Equation', *Trans. Amer. Math. Soc.* **80** (1955), 216–280.
10. Dynkin, E. B.: 'Markoff Processes and Semigroups of Operators, and Infinitesimal Operators of Markoff Processes', *Teorya Veroyatn* **1** (1956), 25–37.

11. Dynkin, E. B.: *Markov Processes*, vols I & II, Springer-Verlag, Berlin, Heidelberg, New York, 1965. (Translation of the Russion edition, 1962.)
12. Feller, W.: 'The Parabolic Differential Equation and the Associated Semi-groups of Transformations', *Ann. Math.* (2), **55** (1952), 468–519.
13. Feller, W.: 'On the Generation of Unbounded Semigroups of Bounded Linear Operators', *Ann. Math* (2) **58** (1953), 166–174.
14. Friedman, A.: *Generalized Functions and Partial Differential Equations*, Prentice-Hall, Englewood Cliffs, 1963.
15. Friedman, A. and Pinsky, M. (eds): *Stochastic Analysis*, Academic Press, New York, 1978.
16. Hida, T.: *Stationary Stochastic Processes*, Princeton University Press, Princeton, NJ, 1970.
17. Hida, T.: *Brownian Motion*, Appl. Math. vol. 11, Springer-Verlag, Berlin, Heidelberg, New York, 1980.
18. Hille, E.: 'Notes on Linear Transformations, I'. *Trans . Amer. Math. Soc.* **39** (1936), 131–153.
19. Hille, E.: *Functional Analysis and Semi-Groups*, Amer. Math. Soc. Colloq. Pub. 31, New York, 1948.
20. Hille, E.: 'On the Integration Problem for Fokker–Planck's Equation in the Theory of Stochastic Processes', C.R. Onzieme Cong. Math. Scand., Trondheim, 1949, 183–194, also in, *Einar Hille, Selected Papers*, MIT Press, Cambridge, Mass., London, 1975.
21. Hille, E. and Phillips, R. S.: *Functional Analysis and Semi-Groups* (revised edn), Amer. Math. Soc. Colloq. Pub. 31 Providence, RI, 1957.
22. Hunt, G. A.: 'Semigroups of Measures on Lie Groups', *Trans. Amer. Math. Soc.* **81** (1956), 264–293.
23. Hunt, G. A.: 'Markoff Processes and Potentials, I-III', *Ill. J. Math.* **1** (1957), 44–93, 316–369: **2** (1958), 151–215.
24. Itô, K.: 'Lectures on Stochastic Processes', Tata Institute of Fundamental Research, Bombay, 1960, 1967.
25. Itô, K. and McKean, H. P.: *Diffusion Processes and their Sample Paths*, Springer-Verlag, Berlin, Heidelberg, New York, 1965.
26. Kakutani, S.: Concrete Representation of Abstract (L)-space and the Mean Ergodic Theorem', *Ann. Math.* **42** (1941), 523–537.
27. Kakutani, S. and Yosida, K.: Operator-Theoretic Treatment of Markoff Process and Mean Ergodic Theorems', *Ann. Math.* **42** (1941), 188-228.
28. Kato, T.: 'Remarks on Pseudo-resolvents and Infinitesimal Generators of Semi-groups', *Proc. Japan Acad.* **35** (1959), 467–468.
29. Kato, T.: *Perturbation Theory for Linear Operators*, Springer-Verlag, Berlin, Heidelberg, New York, 1966.
30. Kato, T.: 'Nonlinear Semigroups and Evolution Equations', *J. Math. Soc. Japan* **19** (1967), 508–520.
31. Kurtz, T. G.: *Approximation of Population Processes*, CBMS, Applied Math. vol. 36, SIAM, Philadelphia, 1981.
32. Lax, P. D. and Milgram, A. N.: 'Parabolic Equations', in *Contributions to the Theory of Partial Differential Equations*, Princeton University Press, Princeton, NJ, 1954.
33. Lax, P. D. and Phillips, R. S.: 'Local Boundary Conditions for Dissipative Systems of Linear Partial Differential Operators', *Comm. Pure Appl. Math.* **13** (1960), 427–455.
34. Lax, P. D. and Phillips, R. S.: *Scattering Theory*, Academic Press, New York, 1967.
35. Lions, J. L.: 'Une remarque sur les applications du theoreme de Hille-Yosida', *J. Math. Soc. Japan* **9** (1957), 62–70.
36. Lions, J. L.: *Optimal Control of Systems Governed by Partial Differential Equations*, Die Grundlehren, bd. 170, Springer-Verlag, Berlin, Heidelberg, New York, 1971 (translated by Dr. S. K. Mitter).
37. Lumer, G. and Phillips, R. S.: 'Dissipative Operators in Banach Space', *Pacific J. Math.* **11** (1961), 679–698.
38. Nelson, E.: 'An Existence Theorem for Second-Order Parabolic Equations', *Trans. Amer. Math. Soc.* **88** (1958), 414–429. See also *Dynamical Theories of Brownian Motion*, Princeton University Press, Princeton, NJ, 1967 and 1972.
39. Nelson, E.: 'Feynman Integrals and the Schrödinger Equation', *J. Math. Phys.* **5** (1964), 332–343.
40. Nelson, E.: 'The Free Markoff Field', *J. Funct. Anal.* **12** (1973), 211–227.
41. Phillips, R. S.: 'Semi-groups of Positive Contraction Operators', *Czech. Math. J.* **12** (1962), 294–313.
42. Schwartz, L.: 'Lectures on Mixed Problems in Partial Differential Equations and the Representation of Semi-Groups', Tata Inst. of Fund. Research, Bombay, 1958.
43. Segal, I. E.: 'Nonlinear Semigroups', *Ann. Math.* (2) **78** (1963), 339–364.
44. Segal, I. E.: 'Nonlinear Functions of Weak Processes, II', *J. Funct. Anal.* **6** (1970), 29–75.

45. Yau, S.-T.: 'Some Function-theoretic Properties of Complete Riemannian Manifolds and their Applications to Geometry', *Indiana U. Math. J.* **25** (1976), 659–670.
46. Yau, S.-T.: 'On the Heat Kernel of a Compete Riemannian Manifold', *J. Math. pure et appl.* **57** (1978), 191–201.
47. Yosida, K.: 'On the Differentiability and the Representation of One-parameter Semi-groups of Linear Operators', *J. Math. Soc. Japan* **1** (1948), 15–21.
48. Yosida, K.: 'On the Integration of Diffusion Equations in Riemannian Spaces', *Proc. Amer. Math. Soc.* **3** (1952), 864–873.
49. Yosida, K.: 'An Extension of Fokker–Planck's Equation', *Proc. Japan Acad.* **25** (1949), 1–3.
50. Yosida, K.: 'An Operator-theoretic Treatment of Temporally Homogeneous Markoff Processes', *J. Math. Soc. Japan* **1** (1949), 244–253.
51. Yosida, K.: *Functional Analysis*, Grundlehren bd. 123, Fifth Edn, Springer-Verlag, Berlin, Heidelberg, New York, 1978.
52. Arendt, W., Chernoff, P. R., and Kato, T.: 'A Generalization of Dissipativity and Positive Semigroups', *J. Operator Theory* **8** (1982), 167–180.
53. Glimm, J. and Jaffe, A.: *Quantum Physics, A Functional Integral Point of View*, Springer-Verlag, Berlin, Heidelberg, New York, 1981.
54. Bratteli, O. and Robinson, D. W.: *Operator Algebras and Quantum Statistical Mechanics*, vols. I-II, Springer-Verlag, Berlin, Heidelberg, New York, 1979 & 1981.
55. Guerra, F., Rosen, L., and Simon, B.: 'The $P(\phi)_2$ Euclidean, Quantum Field Theory as Classical Statistical Mechanics', *Ann. Math.* **101** (1975), 111–259.

Acta Applicandae Mathematicae **1**, 221–296. 0167–8019/84/0023–0221$11.40.

Positive One-Parameter Semigroups on Ordered Banach Spaces

CHARLES J. K. BATTY* and DEREK W. ROBINSON
Department of Mathematics, Institute of Advanced Studies, Australian National University, Canberra, Australia

(Received: 9 May 1983)

Abstract. In this review we describe the basic structure of positive continuous one-parameter semigroups acting on ordered Banach spaces. The review is in two parts.

First we discuss the general structure of ordered Banach spaces and their ordered duals. We examine normality and generation properties of the cones of positive elements with particular emphasis on monotone properties of the norm. The special cases of Banach lattices, order-unit spaces, and base-norm spaces, are also examined.

Second we develop the theory of positive strongly continuous semigroups on ordered Banach spaces, and positive weak*-continuous semigroups on the dual spaces. Initially we derive analogues of the Feller–Miyadera–Phillips and Hille–Yosida theorems on generation of positive semigroups. Subsequently we analyse strict positivity, irreducibility, and spectral properties, in parallel with the Perron–Frobenius theory of positive matrices.

AMS (MOS) subject classifications (1980). 46A40, 15A48, 06F20, 46L05, 46L10, 54C40, 54C45, 47B55, 47D05, 47D07, 47B44, 46L55, 46L60, 46B20.

Key words. Ordered Banach space, normal cone, generating cone, monotone norm, Riesz norm, order-unit, Banach lattice, C^*-algebra, half-norm, dissipative, C_0-semigroup, C_0^*-semigroup, Perron–Frobenius theory, irreducible semigroup.

Contents

*Permanent address: Dept. of Mathematics, University of Edinburgh, King's Buildings, Edinburgh, EH9 3JZ, U.K.

1. Ordered Banach Spaces

1.0. INTRODUCTION

The theory of ordered Banach spaces is a development of the structure associated with the classical Banach spaces of real functions. Each of these function spaces, e.g., $C(X)$ or $L^p(X; d\mu)$, can be ordered by setting $f \geqslant g$ whenever the function $f - g$ is pointwise positive. This ordering can, however, be described in a more geometric manner which is more convenient for the introduction of other order relations.

The pointwise positive functions in each of the classical real Banach spaces form a convex cone P because if $f, g \geqslant 0$ then $\lambda f + \mu g \geqslant 0$ for all positive λ, μ. In terms of this cone the ordering $f \geqslant g$ is equivalent to the statement that $f - g \in P$. More generally any cone P induces an order by setting $f \geqslant g$ whenever $f - g \in P$. Properties of the order relation are then determined by the geometric and topological properties of the cone P. Duality properties are also important.

If P is a convex cone in a real Banach space $\mathscr{B}$ one can define a cone P^* in the dual $\mathscr{B}^*$ as the elements of $\mathscr{B}^*$ which are positive on P, and then $\mathscr{B}^*$ is ordered by P^*. If, for example, P is the cone of positive functions in $L^p(X; d\mu)$ then P^* can be identified with the positive functions in $L^q(X; d\mu)$ whenever $1 \leqslant p < +\infty$ and $q^{-1} = 1 - p^{-1}$. The dual P^* of the cone P of positive functions in $C(X)$, or $L^\infty(X; d\mu)$, is, however, a cone of positive measures. Thus duality leads one away from the simplest form of positivity and provides one motivation for the general geometric approach to ordering.

There are two kinds of property of a cone P which are essential for an interesting order structure. First the cone must not be too large. Typically this is expressed by some sort of pointedness condition, e.g., $\pm f \in P$ if, and only if, $f = 0$. Second the cone must not be too small. For example a general f should be decomposable as a difference, $f = g - h$, of a positive g and a positive h, at least approximately. These two types of restraint on P are referred to as normality and generation properties, respectively. In the next sections we discuss a whole hierarchy of such conditions.

It is a simple observation that the larger the cone P the smaller the cone P^*. Alternatively stated a normality restriction on P is equivalent to a generation condition on P^*, and vice-versa. Duality results of this nature play a major role in the general analysis of ordered spaces.

There are other more detailed properties of order relations which are also of interest. For example the classical function spaces are lattices with respect to the order defined by pointwise positivity. Moreover, the norm of a function and the norm of its modulus coincide, and as the modulus increases so does the norm. Spaces with these properties are called Banach lattices. They have been extensively studied and possess a rich, well understood, structure. Unfortunately they do not describe all the commonly encountered examples of ordered spaces.

Consider the Banach space $\mathscr{B}$ of all bounded self-adjoint operators on the Hilbert space $\mathscr{H}$. Furthermore let $\mathscr{B}_+$ denote those operators with non-negative specturm, i.e., the positive (semi-) definite operators on $\mathscr{H}$. Then $\mathscr{B}_+$ is a convex cone, with strong normality and generation properties, but the associated order on $\mathscr{B}$ is a lattice ordering if, and only if, $\mathscr{H}$ has dimension one. More generally if $\mathscr{C}$ is the self-adjoint part of a Banach *-subalgebra of all bounded operators on $\mathscr{H}$, ordered by $\mathscr{C}\cap\mathscr{B}_+$, then the order has the lattice property if, and only if, the elements of $\mathscr{C}$ commute. Since each commutative $\mathscr{C}$ can be identified with an algebra of continuous functions, these operator examples can be viewed as non-commutative versions of the continuous functions.

It is naturally of interest to develop the theory of ordered Banach spaces in a way which unifies the classical function spaces and their non-commutative analogues. A useful notion for this unification is the Riesz norm discussed in Section 1.3, and it is of particular interest that this notion is self-dual. Therefore the theory of Riesz norm spaces also covers the duals, or preduals, of the function spaces, or operator spaces.

A more specific concept which occurs both in the commutative and non-commutative setting is that of an order-unit. The constant function 1 is an order-unit of $L^\infty(X;d\mu)$ and the identity operator is an order-unit of the space of all bounded self-adjoint operators on a Hilbert space $\mathscr{H}$, i.e., each element of the space is bounded by a multiple of the order-unit. In geometric terms order-units are identifiable as interior points of the cone of positive elements. Interior points and the dual concept of base are discussed in detail in Section 1.4.

Finally if $\mathscr{B}$ is a Banach space ordered by a positive cone $\mathscr{B}_+$ then the Banach space $\mathscr{L}$ of bounded operators on $\mathscr{B}$ can be ordered by specifying that an operator $A\in\mathscr{L}$ is positive if $A\mathscr{B}_+\subseteq\mathscr{B}_+$. The study of the positive bounded operators $\mathscr{L}_+$ is of interest for several reasons and in Section 1.7 we discuss relationships between order properties of $\mathscr{B}$ and $\mathscr{L}$.

1.1. NORMAL GENERATING CONES

An *ordered Banach space* $(\mathscr{B}, \mathscr{B}_+, \|\cdot\|)$ consists of a real Banach space $\mathscr{B}$ with norm $\|\cdot\|$ and a *positive cone* $\mathscr{B}_+$ which is defined as a norm closed subset of $\mathscr{B}$ satisfying

$$\lambda\mathscr{B}_+ + \mu\mathscr{B}_+ \subseteq \mathscr{B}_+$$

for all λ, $\mu \geqslant 0$. Elements of $\mathscr{B}$ will be denoted by a, b, c, ... and $\mathscr{B}_\alpha$ will denote the closed ball of radius α, i.e.,

$$\mathscr{B}_\alpha = \{a; a\in\mathscr{B}, \|a\| \leqslant \alpha\}.$$

Associated to each $(\mathscr{B}, \mathscr{B}_+, \|\cdot\|)$ there is an ordered *dual space* $(\mathscr{B}^*, \mathscr{B}^*_+, \|\cdot\|^*)$ consisting of the real linear functionals $\mathscr{B}^* = \{\omega, \xi, \eta \ldots\}$ over $\mathscr{B}$ which are bounded, so that the dual-norm

$$\|\omega\|^* = \sup\{|\omega(a)|; a\in\mathscr{B}_1\}$$

is finite, and of the weak*-closed *dual cone* $\mathscr{B}_+$ which is defined by

$$\mathscr{B}^*_+ = \{\omega; \omega\in\mathscr{B}^*, \omega(a) \geqslant 0 \text{ for all } a\in\mathscr{B}_+\}.$$

Note that the dual cone automatically satisfies

$$\lambda\mathscr{B}^*_+ + \mu\mathscr{B}^*_+ \subseteq \mathscr{B}^*_+, \quad \lambda, \mu \geqslant 0.$$

The norm closed ball of radius α in $\mathscr{B}^*$ is denoted by $\mathscr{B}^*_\alpha$.

Order relations are defined on $\mathscr{B}$, and on $\mathscr{B}^*$, by setting $a \geqslant b$ whenever $a - b\in\mathscr{B}_+$, and $\xi \geqslant \eta$ whenever $\xi - \eta\in\mathscr{B}^*_+$. Thus $a \geqslant 0$ is equivalent to $a\in\mathscr{B}_+$, and $\xi \geqslant 0$ is equivalent to $\xi\in\mathscr{B}^*_+$.

There are two deficiencies in this structure. First there is no condition which ensures that $\mathscr{B}_+$ is large enough to introduce an interesting order relation and there is no condition which ensures that $\mathscr{B}_+$ is not too large. The purpose of this section is to introduce and analyze such conditions. Further refinements of these conditions are discussed in the subsequent sections. We begin with the weakest possible form of such conditions.

First the cone $\mathscr{B}_+$ is defined to be *weakly generating* if $\mathscr{B}_+ - \mathscr{B}_+$ is norm dense in $\mathscr{B}$, i.e., if each $a\in\mathscr{B}$ is the norm limit of a sequence $\{b_n - c_n\}_{n\geqslant 1}$ of differences of elements $b_n, c_n\in\mathscr{B}_+$. Similarly the dual cone $\mathscr{B}^*_+$ is *-weakly generating if $\mathscr{B}^*_+ - \mathscr{B}^*_+$ is weak*-dense in $\mathscr{B}^*$.

Second the cone $\mathscr{B}_+$ is defined to be *proper*, or *pointed*, if

$$\mathscr{B}_+ \cap -\mathscr{B}_+ = \{0\}.$$

This is equivalent to antisymmetry of the order $\geqslant$, i.e., $a \geqslant b$ and $a \leqslant b$ if, and only if, $a = b$. Similarly $\mathscr{B}^*_+$ is proper, or pointed, if

$$\mathscr{B}^*_+ \cap -\mathscr{B}^*_+ = \{0\}.$$

The generation property has the tendency to make $\mathscr{B}_+$ 'large' and the dual cone $\mathscr{B}^*_+$ 'small'. Conversely, the pointedness property requires $\mathscr{B}_+$ to be 'small' and $\mathscr{B}^*_+$ to be 'large'. This elementary observation is at the root of a series of duality properties of which the following proposition is the simplest.

PROPOSITION 1.1.1. *The following pairs of conditions are equivalent*:

1. (1'.) $\mathscr{B}_+$ ($\mathscr{B}^*_+$) *is* (*)-*weakly generating,*
2. (2'.) $\mathscr{B}^*_+$ ($\mathscr{B}_+$) *is proper,*

i.e., 1 ⇔ 2 *and* 1' ⇔ 2'.

The proof of this statement is a straightforward application of the Hahn–Banach theorem on which we will not elaborate. Instead we turn to the analysis of stronger properties of generation and pointedness.

First the cone $\mathscr{B}_+$ is defined to be *generating* if $\mathscr{B} = \mathscr{B}_+ - \mathscr{B}_+$, i.e., if each $a \in \mathscr{B}$ has a decomposition $a = b - c$ with $b, c \in \mathscr{B}_+$. This is equivalent to requiring that for each $a \in \mathscr{B}$ there is a $b \in \mathscr{B}_+$ such that $b \geqslant a$, in which case a has the decomposition $a = b - (b - a)$ into components $b, b - a \in \mathscr{B}_+$.

Second the cone $\mathscr{B}_+$ is defined to be *normal* if there is an $\alpha \geqslant 1$ such that $c \leqslant a \leqslant b$ always implies $\|a\| \leqslant \alpha\{\|b\| \vee \|c\|\}$. This property implies that $\mathscr{B}_+$ is proper. Moreover, if $a, b \in \mathscr{B}_+$ and $\|a\| = 1 = \|b\|$ then it implies that $\|a + b\| \geqslant \alpha^{-1} > 0$. Thus α^{-1} is a positive measure of the 'pointedness' of the cone.

Normality is a condition of compatibility of the order and the topology. It is equivalent to the requirement that order-bounded sets are norm-bounded. This will be established in Proposition 1.4.3.

Normality and generation for $\mathscr{B}^*_+$ are defined analogously. Again there is a duality between these properties for $\mathscr{B}_+$ and $\mathscr{B}^*_+$. But before giving this we first establish a uniformity of the generation property which allows a more precise indexation of generation and a subsequent closer comparison with normality.

PROPOSITION 1.1.2. *The following conditions are equivalent:*

1. $\mathscr{B}_+$ *is generating,*
2. *there is an* $\alpha \geqslant 1$ *such that each* $a \in \mathscr{B}$ *has a decomposition* $a = b - c$ *with* $b, c \in \mathscr{B}_+$ *and*

$$\|b\| \vee \|c\| \leqslant \alpha \|a\|.$$

Proof. $2 \Rightarrow 1$. By definition.

$1 \Rightarrow 2$. Condition 1 can be expressed as

$$\mathscr{B}_1 \subseteq \bigcup_{\alpha \geqslant 1} (\mathscr{B}_\alpha \cap \mathscr{B}_+ - \mathscr{B}_\alpha \cap \mathscr{B}_+).$$

Hence by the Baire category theorem there exists a $\beta \geqslant 1$ such that

$$\mathscr{B}_1 \subseteq \overline{(\mathscr{B}_\beta \cap \mathscr{B}_+ - \mathscr{B}_\beta \cap \mathscr{B}_+)}.$$

The proof is then completed by estimating that

$$\mathscr{B}_1 \subseteq (\mathscr{B}_\alpha \cap \mathscr{B}_+ - \mathscr{B}_\alpha \cap \mathscr{B}_+)$$

for all $\alpha > \beta$.

Since this last type of estimate will be used several times in the sequel we present it in a suitably general form. But first recall that a subset $\mathscr{C} \subset \mathscr{B}$ is called *σ-convex* if the conditions $c_n \in \mathscr{C}$, $\lambda_n \geqslant 0$, $\sum_{n \geqslant 1} \lambda_n = 1$, together with the existence of

$$c = \sum_{n \geqslant 1} \lambda_n c_n,$$

always imply that $c \in \mathscr{C}$. For example $\mathscr{B}_\beta, \mathscr{B}_\beta \cap \mathscr{B}_+, \mathscr{B}_\beta \cap \mathscr{B}_+ - \mathscr{B}_\beta \cap \mathscr{B}_+$ are all σ-convex sets.

LEMMA 1.1.3. *If $\mathscr{C} \subseteq \mathscr{B}$ is a σ-convex subset such that $\mathscr{B}_1 \subseteq \bar{\mathscr{C}}$ then $\mathscr{B}_1 \subseteq \alpha\mathscr{C}$ for all $\alpha > 1$.*

Proof. For $a \in \mathscr{B}_1$ and $0 < \delta < 1$ choose $a_1 \in \mathscr{C}$ such that $\|a - a_1\| < \delta$. Hence $\delta^{-1}(a - a_1) \in \mathscr{B}_1$ and one can choose $a_2 \in \mathscr{C}$ such that $\|\delta^{-1}(a - a_1) - a_2\| < \delta$. By iteration one finds $a_n \in \mathscr{C}$ such that

$$\left\| \delta^{-n+1} a - \sum_{n=1}^{n} \delta^{-n+m} a_m \right\| < \delta.$$

Therefore setting $\lambda_m = (1 - \delta)\delta^{m-1}$ one has

$$\left\| a - (1 - \delta)^{-1} \sum_{m=1}^{n} \lambda_m a_m \right\| < \delta^n$$

and $a \in (1 - \delta)^{-1}\mathscr{C}$ by σ-convexity of $\mathscr{C}$. □

In comparing normality and generation properties of $\mathscr{B}_+$ it is of interest to keep track of the index α of uniformity. This can be done in many ways, but we concentrate on the following pair. The cone $\mathscr{B}_+$ is defined to be *α_+-generating* if each $a \in \mathscr{B}$ has a decomposition $a = b - c$ with $b, c \in \mathscr{B}_+$ and

$$\|b\| + \|c\| \leqslant \alpha \|a\|$$

and to be *$\beta_\vee$-generating* if the decomposition can be chosen such that

$$\|b\| \vee \|c\| \leqslant \beta \|a\|.$$

Clearly $\alpha \geqslant 1$, by the triangle inequality, and a simple iterative argument establishes that one must have $\beta \geqslant 1$. Next $\mathscr{B}_+$ is defined to be *approximately α_+-generating* if it is α'_+-generating for all $\alpha' > \alpha$ and *approximately $\beta_\vee$-generating* if it is $\beta'_\vee$-generating for all $\beta' > \beta$. Similarly we introduce two types of normaliy.

The cone is defined to be *$\alpha_\vee$-normal* if $c \leqslant a \leqslant b$ always implies $\|a\| \leqslant \alpha(\|b\| \vee \|c\|)$ and to be *β_+-normal* if $c \leqslant a \leqslant b$ implies $\|a\| \leqslant \beta(\|b\| + \|c\|)$.

The following statement now generalizes the duality properties contained in Proposition 1.1.1.

THEOREM 1.1.4. *The following pairs of conditions are equivalent:*

$1_\vee$. (1_+.) *$\mathscr{B}_+$ is $\alpha_\vee$-normal (β_+-normal),*

2_+. ($2_\vee$.) *$\mathscr{B}^*_+$ is α_+-generating ($\beta_\vee$-generating).*

Moreover the following pairs of conditions are also equivalent:

$1'_+$. ($1'_\vee$.) *$\mathscr{B}_+$ is approximately α_+-generating (approximately $\beta_\vee$-generating),*

$2'_\vee$. ($2'_+$.) *$\mathscr{B}^*_+$ is $\alpha_\vee$-normal (β_+-normal).*

Proof. We concentrate on proving $1_+ \Leftrightarrow 2_\vee$ and $1'_\vee \Leftrightarrow 2'_+$. A proof of $1_\vee \Leftrightarrow 2_+$ and $1'_+ \Leftrightarrow 2'_\vee$ can be constructed on similar lines and alternative proofs of these equivalences can be found in [6].

One main ingredient of the proof is the Hahn–Banach theorem. The essential information is contained in the following general lemma which will subsequently be used several times in similar contexts.

LEMMA 1.1.5. *If* $\omega \in \mathscr{B}^*$ *and* $\lambda, \mu \in \mathbb{R}$ *then*

$$\sup\{\lambda\omega(a) + (1-\lambda)\omega(b) + (\mu-\lambda)\omega(c); b, c \geqslant 0, \|a\| + \|a-b-c\| \leqslant 1\} = \inf\{\|\eta\|^* \vee \|\eta - \lambda\omega\|^*; \eta \in \mathscr{B}^*, \eta \geqslant \omega, \eta \geqslant \mu\omega\}$$

and the infimum is attained, whenever it is finite.

Proof. Let S denote the value of the supremum and I the value of the infimum. Now if $b, c \geqslant 0$, $\|a\| + \|a-b-c\| \leqslant 1$, $\eta \geqslant \omega$, and $\eta \geqslant \mu\omega$, then

$$\lambda\omega(a) + (1-\lambda)\omega(b) + (\mu-\lambda)\omega(c) \leqslant (\lambda\omega - \eta)(a-b-c) + \eta(a) \leqslant \|\eta - \lambda\omega\|^* \vee \|\eta\|^*.$$

Thus $S \leqslant I$.

The converse inequality follows by application of the multiple Hahn–Banach theorem, Theorem A2 of the appendix, with $n = 4$ and with the p_i chosen such that

$$p_1(b) = S\|b\|, \qquad p_2(b) = S\|b\| + \lambda\omega(b)$$

$$p_3(b) = \begin{cases} \omega(b) & \text{if } -b \in \mathscr{B}_+, \\ +\infty & \text{if } -b \notin \mathscr{B}_+, \end{cases} \qquad p_4(b) = \begin{cases} \mu\omega(b) & \text{if } -b \in \mathscr{B}_+. \\ +\infty & \text{if } -b \notin \mathscr{B}_+ \end{cases}$$

Thus if $b_1 + b_2 + b_3 + b_4 = 0$ one has

$$\sum_{i=1}^{4} p_i(b_i) = +\infty$$

for $-b_3 \notin \mathscr{B}_+$ or $-b_4 \notin \mathscr{B}_+$ and

$$\sum_{i=1}^{4} p_i(b_i) = S(\|b_1\| + \|b_2\|) - \{\lambda\omega(b_1) + (1-\lambda)\omega(-b_3) + (\mu-\lambda)\omega(-b_4)\} \geqslant 0$$

for $-b_3, -b_4 \in \mathscr{B}_+$. Hence by Theorem A2 there exists a linear functional $\eta; \mathscr{B} \mapsto \mathbb{R}$ satisfying $\eta \leqslant p_i$ for $i = 1, 2, 3, 4$. In particular $\eta \in \mathscr{B}^*$, $\|\eta\|^* \leqslant S$, $\|\eta - \lambda\omega\|^* \leqslant S$, $\eta \geqslant \omega$, and $\eta \geqslant \mu\omega$. Thus $I \leqslant S$. But this implies the equality $I = S$ and also establishes that η attains the infimum. □

Now let us return to the proof of Theorem 1.1.4.

$1_+ \Leftrightarrow 2_\vee$. Choosing $\lambda = 1$ and $\mu = 0$ in Lemma 1.1.5 one finds, after a slight rearrangement of notation, that

$$\sup\{\omega(a); b \leqslant a \leqslant c, \|b\| + \|c\| \leqslant 1\} = \inf\{\|\xi\|^* \vee \|\eta\|^*; \xi, \eta \geqslant 0, \omega = \xi - \eta\}$$

and the infimum is attained. But this is just a statement of the equivalence of β_+-normality of $\mathscr{B}_+$ and $\beta_\vee$-generation of $\mathscr{B}^*_+$. For example if $\mathscr{B}_+$ is β_+-normal, the identity gives $\beta\|\omega\|^* \geqslant S = I$ and, since the infimum is attained, $\mathscr{B}^*_+$ is $\beta_\vee$-generating. Conversely if $\mathscr{B}^*_+$ is $\beta_\vee$-generating the identity gives

$$\beta\|\omega\|^* \geqslant \sup\{\omega(a); b \leqslant a \leqslant c, \|b\| + \|c\| \leqslant 1\}$$

and this implies that $\mathscr{B}_+$ is β_+-normal.

$1'_\vee \Rightarrow 2'_+$. Let $a = b - c$ with $b, c \in \mathscr{B}_+$ and $\|b\| \vee \|c\| \leqslant \beta' \|a\|$. Thus, if $\xi \leqslant \omega \leqslant \eta$, then

$$\xi(b) - \eta(c) \leqslant \omega(a) \leqslant \eta(b) - \xi(c).$$

Therefore

$$|\omega(a)| \leqslant \beta'(\|\xi\|^* + \|\eta\|^*)\|a\|,$$

which implies that

$$\|\omega\|^* \leqslant \beta'(\|\xi\|^* + \|\eta\|^*).$$

Since this is valid for all $\beta' > \beta$ the cone $\mathscr{B}^*_+$ is β_+-normal.

$2'_+ \Rightarrow 1'_\vee$. Condition $1'_\vee$ is equivalent to

$$\mathscr{B}_1 \subseteq (\mathscr{B}_{\beta'} \cap \mathscr{B}_+ - \mathscr{B}_{\beta'} \cap \mathscr{B}_+)$$

for all $\beta' > \beta$. But the set $\mathscr{B}_\beta \cap \mathscr{B}_+ - \mathscr{B}_\beta \cap \mathscr{B}_+$ is σ-convex and hence, by Lemma 1.1.3, it suffices to establish that

$$\mathscr{B}_1 \subseteq \overline{(\mathscr{B}_\beta \cap \mathscr{B}_+ - \mathscr{B}_\beta \cap \mathscr{B}_+)}.$$

But taking polars and using the bipolar theorem, Theorem A3 of the appendix, this is equivalent to

$$\mathscr{B}^*_1 \supseteq (\mathscr{B}_\beta \cap \mathscr{B}_+ - \mathscr{B}_\beta \cap \mathscr{B}_+)^0. \tag{*}$$

Now if $\omega \in (\mathscr{B}_\beta \cap \mathscr{B}_+ - \mathscr{B}_\beta \cap \mathscr{B}_+)^0$ then $\beta\omega(a - b) \leqslant 1$ for all $a, b \in \mathscr{B}_1 \cap \mathscr{B}_+$ and hence

$$\beta \sup\{\omega(a); a \in \mathscr{B}_1 \cap \mathscr{B}_+\} - \beta \inf\{\omega(b); b \in \mathscr{B}_1 \cap \mathscr{B}_+\} \leqslant 1.$$

But Lemma 1.1.5 with $\lambda = 0$ and $\mu = 1$ gives

$$\sup\{\omega(a); a \in \mathscr{B}_1 \cap \mathscr{B}_+\} = \inf\{\|\xi\|^*; \xi \in \mathscr{B}^*, \xi \geqslant \omega\}$$

and replacing ω by $-\omega$ it also gives

$$\inf\{\omega(b); b \in \mathscr{B}_1 \cap \mathscr{B}_+\} = -\inf\{\|\eta\|^*; \eta \in \mathscr{B}^*, \eta \geqslant -\omega\}.$$

Combining these identities with the foregoing inequality one deduces that

$$\beta \inf\{\|\xi\|^* + \|\eta\|^*; \xi, \eta \in \mathscr{B}^*, \eta \leqslant \omega \leqslant \xi\} \leqslant 1.$$

Therefore $\omega \in \mathscr{B}^*_1$, by Condition $2'_+$ and (*) is valid. □

The second statement in Theorem 1.1.4 is optimal in the sense that approximate α-generation does not necessarily imply α-generation. Example 1.1.9 below provides a counterexample. This asymmetry between $\mathscr{B}$ and $\mathscr{B}^*$, i.e., the appearance of approximate properties in $\mathscr{B}$, arises because $\mathscr{B}^*_1$ is always weak*-compact but $\mathscr{B}_1$ has no general compactness property. In special cases, e.g., if $\mathscr{B}$ is reflexive, one can deduce from compactness that approximate α-generation implies α-generation but not in general.

EXAMPLE 1.1.6 (Classical function spaces). Let $\mathscr{B} = L^p(X; d\mu)$ for some measure space $(X; d\mu)$ and some $p \in [1, \infty]$. Define q such that $1/p + 1/q = 1$. Next let $\mathscr{B}_+$ be the cone of pointwise positive functions in $\mathscr{B}$. It follows that $\mathscr{B}_+$ is 1_+-, or $(2^{1/p})_\vee$-, normal and $1_\vee$-, or $(2^{1/q})_+$-, generating. The result for $C(X)$ is identical to that for $L^\infty(X; d\mu)$, i.e., $\mathscr{B}_+$ is 1_+-, or $1_\vee$-, normal and $1_\vee$-, or 2_+-, generating. □

EXAMPLE 1.1.7 (C^*-algebras). Let $\mathscr{B}$ be the self-adjoint part of a C^*-algebra and $\mathscr{B}_+$ the cone of positive elements of the algebra. This is a non-abelian version of $C(X)$ and the conclusions are identical, $\mathscr{B}_+$ is 1_+-, or $1_\vee$-, normal and $1_\vee$-, or 2_+-, generating. □

EXAMPLE 1.1.8 (Affine functions). Let K be a compact convex subset of a locally convex Hausdorff space and $A(K)$ the continuous affine functions equipped with the supremum norm. If $A(K)_+$ is the set of pointwise positive functions it is again 1_+-, or $1_\vee$-, normal and $1_\vee$-, or 2_+-generating. □

EXAMPLE 1.1.9 (Approximate α-generation). Let $\mathscr{B}$ be the Banach space of sequences $a = \{a_n\}_{n \geq 1}$ satisfying

1. $a_n \to 0$ as $n \to \infty$

2. $a_1 + a_2 = \sum_{n \geq 1} a_{n+2}/2^n$

equipped with the supremum norm and let $\mathscr{B}_+$ consist of the $a \in \mathscr{B}$ with $a_n \geq 0$. It follows that $\mathscr{B}_+$ is approximately $1_\vee$-generating but not $1_\vee$-generating, and approximately 2_+-generating but not 2_+-generating. □

EXAMPLE 1.1.10 (A non-normal cone). Let $\mathscr{H} = L^2(\mathbb{R}^\nu; d^\nu x)$ with norm $\|\cdot\|_2$ and let $H = -\nabla^2$ denote the positive self-adjoint Laplace operator on $\mathscr{H}$. Define $\mathscr{B} = D(H^{1/2})$ and

$$\begin{aligned}\|f\| &= (\|f\|_2^2 + \|H^{1/2}f\|_2^2)^{1/2} \\ &= \left(\int d^\nu x \left\{|f(x)|^2 + \sum_{i=1}^{\nu} \left|\frac{\partial f(x)}{\partial x_i}\right|^2\right\}\right)^{1/2}\end{aligned}$$

for $f \in \mathscr{B}$. If $\mathscr{B}_+$ is the cone of pointwise positive functions in $\mathscr{B}$, then $\mathscr{B}_+$ is not normal. □

In the next sections we examine variants of the normality and generation conditions which are directly related to monotonicity properties of the norm.

1.2. MONOTONE NORMS

Let $(\mathscr{B}, \mathscr{B}_+, \|\cdot\|)$ be an ordered Banach space. The norm is defined to be *α-monotone* if $0 \leq a \leq b$ always implies that $\|a\| \leq \alpha \|b\|$. If $\mathscr{B}_+ \neq \{0\}$ one must have $\alpha \geq 1$, and if $\alpha = 1$, we simplify the terminology by saying that the norm is *monotone*.

If $\mathscr{B}_+$ is α_+-, or $\alpha_\vee$-, normal, then $\|\cdot\|$ is α-monotone and conversely if $\|\cdot\|$ is α-

monotone, then $\mathscr{B}_+$ is $(\alpha + \frac{1}{2})_+$-, or $(2\alpha + 1)_\vee$-, normal. Another relationship between normality and monotonicity is given in terms of equivalent norms.

PROPOSITION 1.2.1. *The following conditions are equivalent:*

1. *$\mathscr{B}_+$ is normal,*
2. *there exists an equivalent monotone norm.*

In particular if $\mathscr{B}_+$ is normal it is 1_+-normal with respect to the equivalent monotone norm

$$\|a\|_+ = \inf\{\|b\| + \|c\|; c \leqslant a \leqslant b\}$$

and it is $1_\vee$-normal with respect to the equivalent monotone norm

$$\|a\|_\vee = \inf\{\|b\| \vee \|c\|; c \leqslant a \leqslant b\}.$$

The proof that $1 \Rightarrow 2$ follows by verifying that $\|\cdot\|_+$, or $\|\cdot\|_\vee$, is an equivalent monotone norm. This is straightforward, as are the other statements, and hence we omit the details. We note in passing that duals to $\|\cdot\|_+$ and $\|\cdot\|_\vee$ are given by

$$\|\omega\|_+^* = \inf\{\|\xi\|^* \vee \|\eta\|^*; \omega = \xi - \eta, \xi, \eta \in \mathscr{B}_+^*\},$$

$$\|\omega\|_\vee^* = \inf\{\|\xi\|^* + \|\eta\|^*; \omega = \xi - \eta, \xi, \eta \in \mathscr{B}_+^*\},$$

e.g., the first identity is a consequence of Lemma 1.1.5 with $\lambda = 1, \mu = 0$.

Next we examine dual characterizations of α-monotonicity. For this it is appropriate to introduce a different index of generation.

The cone $\mathscr{B}_+$ is defined to be *α-dominating* if each $a \in \mathscr{B}$ has a decomposition $a = b - c$ with $b, c \in \mathscr{B}_+$ and $\|b\| \leqslant \alpha \|a\|$. This is equivalent to requiring that to each $a \in \mathscr{B}$ there is a $b \in \mathscr{B}_+$ with $b \geqslant a$ and $\|b\| \leqslant \alpha \|a\|$. More generally $\mathscr{B}_+$ is defined to be *approximately α-dominating* if it is α'-dominating for all $\alpha' > \alpha$. Subsequently we also use the terminology (approximately) dominating in place of (approximately) 1-dominating.

THEOREM 1.2.2. *The following conditions are equivalent:*

1. *$\|\cdot\|$ is α-monotone,*
2. *$\mathscr{B}_+^*$ is α-dominating.*

Moreover the following are also equivalent:

1'. *$\mathscr{B}_+$ is approximately α-dominating,*

2'. *$\|\cdot\|^*$ is α-monotone.*

Proof. The proof is very similar to that of Theorem 1.1.4 and so we only sketch the outlines.

$1 \Leftrightarrow 2$. This follows by another application of Lemma 1.1.5, but this time with $\lambda = 0 = \mu$. After a slight rearrangement of notation the lemma gives

$$\sup\{\omega(a); 0 \leqslant a \leqslant b, \|b\| \leqslant 1\} = \inf\{\|\eta\|^*; \eta \in \mathscr{B}_+^*, \eta \geqslant \omega\}$$

and the required equivalence follows straightforwardly.

$1' \Rightarrow 2'$. This is again an easy estimate.

$2' \Rightarrow 1'$. Property $1'$ is equivalent to the statement that

$$\mathscr{B}_1 \subseteq (\mathscr{B}_{\alpha'} \cap \mathscr{B}_+ - \mathscr{B}_+)$$

for all $\alpha' > \alpha$. But we next argue that $\mathscr{C} = (\mathscr{B}_\alpha \cap \mathscr{B}_+ - \mathscr{B}_+)$ is a σ-convex set and hence, by Lemma 1.1.3, it suffices to prove that

$$\mathscr{B}_1 \subseteq \overline{(\mathscr{B}_\alpha \cap \mathscr{B}_+ - \mathscr{B}_+)}. \tag{*}$$

In order to establish the σ-convexity of $\mathscr{C}$, assume $c_n \in \mathscr{C}$ and $c = \Sigma \lambda_n c_n$ exists in $\mathscr{B}$ for $\lambda_n \geqslant 0$ and $\Sigma \lambda_n = 1$. Then $c_n = a_n - b_n$ with $a_n \in \mathscr{B}_\alpha \cap \mathscr{B}_+$ and $b_n \in \mathscr{B}_+$. But $a = \Sigma \lambda_n a_n$ automatically exists in $\mathscr{B}_\alpha \cap \mathscr{B}_+$ and $\Sigma \lambda_n b_n$ must exist in $\mathscr{B}_+$, because $b_n = a_n - c_n$.

Finally $(*)$ can be established in its polar form

$$\mathscr{B}_1^* \supseteq (\mathscr{B}_\alpha \cap \mathscr{B}_+ - \mathscr{B}_+)^\circ$$

by a slight variation of the argument used to prove the analogous statement in Theorem 1.1.4. □

The case $\alpha = 1$, corresponding to monotonicity of the norm, has an alternative dual characterization of a completely different nature. First note that for any $a \in \mathscr{B}$ there exists, by the Hahn–Banach theorem, an $\omega \in \mathscr{B}_1^*$ such that $\omega(a) = \|a\|$. If, however, a is positive, it is not generally possible to conclude anything about the positivity of ω. For this one needs monotonicity of the norm.

THEOREM 1.2.3. *The following conditions are equivalent:*

1. $\|\cdot\|$ *is monotone on* $\mathscr{B}$,
2. *for each* $a \in \mathscr{B}_+$ *there is an* $\omega \in \mathscr{B}_+^* \cap \mathscr{B}_1^*$ *such that* $\omega(a) = \|a\|$.

Moreover the following conditions are also equivalent:

1*. $\|\cdot\|^*$ *is monotone on* $\mathscr{B}^*$,

2*. *if* $\omega \in \mathscr{B}^*$ *then*

$$\|\omega\|^* = \sup\{\omega(a);\, a \in \mathscr{B}_+ \cap \mathscr{B}_1\}.$$

Proof. $1 \Rightarrow 2$. If $\omega \in \mathscr{B}_1^*$ then since monotonicity of the norm $\|\cdot\|$ is equivalent to domination of $\mathscr{B}_+^*$, by Theorem 1.2.2, there exists an $\eta \in \mathscr{B}_+^* \cap \mathscr{B}_1^*$ such that $\eta \geqslant \omega$. Therefore if $a \in \mathscr{B}_+$ and $\omega(a) = \|a\|$ one has

$$\|a\| = \omega(a) \leqslant \eta(a) \leqslant \|a\|$$

and η satisfies Condition 2.

$2 \Rightarrow 1$. If $0 \leqslant a \leqslant b$ and $\omega \in \mathscr{B}_+^* \cap \mathscr{B}_1^*$ satisfies $\omega(a) = \|a\|$ then

$$\|a\| = \omega(a) \leqslant \omega(b) \leqslant \|b\|$$

and the norm is monotone.

The equivalence of Conditions 1* and 2* is proved in a similar fashion, but an approximation technique is necessary. □

EXAMPLE 1.2.4 (Function spaces). If $\mathscr{B} = L^p(X; d\mu)$ or $C(X)$ with $\mathscr{B}_+$ the pointwise positive functions, then the norm is monotone. Moreover for $p\in\langle 1, \infty\rangle$ the norm of each positive $f\in L^p$ is attained by the unique positive normalized element $f^{p-1}/\|f\|_p^{p-1}$ of the dual. If $p = 1$ or ∞, or if $\mathscr{B} = C(X)$, uniqueness fails in general. □

EXAMPLE 1.2.5 (C^*-algebras). If $\mathscr{B}$ is the self-adjoint part of a C^*-algebra, and $\mathscr{B}_+$ the positive elements, then the norm is monotone because $\mathscr{B}_+$ is $1_\vee$-normal (Example 1.1.7) and the dual norm is monotone because $\mathscr{B}_+$ is $1_\vee$-generating, and hence $\mathscr{B}^*_+$ is 1_+-normal by duality. Moreover, if $a\in\mathscr{B}_+\backslash\{0\}$ and $\omega\in\mathscr{B}^*$ satisfies $\omega(a) = \|\omega\|^*\|a\|$, then ω is automatically positive. This follows because $\mathscr{B}^*_+$ is 1_+-generating and if $\omega = \eta - \xi$ with $\eta, \xi\in\mathscr{B}^*_+$ and $\|\omega\|^* = \|\eta\|^* + \|\xi\|^*$ one has

$$\omega(a) \leqslant \eta(a) \leqslant \|\eta\|^*\|a\| \leqslant \|\omega\|^*\|a\|$$

which implies $\|\eta\|^* = \|\omega\|^*, \|\xi\|^* = 0$, and ω is positive. □

EXAMPLE 1.2.6 (Affine functions). Consider the space $\mathscr{B} = A(K)$ of affine functions of Example 1.1.8. Then the norm and dual norm are both monotone and if $a\in\mathscr{B}_+\backslash\{0\}$, $\omega\in\mathscr{B}^*_1$, and $\omega(a) = \|\omega\|^*\|a\|$, then $\omega\in\mathscr{B}^*_+$. These statements follow from Example 1.1.8 and the reasoning used in Example 1.2.5. □

EXAMPLE 1.2.7 (Two-dimensional spaces). Let $\mathscr{B} = \mathbb{R}^2$ with $\mathscr{B}_+ = \mathbb{R}^2_+$. The unit ball is convex but the norm is monotone if, and only if, the left derivative of the unit sphere at each point in the first quadrant is negative. The unit sphere in the third quadrant is determined by symmetry from the first quadrant, but monotonicity places no further restraint on the sphere in the second and fourth quadrants. □

1.3. ABSOLUTELY MONOTONE NORMS

In this section we consider a more stringent notion of monotonicity for the norm, absolute monotonicity. This is of interest for several reasons. First, the norms and dual norms of the classical function spaces, and of C^*-algebras, have this property. Second, absolute monotonicity of an equivalent norm and its dual norm are characteristic of normal generating cones. Third, the concept is useful in the theory of semigroup generators, the topic of Part 2.

Let $(\mathscr{B}, \mathscr{B}_+, \|\cdot\|)$ be an ordered Banach space. The norm $\|\cdot\|$ is defined to be *α-absolutely monotone* if $-b \leqslant a \leqslant b$ always implies $\|a\| \leqslant \alpha\|b\|$. Note that $-b \leqslant a \leqslant b$ requires $b \geqslant 0$ and taking $a = b\in\mathscr{B}_+\backslash\{0\}$ one must have $\alpha \geqslant 1$.

Next we define the dual concept. The cone $\mathscr{B}_+$ is defined to be *α-absolutely dominating* if for each $a\in\mathscr{B}$ there is a $b \geqslant 0$ such that $-b \leqslant a \leqslant b$ and $\|b\| \leqslant \alpha\|a\|$. More generally $\mathscr{B}_+$ is defined to be *approximately α-absolutely dominating* if it is α'-absolutely dominating for all $\alpha' > \alpha$.

Subsequently we use the simplified terminology absolutely monotone for 1-absolutely monotone and (approximately) absolutely dominating for (approximately) 1-absolutely dominating.

The duality between these concepts is as follows.

THEOREM 1.3.1. *The following conditions are equivalent:*

1. $\|\cdot\|$ *is* α*-absolutely monotone,*
2. $\mathscr{B}^*_+$ *is* α*-absolutely dominating.*

Moreover the following conditions are also equivalent:

1′. $\mathscr{B}_+$ *is approximately* α*-absolutely dominating,*
2′. $\|\cdot\|^*$ *is* α*-absolutely monotone.*

Proof. The proof is analogous to the proofs of Theorem 1.1.4 and 1.2.2. Hence we only sketch the outlines.

$1 \Leftrightarrow 2$. This follows from Lemma 1.1.5 by setting $\lambda = 0$ and $\mu = -1$. After a slight rearrangement of notation this gives

$$\sup\{\omega(a); b \pm a \geqslant 0, \|b\| \leqslant 1\} = \inf\{\|\eta\|^*; \eta \in \mathscr{B}^*, \eta \geqslant \pm\omega\} \tag{*}$$

and the infimum is attained. The required equivalence follows immediately.

$1' \Rightarrow 2'$. If $-b \leqslant a \leqslant b$ and $-\eta \leqslant \omega \leqslant \eta$ then $|\omega(a)| \leqslant \eta(b)$. Hence if $\|b\| \leqslant \alpha' \|a\|$ one concludes that $\|\omega\|^* \leqslant \alpha' \|\eta\|^*$.

$2' \Rightarrow 1'$. Define $\mathscr{C}$ by

$$\mathscr{C} = \{a; a \in \mathscr{B}, -b \leqslant a \leqslant b \text{ for some } b \in \mathscr{B}_1\}.$$

It follows immediately that $\mathscr{C}$ is σ-convex. Hence by Lemma 1.1.3 it suffices to show that $\mathscr{B}_1 \subseteq \alpha\overline{\mathscr{C}}$ or, equivalently $\mathscr{C}^0 \subseteq \mathscr{B}^*_\alpha$. But this follows directly from (*). □

The most interesting monotonicity and domination properties are those with $\alpha = 1$ and the most interesting situation is when both such properties are satisfied. We define the norm of an ordered Banach space $(\mathscr{B}, \mathscr{B}_+, \|\cdot\|)$ to be a *Riesz norm* (*a strong Riesz norm*) if $\|\cdot\|$ is absolutely monotone and $\mathscr{B}_+$ is approximately absolutely dominating (absolutely dominating).

There are several implications of Theorem 1.3.1 for Riesz norms which are worth noting. First $\|\cdot\|$ is a Riesz norm if, and only if, it is absolutely monotone and the dual norm is also absolutely monotone. Second if $\|\cdot\|$ is a Riesz norm then the dual norm $\|\cdot\|^*$ is a strong Riesz norm and conversely if $\|\cdot\|^*$ is a Riesz norm then $\|\cdot\|$ is a Riesz norm. Finally, by combination of the various definitions, one concludes that $\|\cdot\|$ is a Riesz norm if, and only if, it has the representation

$$\|a\| = \inf\{\|b\|; b \in \mathscr{B}, -b \leqslant a \leqslant b\},$$

and if $\|\cdot\|$ is a strong Riesz norm then the infimum is attained.

The first point of interest of Riesz norms is the following result.

THEOREM 1.3.2. *The following conditions are equivalent:*

1. $\mathscr{B}_+$ *is normal and generating,*
2. *there exists an equivalent Riesz norm.*

Moreover if $\mathscr{B}_+$ *is normal and generating the equivalent Riesz norm* $\|\cdot\|_r$ *can be defined by*

$$\|a\|_r = \inf\{\|b\|; b \in \mathscr{B}, -b \leqslant a \leqslant b\}$$

and then the dual Riesz norm is given by

$$\|\omega\|_r^* = \inf\{\|\eta\|^*; \eta\in\mathscr{B}^*, -\eta\leqslant\omega\leqslant\eta\}.$$

Proof. 1 ⇒ 2. Since $\mathscr{B}_+$ is generating $a\in\mathscr{B}\mapsto\|a\|_r$ is everywhere defined. Now $\mathscr{B}_+$ is $\alpha_\vee$-normal for some $\alpha\geqslant 1$ and hence $-b\leqslant a\leqslant b$ implies $\|a\|\leqslant\alpha\|b\|$. Therefore $\|a\|\leqslant\alpha\|a\|_r$. But $\mathscr{B}_+$ is also β_+-generating for some $\beta\geqslant 1$ and hence a has a decomposition $a=b-c$ with $b,c\geqslant 0$ and $\|b\|+\|c\|\leqslant\beta\|a\|$. Since $-(b+c)\leqslant a\leqslant$ $\leqslant(b+c)$ and $\|b+c\|\leqslant\|b\|+\|c\|$ it follows that $\|a\|_r\leqslant\beta\|a\|$. Consequently $\|\cdot\|_r$ is equivalent to $\|\cdot\|$.

2 ⇒ 1. Since an equivalent Riesz norm exists $\mathscr{B}_+$ is generating, by definition. Let $\|\cdot\|_r$ denote the equivalent norm and chose $\gamma,\delta>0$ such that $\gamma\|a\|\leqslant\|a\|_r\leqslant\delta\|a\|$ for all $a\in\mathscr{B}$. Then if $0\leqslant a\leqslant b$ one has $\|a\|_r\leqslant\|b\|_r$ by absolute monotonicity and hence $\|a\|\leqslant\gamma^{-1}\delta\|b\|$. Thus $\|\cdot\|$ is $\gamma^{-1}\delta$-monotone and consequently $\mathscr{B}_+$ is normal.

The representation of the dual Riesz norm $\|\cdot\|_r^*$ follows from Lemma 1.1.5 with $\lambda=0$ and $\mu=-1$. Explicitly one obtains

$$\begin{aligned}\inf\{\|\eta\|^*;\eta\in\mathscr{B}^*, -\eta\leqslant\omega\leqslant\eta\} &= \sup\{\omega(a); -b\leqslant a\leqslant b, \|b\|\leqslant 1\}\\ &= \sup\{|\omega(a)|; \|a\|_r\leqslant 1\} = \|\omega\|_r^*.\end{aligned}$$

□

There is another way of rephrasing the representation of a Riesz norm which is illuminating. If $d\pm a\geqslant 0$ and one sets $b=(d+a)/2$, $c=(d-a)/2$, then $a=b-c$, $d=b+c$, and $b,c\geqslant 0$. Therefore the Riesz norm is given by

$$\|a\| = \inf\{\|b+c\|; a=b-c, b,c\geqslant 0\},$$

and if $\|\cdot\|$ is a strong Riesz norm the infimum is attained. This states that a has a decomposition $a=b-c$ as the difference of positive components b and c, such that $\|a\|\simeq\|b+c\|$. But $b+c$ corresponds to a 'modulus' of a and the fact that a and its modulus have the same norm is a form of orthogonality of b and c. In particular examples, such as function spaces and C^*-algebras, the existence of an appropriate norm conserving modulus allows one to verify the Riesz norm property.

Finally this description of the Riesz norm in terms of positive and negative components can be amplified; absolute monotonicity of $\|\cdot\|$ is equivalent to the bound

$$\|a\|\leqslant\inf\{\|b+c\|; a=b-c, b,c\geqslant 0\}$$

and approximate absolute domination of $\mathscr{B}_+$ is equivalent to

$$\|a\|\geqslant\inf\{\|b+c\|; a=b-c, b,c\geqslant 0\}.$$

EXAMPLE 1.3.3 (Function spaces). If $\mathscr{B}=L^p(X; d\mu)$ with $\mathscr{B}_+$ the pointwise positive functions then $-f\leqslant g\leqslant f$ implies $|g|\leqslant|f|$ and hence $\|g\|_p\leqslant\|f\|_p$, for all $p\in[1,\infty]$. Thus the norms $\|\cdot\|_p$ are all absolutely monotone. But given f one has $-|f|\leqslant f\leqslant|f|$ and $\||f|\|_p=\|f\|_p$. Thus $\mathscr{B}_+$ is absolutely dominating for all $p\in[1,\infty]$. Therefore all the L^p-norms are strong Riesz norms. Similarly the supremum norm on $C(X)$ is a strong Riesz norm. □

EXAMPLE 1.3.4 (C^*-algebras). Let $\mathscr{B}$ be the self-adjoint part of a C^*-algebra ordered by the positive elements, then the C^*-norm is a strong Riesz norm. This follows because each $a \in \mathscr{B}$ has a modulus $|a|$ satisfying $|a| \geqslant \pm a$ and $\| |a| \| = \| a \|$. In particular $\| \cdot \|$ is absolutely dominating. But $\| \cdot \|$ is absolutely monotone because $\mathscr{B}_+$ is $1_\vee$-normal. □

EXAMPLE 1.3.5 (Two-dimensional spaces). Let $\mathscr{B} = \mathbb{R}^2$ with $\mathscr{B}_+ = \mathbb{R}^2_+$. Monotonicity of the norm places restraints on S_1, and S_3, the segments of the unit sphere in the first, and third, quadrant (see Example 1.2.7). Now for $a = (a_1, a_2) \in S$, consider the closed rectangle $\square_a$ with vertices $(\pm a_1, \pm a_2)$. The norm on $\mathscr{B}$ is absolutely monotone if

$$\mathscr{B}_1 \supseteq \bigcup_{a \in S_1} \square_a$$

and $\mathscr{B}_+$ is absolutely dominating if

$$\mathscr{B}_1 \subseteq \bigcup_{a \in S_1} \square_a .$$

Thus $\| \cdot \|$ is a (strong) Riesz norm if, and only if,

$$\mathscr{B}_1 = \bigcup_{a \in S_1} \square_a .$$

These characterizations can also be stated in terms of symmetry properties of $\mathscr{B}_1$, e.g., the norm is a Riesz norm if, and only if, $\mathscr{B}_1$ is symmetric under reflections $(a_1, a_2) \mapsto (-a_1, a_2)$. □

In the next two sections we consider two special classes of ordered Banach space which extend these examples, spaces with order units and spaces with a lattice ordering.

1.4. INTERIOR POINTS AND BASES

An element u of the cone $\mathscr{B}_+$ is defined to be an *interior point* if $\mathscr{B}_+$ contains an open neighbourhood of u, i.e., if there exists an $\varepsilon > 0$ such that $\{a; \| u - a \| < \varepsilon\} \subseteq \mathscr{B}_+$. The set of interior points of $\mathscr{B}_+$ is denoted by int $\mathscr{B}_+$. Note that if $u \in \operatorname{int} \mathscr{B}_+$, $b \in \mathscr{B}_+$, and $\lambda > 0$, then $\lambda u + b \in \operatorname{int} \mathscr{B}_+$. Therefore int $\mathscr{B}_+$ is either empty or norm dense in $\mathscr{B}_+$.

An element $u \in \mathscr{B}_+$ is defined to be an *order-unit* if

$$\mathscr{B} = \bigcup_{\lambda \geqslant 0} [-\lambda u, \lambda u]$$

where the order-interval $[c, b]$ is defined by

$$[c, b] = \{a; c \leqslant a \leqslant b\}.$$

But if $u \in \operatorname{int} \mathscr{B}_+$ then $a \in [-\lambda u, \lambda u]$ for all $\lambda > \| a \| / \varepsilon$, because $\| u - (u \pm a/\lambda) \| =$

$= \|a\|/\lambda < \varepsilon$, and hence each interior point is an order unit. Conversely, if u is an order-unit then $\mathscr{B}_1 \subseteq [-\lambda_0 u, \lambda_0 u]$ for some $\lambda_0 > 0$, by the Baire category theorem, and this implies $\{a; \|u - a\| < 1/\lambda_0\} \subseteq \mathscr{B}_+$. Thus each order-unit is an interior point of $\mathscr{B}_+$ and the two concepts, interior point and order-unit, coincide.

If int $\mathscr{B}_+$ is non-empty then $\mathscr{B}_+$ is generating and, for each $u \in \operatorname{int} \mathscr{B}_+$,

$$\|a\|_u = \inf\{\lambda > 0; a \in [-\lambda u, \lambda u]\}$$

defines a semi-norm on $\mathscr{B}$. Moreover $\mathscr{B}_1 \subseteq [-\lambda_0 u, \lambda_0 u]$ implies $\|a\|_u \leqslant \lambda_0 \|a\|$. The semi-norm $\|\cdot\|_u$ is a norm if, and only if, $\mathscr{B}_+$ is proper in which case it is a strong Riesz norm for which $\mathscr{B}_+$ is $1_\vee$-normal. It also follows by a series of simple estimates that $\mathscr{B}_+$ is normal with respect to $\|\cdot\|$ if, and only if, $\|\cdot\|$ and $\|\cdot\|_u$ are equivalent norms.

Now an ordered Banach space $(\mathscr{B}, \mathscr{B}_+, \|\cdot\|)$ is defined to be an *order-unit space* if $\mathscr{B}_+$ contains an interior point u and $\|\cdot\| = \|\cdot\|_u$. Equivalently $(\mathscr{B}, \mathscr{B}_+, \|\cdot\|)$ is an order unit space if $\mathscr{B}_1 = [-u, u]$ for some $u \in \mathscr{B}_+$. The first definition clearly implies the second, but $\mathscr{B}_1 = [-u, u]$ implies $u \in \operatorname{int} \mathscr{B}_+$, $\|a\| \leqslant \|a\|_u$, and conversely $-u\|a\|_u \leqslant a \leqslant u\|a\|_u$ implies $a/\|a\|_u \in \mathscr{B}_1$, i.e., $\|a\|_u \leqslant \|a\|$.

There is a dual concept defined in terms of bases.

A *base* for $\mathscr{B}_+$ is a norm-closed, convex, bounded, subset K of $\mathscr{B}_+$ such that for each $a \in \mathscr{B}_+$ there is a unique $\lambda_K(a) \geqslant 0$ such that $a \in \lambda_K(a)K$. If $\mathscr{B}_+$ has a base it is not difficult to show that it is normal, but not necessarily generating. If, however, $\mathscr{B}_+$ is generating and has a base K

$$\|a\|_K = \inf\{\lambda \geqslant 0; a \in \lambda \operatorname{co}(K \cup -K)\}$$

defines a norm on $\mathscr{B}$. Again it is readily verified that $\|\cdot\|_K$ is a Riesz norm which is equivalent to $\|\cdot\|$. Note that $\operatorname{co}(K \cup -K)$ is σ-convex and hence the $\|\cdot\|_K$-unit ball is the closed convex ball $\overline{\operatorname{co}}(K \cup -K)$, by Lemma 1.1.3.

Now an ordered Banach space $(\mathscr{B}, \mathscr{B}_+, \|\cdot\|)$ is defined to be a *base-norm space* if $\mathscr{B}_+$ is generating and has a base K and $\|\cdot\| = \|\cdot\|_K$. Equivalently $(\mathscr{B}, \mathscr{B}_+, \|\cdot\|)$ is a base-norm space if $\mathscr{B}_1 = \overline{\operatorname{co}}(K \cup -K)$ for some base K. Similarly if $\|\cdot\|^* = \|\cdot\|_K$ for some weak*-compact base K of $\mathscr{B}^*_+$, or if $\mathscr{B}^*_1 = \operatorname{co}(K \cup -K)$, then $(\mathscr{B}^*, \mathscr{B}^*_+, \|\cdot\|^*)$ is defined to be a *dual base-norm space*.

The duality between these concepts and the general representation of order-unit spaces is given by the following.

THEOREM 1.4.1. *The following conditions are equivalent:*

1. $(\mathscr{B}, \mathscr{B}_+, \|\cdot\|)$ *is an order-unit space,*
2. $(\mathscr{B}^*, \mathscr{B}^*_+, \|\cdot\|^*)$ *is a dual base-norm space,*
3. $(\mathscr{B}, \mathscr{B}_+, \|\cdot\|)$ *is isometrically order-isomorphic to the space* $(A(K), A(K)_+, \|\cdot\|_\infty)$ *of continuous affine functions over some compact convex set K, ordered by the positive functions and equipped with the supremum norm.*

Moreover the following conditions are equivalent:

1′. $(\mathscr{B}, \mathscr{B}_+, \|\cdot\|)$ *is a base-norm space,*

2′. $(\mathscr{B}^*, \mathscr{B}^*_+, \|\cdot\|^*)$ *is an order-unit space.*

Proof. 1 ⇒ 2. Let u be the interior element with $\|\cdot\|_u = \|\cdot\|$. For $\omega \in \mathscr{B}^*_+$

$$\begin{aligned}\|\omega\|^* &= \sup\{\omega(b);\ -u \leqslant b \leqslant u\}\\ &= \omega(u).\end{aligned}$$

Therefore

$$K = \{\omega; \omega \in \mathscr{B}^*_+, \|\omega\|^* = 1\}$$

is a weak*-compact base for $\mathscr{B}^*_+$. Since $\mathscr{B}_+$ is $1_\vee$-normal, $\mathscr{B}^*_+$ is 1_+-generating, by Theorem 1.1.4, and hence

$$\mathscr{B}^*_1 = \mathrm{co}(K \cup -K) = \overline{\mathrm{co}}(K \cup -K).$$

2 ⇒ 3. Let K be a weak*-compact base for $\mathscr{B}^*_+$ with $\mathscr{B}^*_1 = \mathrm{co}(K \cup -K)$. For each $b \in \mathscr{B}$ define $\theta(b)$ on K by $(\theta(b))(\omega) = \omega(b)$. Then θ is an isometric order-isomorphism of $\mathscr{B}$ into $A(K)$. Now for $a \in A(K)$ define $\tilde{a}$ on $\mathscr{B}^*$ by

$$\tilde{a}(\lambda\eta - \mu\xi) = \lambda a(\eta) - \mu a(\xi)$$

when $\lambda, \mu \geqslant 0$ and $\eta, \xi \in K$. Since K is a base and a is affine this gives a well-defined linear functional $\tilde{a}$ on $\mathscr{B}^*$. But $\tilde{a}$ is weak*-continuous on $\mathscr{B}^*_1 = \mathrm{co}(K \cup -K)$ by weak*-compactness and therefore $\tilde{a} \in \mathscr{B}$ by the Krein–Smulian theorem. Since $\theta(\tilde{a}) = a$ the mapping θ is surjective.

3 ⇒ 1. This is a simple verification with $u = 1$.

1′ ⇒ 2′. Let K be a base of $\mathscr{B}_+$ with $\mathscr{B}_1 = \overline{\mathrm{co}}(K \cup -K)$. The associated functional $a \in \mathscr{B}_+ \mapsto \lambda_K(a) \in \mathbb{R}_+$ extends to a linear functional on $\mathscr{B}$ which is an order-unit for $\mathscr{B}^*$ with $\|\cdot\|_{\lambda_K} = \|\cdot\|^*$.

2′ ⇒ 1′. Let ρ be the order-unit of $\mathscr{B}^*$ for which $\|\cdot\|_\rho = \|\cdot\|^*$ and define $K = \{b; b \in \mathscr{B}_+, \rho(b) = 1\}$. It follows from the bipolar theorem, Theorem A3 of the appendix, that K is a base for $\mathscr{B}_+$ and $\mathscr{B}_1 = \overline{\mathrm{co}}(K \cup -K)$. □

Theorem 1.4.1 is an isometric version of part of the following result.

PROPOSITION 1.4.2. 1. int $\mathscr{B}_+ \neq \emptyset$ *if, and only if,* $\mathscr{B}^*_+$ *has a weak*-compact base K. In this case,*

$$\mathrm{int}\,\mathscr{B}_+ = \{u; u \in \mathscr{B}_+, \inf_{\omega \in K} \omega(u) > 0\}.$$

If $\mathscr{B}_+$ *is normal, the correspondence between interior points of* $\mathscr{B}_+$ *and weak*-compact bases for* $\mathscr{B}^*_+$ *is bijective.*

2. $\mathscr{B}_+$ *has a base K if, and only if,* int $\mathscr{B}^*_+ \neq \emptyset$. *In this case,*

$$\mathrm{int}\,\mathscr{B}^*_+ = \{\rho; \rho \in \mathscr{B}^*_+, \inf_{a \in K} \rho(a) > 0\}.$$

If $\mathscr{B}_+$ *is generating, the correspondence between bases for K and interior points of* $\mathscr{B}^*_+$ *is bijective.*

3. *$\mathscr{B}_+$ has a base K if, and only if, there is a constant α such that*

$$\alpha\left\|\sum_{a\in S} a\right\| \geqslant \sum_{a\in S} \|a\|$$

for all finite subsets S of $\mathscr{B}_+$.

Proof. 1. If $u\in\operatorname{int}\mathscr{B}_+$, then $K=\{\omega;\omega\in\mathscr{B}^*_+,\omega(u)=1\}$ is a weak*-compact base for $\mathscr{B}^*_+$. If $\mathscr{B}^*_+$ has a weak*-compact base K, elementary estimates show that $\operatorname{int}\mathscr{B}_+=\{u;u\in\mathscr{B}_+,\inf_{\omega\in K}\omega(u)>0\}$, and the Hahn–Banach separation theorem, Theorem A3 of the appendix, shows that $\operatorname{int}\mathscr{B}_+\neq\emptyset$. If, moreover, $\mathscr{B}_+$ is normal, so $\mathscr{B}^*_+$ is generating, λ_K extends to a linear functional on $\mathscr{B}^*$, which is weak*-continuous on $\operatorname{co}(K\cup -K)$, a (norm) neighbourhood of 0 in $\mathscr{B}^*$. By the Krein–Smulian theorem. λ_K belongs to $\mathscr{B}$ and hence to $\operatorname{int}\mathscr{B}_+$.

2. The proof is similar to part 1. If $\rho\in\operatorname{int}\mathscr{B}^*_+$, then $K=\{a;a\in\mathscr{B}_+,\rho(a)=1\}$ is a base for $\mathscr{B}_+$. If $\mathscr{B}_+$ is generating and has a base K, then λ_K extends to a linear functional in $\operatorname{int}\mathscr{B}^*_+$.

3. If $\mathscr{B}_+$ has a base K, and $\alpha=\sup\{\|a\|/\|b\|;a,b\in K\}$, it is readily verified that $\alpha\|\Sigma_{a\in S}a\|\geqslant\Sigma_{a\in S}\|a\|$. Conversely, if this inequality is always satisfied, then $\|b\|\geqslant 1$ whenever $b\in\operatorname{co}\{a\in\mathscr{B}_+;\|a\|=\alpha\}$, so by the Hahn–Banach separation theorem, there exists $\omega\in\mathscr{B}^*_1$ such that $\omega(a)\geqslant 1$ whenever $a\in\mathscr{B}_+$, $\|a\|=\alpha$. Then $\eta\in\mathscr{B}^*_+$ whenever $\|\omega-\eta\|^*\leqslant\alpha^{-1}$, so $\omega\in\operatorname{int}\mathscr{B}^*_+$. □

If $\alpha\|\Sigma_{a\in S}a\|\geqslant\|\Sigma_{a\in S}a\|$ for all finite $S\subseteq\mathscr{B}_+$, then $\|\cdot\|$ is said to be *α-additive*. Alternatively, if for each finite $S\subseteq\mathscr{B}_1$, there exists $b\in\mathscr{B}_\alpha$ such that $a\leqslant b$ for all $a\in S$, then $\mathscr{B}$ is said to be *α-directed*. The above result, together with weak*-compactness, shows that $\|\cdot\|$ is α-additive if, and only if, $\mathscr{B}^*$ is α-directed, It may also be shown that $\mathscr{B}$ is α′-directed for all $\alpha'>\alpha$ if, and only if, $\|\cdot\|^*$ is α-additive (see [6]).

The existence of interior points in $\mathscr{B}_+$ and normality of $\mathscr{B}_+$ are in a sense complementary properties. This complementarity can be summarized in terms of the set $b(\mathscr{B})$ of norm-bounded sets in $\mathscr{B}$ and the set $o(\mathscr{B})$ of order-bounded sets.

PROPOSITION 1.4.3. 1. $\operatorname{int}\mathscr{B}_+\neq\emptyset$ *if, and only if,* $b(\mathscr{B})\subseteq o(\mathscr{B})$.

2. *$\mathscr{B}_+$ is normal if, and only if,* $o(\mathscr{B})\subseteq b(\mathscr{B})$.

Proof. 1. If $\{a;\|u-a\|<\varepsilon\}\subseteq\mathscr{B}_+$ then $\mathscr{B}_1\subseteq[-\varepsilon^{-1}u,\varepsilon^{-1}u]$ and conversely if $\mathscr{B}_1\subseteq[c,b]$ then $\{a;\|a-b\|<1\}\subseteq\mathscr{B}_+$.

2. If $\mathscr{B}_+$ is normal there is an $\alpha\geqslant 1$ such that $[c,b]\subseteq\mathscr{B}_\beta$ for $\beta=\alpha(\|b\|\vee\|c\|)$. Conversely if $\mathscr{B}_+$ is not normal there exist sequences a_n,b_n such that $a_n\in[0,b_n]$ and $\|b_n\|\leqslant 1$, $\|a_n\|>4^n$. Thus defining $b=\Sigma 2^{-n}b_n$ one has $2^{-n}a_n\in[0,b]$ but $\|2^{-n}a_n\|>2^n$. Thus $[0,b]\notin b(\mathscr{B})$. □

One immediate corollary is that $o(\mathscr{B})=b(\mathscr{B})$ if, and only if, $\mathscr{B}_+$ is normal and $\operatorname{int}\mathscr{B}_+\neq\emptyset$.

There are various weaker notions of interior point, or order-unit. For example $u\in\mathscr{B}_+$ is called a *quasi-interior point* if

$$\omega(u)>0$$

for all $\omega\in\mathscr{B}_+^*\setminus\{0\}$. Equivalently u is a quasi-interior point if for each $a\in\mathscr{B}$ and $\varepsilon>0$ there exists a $b\in\mathscr{B}$ and $\lambda\in\mathbb{R}_+$ with $\|a-b\|<\varepsilon$, $b\leqslant\lambda u$. (The first of these definitions states that $\{0\}=(\mathbb{R}_+u-\mathscr{B}_+)^0$ and the second states the polar form $\mathscr{B}=(\mathbb{R}_+u-\mathscr{B}_+)$.)

Quasi-interior points exist only if $\mathscr{B}_+$ is weakly generating, and the converse is true if $\mathscr{B}$ is separable. It can happen, however, that quasi-interior points exist but $\mathscr{B}_+$ is not generating, e.g., this is the situation in the dual of the space given in Example 1.1.10. Nevertheless one has one of two extreme situations, the set qu.int $\mathscr{B}_+$ of quasi-interior points is empty, or norm-dense in $\mathscr{B}_+$. This follows because $u\in$qu.int $\mathscr{B}_+$ implies $\lambda u+a\in$qu.int $\mathscr{B}_+$ for all $\lambda>0$ and $a\in\mathscr{B}_+$. Furthermore int $\mathscr{B}_+\subseteq$ qu.int $\mathscr{B}_+$ with equality whenever int $\mathscr{B}_+\neq\emptyset$. This is a consequence of the Hahn–Banach separation theorem, Theorem A3 of the appendix.

Any proper generating cone in a finite-dimensional space has both interior points and bases. But in infinite dimensions all possibilities occur.

EXAMPLE 1.4.4 (Function spaces). Let $\mathscr{B}=L^p(X;d\mu)$ with $p\in[1,\infty\rangle$ and let $\mathscr{B}_+$ be the cone of pointwise positive functions. If $\mathscr{B}$ is infinite-dimensional then int $\mathscr{B}_+=\emptyset$ and qu.int $\mathscr{B}_+=\{f;f>0\text{ a.e.}\}$; if $p>1$ then $\mathscr{B}$ has no bases and if $p=1$ then $\mathscr{B}_+$ has the base $K=\{f;f\geqslant 0,\int d\mu f=1\}$. Note that if $d\mu$ is not σ-finite qu.int $\mathscr{B}_+=\emptyset$. The case $\mathscr{B}=L^\infty(X;d\mu)$ is somewhat different. This is an order-unit space with order-unit the constant function of value one and if $\mathscr{B}$ is infinite-dimensional it has no bases. □

EXAMPLE 1.4.5 (C^*-algebras). Let $\mathscr{B}$ be the self-adjoint part of an infinite-dimensional C^*-algebra with identity ordered by the positive elements. Then $\mathscr{B}$ is an order-unit space, with order-unit the identity, and int $\mathscr{B}_+=$ qu.int $\mathscr{B}_+=\{a;a\geqslant 0,$ a invertible$\}$. The theory of C^*algebras [59] shows that $\mathscr{B}$ has a subspace isometrically order-isomorphic to $C(X)$ with X infinite. Since $C(X)_+^*$ has no interior points it follows that $\mathscr{B}_+$ has no bases. □

EXAMPLE 1.4.6 (Order-units and bases). Let $\mathscr{B}$ be the subspace of l^∞ consisting of the sequences of $a=\{a_n\}_{n\geqslant 1}$ satisfying $a_{2n-1}+a_{2n}=a_1+a_2$ and let $\mathscr{B}_+=$ $=\{a;a\in\mathscr{B},a_n\geqslant 0\}$. Then $\mathscr{B}$ is an order-unit space with order-unit $a_n=1$, int $\mathscr{B}_+=$ $=$ qu.int $\mathscr{B}_+=\{a;a\in\mathscr{B},\inf a_n>0\}$, and $\mathscr{B}_+$ has the base $K=\{a;a\in\mathscr{B}_+,a_1+a_2=1\}$. □

1.5. BANACH LATTICES

Consider an ordered Banach space $(\mathscr{B},\mathscr{B}_+,\|\cdot\|)$ which is a lattice in the given ordering, i.e., each pair $a,b\in\mathscr{B}$ has a least upper bound $a\vee b$ and a greatest lower bound $a\wedge b$. If $|a|\leqslant|b|$ always implies $\|a\|\leqslant\|b\|$, where $|a|=a\vee -a$ denotes the modulus of a, then $(\mathscr{B},\mathscr{B}_+,\|\cdot\|)$ is defined to be a *Banach lattice*. For example if $\mathscr{B}$ is a lattice and $\|\cdot\|$ is a Riesz norm then $\mathscr{B}$ is a Banach lattice. In particular $C(X)$ and $L^p(X;d\mu)$

are Banach lattices. Moreover if $\mathscr{B}$ is a lattice and either an order-unit space, or a base-norm space, then it is a Banach lattice, and can be identified with $C(X)$ or $L^1(X; d\mu)$ (see Theorems 1.5.1 and 1.5.2). Alternatively the self-adjoint-part of a C^*-algebra is a lattice if, and only if, the algebra is commutative [11].

There is a very detailed theory of Banach lattices and an extensive literature on this subject (see, for example, [75, 31, 82]. We mention only those aspects which are of interest in the sequel.

In any Banach lattice the norm is a strong Riesz norm, e.g., since $-|a| \leqslant a \leqslant |a|$ one has absolute domination of $\mathscr{B}_+$. Moreover $-b \leqslant a \leqslant b$ implies $|a| \leqslant b = |b|$ and absolute monotonicity of $\|\cdot\|$ follows from the Banach lattice property. It can also be established that the dual $(\mathscr{B}^*, \mathscr{B}^*_+, \|\cdot\|^*)$ of a Banach lattice is a Banach lattice. In fact the dual lattice has the extra property of order completeness, i.e., each bounded subset of $\mathscr{B}^*$ has a least upper bound in $\mathscr{B}^*$.

The cone $\mathscr{B}_+$ of a Banach lattice is 1_+-normal and $1_\vee$-generating. For example the generating property follows by introduction of the *canonical decomposition* $a = a_+ - a_-$ of a into *positive* and *negative components* $a_\pm = (|a| \pm a)/2$. Thus

$$\|a_+\| \vee \|a_-\| \leqslant (\||a|\| + \|a\|)/2 = \|a\|.$$

The 1_+-normality can also be verified by a simple calculation, or deduced by duality from the $1_\vee$-generation of $\mathscr{B}^*_+$.

Next we describe the main representation theorems for lattices which are order-unit spaces, or base-norm spaces. First note that if $(\mathscr{B}, \mathscr{B}_+, \|\cdot\|)$ is a lattice and an order-unit space then

$$\|a \vee b\| = \|a\| \vee \|b\|$$

for all $a, b \in \mathscr{B}_+$. Banach lattices which satisfy this latter property are called AM-spaces and the following theorem gives a representation of such spaces.

THEOREM 1.5.1. *Let $(\mathscr{B}, \mathscr{B}_+, \| \ \|)$ be a Banach lattice. The following conditions are equivalent:*

1. $\|a\| = \|a_+\| \vee \|a_-\|$, $a \in \mathscr{B}$,
2. $\|a \vee b\| = \|a\| \vee \|b\|$, $a, b \in \mathscr{B}_+$,
3. $(\mathscr{B}, \mathscr{B}_+, \|\cdot\|)$ *is isometrically order-isomorphic to a sublattice of* $(C(X), C(X)_+, \|\cdot\|_\infty)$ *for some compact Hausdroff space X.*

Moreover the following conditions are also equivalent:

1′. $(\mathscr{B}, \mathscr{B}_+, \|\cdot\|)$ *is an order-unit space,*
2′. $(\mathscr{B}, \mathscr{B}_+, \|\cdot\|)$ *is isometrically order-isomorphic to* $(C(X), C(X)_+, \|\cdot\|_\infty)$.

Next consider base norm spaces. If $(\mathscr{B}, \mathscr{B}_+, \|\cdot\|)$ is a lattice and a base-norm space, then one can verify that

$$\|a + b\| = \|a\| + \|b\|$$

for all $a, b \in \mathscr{B}_+$. Banach lattices with this property are called AL-spaces. The dual of an AM-space is an AL-space, and vice-versa. The analogue of Theorem 1.5.1 for these dual spaces is the following.

THEOREM 1.5.2. *Let* $(\mathscr{B}, \mathscr{B}_+, \|\cdot\|)$ *be a Banach lattice. The following conditions are equivalent:*

1. $\|a\| = \|a_+\| + \|a_-\|, \quad a \in \mathscr{B}$,
2. $\|a+b\| = \|a\| + \|b\|, \quad a, b \in \mathscr{B}_+$,
3. $(\mathscr{B}, \mathscr{B}_+, \|\cdot\|)$ *is isometrically order-isomorphic to* $L^1(X; \mathrm{d}\mu)$ *for some measure space* $(X, \mathrm{d}\mu)$.
4. $(\mathscr{B}, \mathscr{B}_+, \|\cdot\|)$ *is a base-norm space.*

It is now a simple task to strengthen the results of Section 1.4 when $\mathscr{B}$ is a lattice.

COROLLARY 1.5.3. *Let* $(\mathscr{B}, \mathscr{B}_+, \|\cdot\|)$ *be an ordered Banach space which is a lattice in the given order.*

1. $\operatorname{int} \mathscr{B}_+ \neq \emptyset$ *if, and only if,* $(\mathscr{B}, \mathscr{B}_+, \|\cdot\|)$ *is topologically order-isomorphic to* $(C(X), C(X)_+, \|\cdot\|_\infty)$ *for some compact Hausdorff space X.*
2. $\mathscr{B}_+$ *has a base if, and only if,* $(\mathscr{B}, \mathscr{B}_+, \|\cdot\|)$ *is topologically order-isomorphic to* $L^1(X; \mathrm{d}\mu)$ *for some measure space* $(X, \mathrm{d}\mu)$.

Finally we examine two general properties closely related to the lattice property.

An ordered vector space $(\mathscr{B}, \mathscr{B}_+)$ is said to have the *Riesz interpolation property* if, $a_1, a_2, b_1, b_2 \in \mathscr{B}$ and $a_i \leqslant b_j$ for $i, j = 1, 2$ implies the existence of a $c \in \mathscr{B}$ such that $a_i \leqslant c \leqslant b_j$ for $i, j = 1, 2$. This property is equivalent to the *Riesz decomposition property*, i.e., the property that if $0 \leqslant a \leqslant b_1 + b_2$ with $b_1, b_2 \in \mathscr{B}_+$, then there exist $a_1, a_2 \in \mathscr{B}_+$ such that $a = a_1 + a_2$ and $a_i \leqslant b_i$, $i = 1, 2$, [6], Proposition 2.5.4. Any lattice has the Riesz interpolation property.

THEOREM 1.5.4. *Let* $(\mathscr{B}, \mathscr{B}_+, \|\cdot\|)$ *be an ordered Banach space and suppose that* $\mathscr{B}_+$ *is normal and generating. The following conditions are equivalent:*

1. $(\mathscr{B}, \mathscr{B}_+)$ *has the Riesz interpolation property,*
2. $(\mathscr{B}^*, \mathscr{B}^*_+)$ *is a lattice,*
3. $(\mathscr{B}^*, \mathscr{B}^*_+)$ *has the Riesz interpolation property.*

Moreover the following conditions are also equivalent:

1′. $(\mathscr{B}, \mathscr{B}_+)$ *has the Riesz interpolation property and* $\|\cdot\|$ *is a Riesz norm,*

2′. $(\mathscr{B}^*, \mathscr{B}^*_+, \|\cdot\|^*)$ *is a Banach lattice.*

Proof. This is a standard result (see, for example, [6] Theorem 2.5.7). Note that if $(\mathscr{B}^*, \mathscr{B}^*_+)$ is a lattice then for $a \in \mathscr{B}_+$ and $\omega, \eta \in \mathscr{B}^*$ one has

$$(\omega \vee \eta)(a) = \sup\{\omega(b) + \eta(c);\, a = b + c,\, b, c \in \mathscr{B}_+\}. \qquad \square$$

The following is an analogue of Corollary 1.5.3. (For the various definitions of Choquet simplexes see [2, 6].)

THEOREM 1.5.5. *Let* $(\mathscr{B}, \mathscr{B}_+, \|\cdot\|)$ *be an ordered Banach space with the Riesz interpolation property.*

1. *int* $\mathscr{B}_+ \neq \emptyset$ *if, and only if,* $(\mathscr{B}, \mathscr{B}_+, \|\cdot\|)$ *is topologically order-isomorphic to* $(A(K), A(K)_+, \|\cdot\|_\infty)$ *for some Choquet simplex* K.
2. $\mathscr{B}_+$ *has a base if, and only if,* $(\mathscr{B}, \mathscr{B}_+, \|\cdot\|)$ *is topologically order isomorphic to* $L^1(X; \mathrm{d}\mu)$ *for some measure space* $(X, \mathrm{d}\mu)$.

Proof. 1. This is a standard fact in Choquet theory (see, for example, [6] Theorem 2.7.1).

2. It suffices to assume that $\|\cdot\|$ is a base-norm and to show that $\mathscr{B}$ is a lattice. The result then follows from Corollary 1.5.3.

Let $a, b \in \mathscr{B}_+$ and c_n a sequence in $\mathscr{B}$ such that $a \leqslant c_n, b \leqslant c_n$, and $\|c_n\| < \alpha + 2^{-n}$, where

$$\alpha = \inf\{\|c\|, c \in \mathscr{B}, a \leqslant c, b \leqslant c\}.$$

Using the Riesz interpolation property it may be assumed that $c_{n+1} \leqslant c_n$. Then

$$\|c_n - c_{n+1}\| = \|c_n\| - \|c_{n+1}\| < 2^{-n}.$$

Hence c_n converges to a limit c such that $a \leqslant c, b \leqslant c$, and $\|c\| = \alpha$. Next suppose that $c' \in \mathscr{B}, c' \geqslant a$, and $c' \geqslant b$. By the Riesz interpolation property there exists $c'' \in \mathscr{B}$ such that $a \leqslant c'' \leqslant c, b \leqslant c'' \leqslant c'$. Then

$$\alpha = \|c\| = \|c''\| + \|c - c''\| \geqslant \alpha + \|c - c''\|.$$

Hence $c = c'' \leqslant c'$. Thus $c = a \vee b$. □

1.6. HALF-NORMS

The asymmetric nature of the positive cone is reflected in many properties of ordered Banach spaces. It is therefore useful to have an asymmetric version of the norm. The appropriate notion appears to be a half-norm.

A *half-norm* on a Banach space $\mathscr{B}$ is a continuous sublinear functional $p; \mathscr{B} \mapsto \mathbb{R}$. In particular sublinearity implies $0 = p(0) \leqslant p(a) + p(-a)$. The continuity means that there is a constant k such that $p(a) \leqslant k\|a\|$ for all $a \in \mathscr{B}$, and hence by sublinearity $|p(a)| \leqslant k\|a\|$. A half-norm p is defined to be *proper* if $p(a) \vee p(-a) > 0$ for all $a \in \mathscr{B} \backslash \{0\}$. One can then associate with each proper half-norm p a norm $\|\cdot\|_p$ on $\mathscr{B}$ by the definition

$$\|a\|_p = p(a) \vee p(-a),$$

and it follows that $\|a\|_p \leqslant k\|a\|$.

One can also associate with each (proper) half-norm p a (proper) closed convex cone

$$\mathscr{B}^p_+ = \{a; a \in \mathscr{B}, p(-a) \leqslant 0\}.$$

One then has

$$\{a; a \in \mathscr{B}, p(-a) < 0\} \subseteq \operatorname{int} \mathscr{B}^p_+ .$$

by continuity of p, and the definition of $\mathscr{B}^p_+$.

Conversely if $(\mathscr{B}, \mathscr{B}_+, \|\cdot\|)$ is an ordered Banach space with a (proper) positive cone $\mathscr{B}_+$ there is a (proper) canonical half-norm N on $\mathscr{B}$ given by

$$N(a) = \inf\{\|b\|; b \in \mathscr{B}, b \geqslant a\}.$$

One has $0 \leqslant N(a) \leqslant \|a\|$ and N is compatible with the ordering in the sense that $\mathscr{B}_+^N = \mathscr{B}_+$. In fact if $\operatorname{int}\mathscr{B}_+ \neq \emptyset$ one can construct an alternative half-norm which coincides with the canonical half-norm for all $-a \notin \operatorname{int}\mathscr{B}_+$ but which takes strictly negative values on $-a \in \operatorname{int}\mathscr{B}_+$. This will be discussed in more detail below.

Note that the canonical half-norm can be re-expressed as

$$N(a) = \inf\{\|a + c\|; c \in \mathscr{B}_+\}.$$

Hence N measures the distance of $-a$ from $\mathscr{B}_+$.

Since the dual space $(\mathscr{B}^*, \mathscr{B}^*_+, \|\cdot\|^*)$ is an ordered Banach space, one can also define a canonical half-norm associated with $\mathscr{B}^*$. This will be denoted by N^*. Explicitly one has

$$N^*(\omega) = \inf\{\|\xi\|^*; \xi \in \mathscr{B}^*, \xi \geqslant \omega\}$$

and it follows from weak*-compactness that the infimum is attained.

The half-norms N and N^* may be described by dual relations. For this one needs the following.

LEMMA 1.6.1. *Let ω be a linear functional on an ordered Banach space $(\mathscr{B}, \mathscr{B}_+, \|\cdot\|)$ and $\alpha \in \mathbb{R}_+$. The following conditions are equivalent:*

1. $\omega \in \mathscr{B}^*_+ \cap \mathscr{B}^*_\alpha$,
2. $\omega(a) \leqslant \alpha N(a), \quad a \in \mathscr{B}$.

The proof is an elementary consequence of the definition of N and its properties mentioned above.

PROPOSITION 1.6.2. *The canonical half-norms N and N^* satisfy*

$$N(a) = \sup\{\omega(a); \omega \in \mathscr{B}^*_+ \cap \mathscr{B}^*_1\},$$

$$N^*(\omega) = \sup\{\omega(a); a \in \mathscr{B}_+ \cap \mathscr{B}_1\}.$$

Proof. The first statement follows from the Hahn–Banach theorem and Lemma 1.6.1. The second statement is obtained by setting $\lambda = 0, \mu = 1$, in Lemma 1.1.5 and rearranging (see the proof of Theorem 1.1.4). □

Monotonicity of the norm is easily characterized by the canonical half-norms. The following result may be regarded as an extension of Theorem 1.2.3.

THEOREM 1.6.3. *The following conditions are equivalent:*

1. $\|\cdot\|$ *is monotone on* $\mathscr{B}$,
2. $N(a) = \|a\|, a \in \mathscr{B}_+$,
3. $N^*(\omega) = \inf\{\|\xi\|^*; \xi \geqslant 0, \xi \geqslant \omega\}, \omega \in \mathscr{B}^*$
4. *For each $\omega \in \mathscr{B}^*$ there exist $\xi, \eta \in \mathscr{B}^*_+$ with $\omega = \xi - \eta$ and $\|\xi\|^* = N^*(\omega)$.*

Moreover the following conditions are equivalent:

1*. $\|\cdot\|^*$ *is monotone on* $\mathscr{B}^*$,

2*. $N^*(\omega) = \|\omega\|^*$, $\omega \in \mathscr{B}^*_+$,

3*. $N(a) = \inf\{\|b\|; b \geqslant 0, b \geqslant a\}$, $a \in \mathscr{B}$,

4*. *For each* $a \in \mathscr{B}$ *and* $\alpha > 1$ *there exist* $b, c \in \mathscr{B}_+$ *with* $a = b - c$ *and* $\|b\| \leqslant \alpha N(a)$.

Proof. $1 \Leftrightarrow 2$. This follows immediately from the definitions.

$1 \Rightarrow 3$. This is a consequence of the definition of N^* and Theorem 1.2.2.

$3 \Rightarrow 4$. This follows from weak*-compactness.

$4 \Rightarrow 1$. Since $N^*(\omega) \leqslant \|\omega\|^*$ the cone $\mathscr{B}^*_+$ is dominating and the norm on $\mathscr{B}$ is monotone, by Theorem 1.2.2.

The equivalence of the last four conditions is deduced in a similar fashion. □

The norm $\|\cdot\|_N$ associated with the canonical half-norm is called the *order-norm* on $\mathscr{B}$. It can be re-expressed as

$$\|a\|_N = N(a) \vee N(-a) = \inf\{\|b\| \vee \|c\|; c \leqslant a \leqslant b\}$$

and in this latter form it occurred already in Proposition 1.2.1. In particular the cone $\mathscr{B}_+$ is $1_\vee$-normal with respect to $\|\cdot\|_N$. It also follows from Proposition 1.2.1 and the definition of $\alpha_\vee$-normality that $\|\cdot\|_N$ and $\|\cdot\|$ coincide if, and only if, $\mathscr{B}_+$ is $1_\vee$-normal with respect to $\|\cdot\|$, in which case $\|\cdot\|$ is said to be an order-norm. Alternatively $\|\cdot\|_N$ and $\|\cdot\|$ are equivalent if, and only if, $\mathscr{B}_+$ is normal.

EXAMPLE 1.6.4 (Banach lattices). If $\mathscr{B}$ is a Banach lattice then $N(a) = \|a_+\|$ where a_+ is the positive component in the canonical decomposition of a. This follows because $b \geqslant a$ implies $|b| \geqslant a_+$ and hence $\|b\| \geqslant \|a_+\|$. Alternatively it follows from monotonicity of the dual norm and Condition 3* of Theorem 1.6.3. □

EXAMPLE 1.6.5 (C^*-algebras). Let $\mathscr{B}$ be the self-adjoint part of a C^*-algebra ordered by the positive elements. Each $a \in \mathscr{B}$ has a canonical decomposition into positive and negative components $a = a_+ - a_-$ where $a_\pm = (|a| \pm a)/2$ and $|a|$ is the algebraic modulus of a. Again $N(a) = \|a_+\|$. Moreover the C^*-norm satisfies $\|a\| = \|a_+\| \vee \|a_-\|$ and hence it is an order-norm. □

EXAMPLE 1.6.6 (Affine functions). Consider the space $\mathscr{B} = A(K)$ of affine functions of Example 1.1.8. Then $N(a) = \sup\{a(k) \vee 0; k \in K\}$. The norm is an order-norm. □

EXAMPLE 1.6.7 (Two-dimensional spaces). If $\mathscr{B} = \mathbb{R}^2$ and $\mathscr{B}_+ = \mathbb{R}^2_+$ the norm is an order-norm if, and only if, the left derivative of the unit sphere is everywhere negative, e.g., $\|(a_1, a_2)\| = |a_1| \vee |a_2|$. □

To conclude let us consider spaces with order-units and bases.

First, if $\operatorname{int} \mathscr{B}_+ \neq \emptyset$, or if $\operatorname{int} \mathscr{B}^*_+ \neq \emptyset$, there are natural generalizations of N, and N^*, given by

$$\hat{N}(a) = \sup\{\omega(a); \omega \in \mathscr{B}^*_+, \|\omega\|^* = 1\},$$

$$\hat{N}^*(\omega) = \sup\{\omega(a); a \in \mathscr{B}_+, \|a\| = 1\}.$$

Thus it follows from Proposition 1.6.2 that $N(a) = \hat{N}(a) \vee 0$. But if $a \in \operatorname{int} \mathscr{B}_+$ then $a - \varepsilon b \in \operatorname{int} \mathscr{B}_+$ for all $b \in \mathscr{B}_1$, and a sufficiently small $\varepsilon > 0$. Therefore $\hat{N}(-a) \leqslant -\varepsilon$, i.e., $\hat{N}$ is strictly negative on $\operatorname{int} \mathscr{B}_+$. Combining this result with the discussion at the beginning of the section one has

$$\operatorname{int} \mathscr{B}_+ = \{a; \hat{N}(-a) < 0\}.$$

In fact $\hat{N}(-a)$ is minus the distance of a from $\mathscr{B}_+ \backslash \operatorname{int} \mathscr{B}_+$, whenever $a \in \operatorname{int} \mathscr{B}_+$. Similar observations are valid for N^* and $\hat{N}^*$.

Second, suppose $\mathscr{B}_+$ is proper and $u \in \operatorname{int} \mathscr{B}_+$. There are two natural proper half-norms associated with u:

$$N_u(a) = \inf\{\lambda; \lambda \in \mathbb{R}_+, a \leqslant \lambda u\},$$

$$\hat{N}_u(a) = \inf\{\lambda; \lambda \in \mathbb{R}, a \leqslant \lambda u\}$$

and the associated norms coincide with the order-unit norm introduced in Section 1.4., i.e.,

$$\|a\|_u = N_u(a) \vee N_u(-a)$$

$$= \hat{N}_u(a) \vee \hat{N}_u(-a) = \inf\{\lambda > 0; a \in [-\lambda u, \lambda u]\}.$$

Again $N_u(a) = \hat{N}_u(a)$ if either expression is strictly positive, but $\hat{N}_u$ takes negative values and

$$\operatorname{int} \mathscr{B}_+ = \{a; \hat{N}_u(-a) < 0\}.$$

Both N_u and $\hat{N}_u$ have dual representations:

$$N_u(a) = \sup\{\omega(a); \omega \in \mathscr{B}^*_+, \omega(u) \leqslant 1\},$$

$$\hat{N}_u(a) = \sup\{\omega(a); \omega \in \mathscr{B}^*_+, \omega(u) = 1\}.$$

Moreover if $\{\mathscr{B}, \mathscr{B}_+, \|\cdot\|\}$ is an order-unit space with $\|\cdot\|_u = \|\cdot\|$ then $N_u = N$, the canonical half-norm, and $\hat{N}_u = \hat{N}$. This follows, for example, because

$$N(a) = \inf\{\lambda; \lambda \in \mathbb{R}_+, a \leqslant \lambda b \text{ for some } b \in \mathscr{B}_1\}$$

and $b \in \mathscr{B}_1$ is equivalent to $b \in [-u, u]$.

Third, if $\mathscr{B}_+$ is generating and has a base K there is a positive half-norm N_K defined by

$$N_K(a) = \inf\{\lambda; \lambda \in \mathbb{R}_+, a \leqslant \lambda b \text{ for some } b \in K\}.$$

Moreover, if λ_K is the corresponding order-unit of $\mathscr{B}^*$, so that $\lambda_K(a) = 1$ for $a \in K$ (see Proposition 1.4.2), one has

$$N_K(a) = \sup\{\omega(a); 0 \leqslant \omega \leqslant \lambda_K\},$$

$$\hat{N}_{\lambda_K}(\omega) = \sup\{\omega(a); a \in K\}.$$

One also has the connection

$$\|a\|_K = \inf\{\lambda; \lambda \in \mathbb{R}_+, a \in \lambda\, \mathrm{co}(K \cup -K)\}$$
$$= N_K(a) + N_K(-a)$$

between the base-norm $\|\cdot\|_K$ associated with K (see Section 1.4) and the half-norm N_K. Finally if $\mathscr{B}$ has an order-unit u and $K = \{\omega; \omega \in \mathscr{B}^*_+, \omega(u) = 1\}$ the corresponding base of $\mathscr{B}^*_+$ then

$$N_K(\omega) = \sup\{\omega(a); 0 \leqslant a \leqslant u\}.$$

1.7. BOUNDED OPERATORS

Let $(\mathscr{A}, \mathscr{A}_+, \|\cdot\|)$ and $(\mathscr{B}, \mathscr{B}_+, \|\cdot\|)$ be ordered Banach spaces, $\mathscr{L} = \mathscr{L}(\mathscr{A}, \mathscr{B})$ the Banach space of bounded linear operators $S; \mathscr{A} \mapsto \mathscr{B}$ equipped with the operator norm $\|\cdot\|$, and

$$\mathscr{L}_+ = \{S; S \in \mathscr{L}, S\mathscr{A}_+ \subseteq \mathscr{B}_+\}.$$

Since $\mathscr{L}_+$ is a closed convex cone $(\mathscr{L}, \mathscr{L}_+, \|\cdot\|)$ is an ordered Banach space. Operators $S \in \mathscr{L}_+$ will be referred to as *positive operators*.

If $\mathscr{A}_+$ is weakly generating and $\mathscr{B}_+$ is proper, then $\mathscr{L}_+$ is proper. However $\mathscr{L}_+$ can be shown to be (weakly) generating only under very special circumstances (see, for example, [61, 78]). Indeed if $S_1, S_2 \in \mathscr{L}_+$ and $a_1 \leqslant a \leqslant a_2$, then

$$S_1 a_1 - S_2 a_2 \leqslant (S_1 - S_2)a \leqslant S_1 a_2 - S_2 a_1.$$

Thus the difference $S = S_1 - S_2$ of two positive operators is *order-bounded*, it maps order-intervals $[a_1, a_2]$ into order intervals. But there are many examples of bounded operators which are not order-bounded, even under fairly stringent conditions on $\mathscr{A}$ and $\mathscr{B}$. One weak result in this direction is the following.

PROPOSITION 1.7.1. *If $\mathscr{A}_+$ is normal and* int $\mathscr{B}_+ \neq \emptyset$ *then every $S \in \mathscr{L}(\mathscr{A}, \mathscr{B})$ is order bounded.*

Proof. Using the notation of Proposition 1.4.3 one has

$$S(o(\mathscr{A})) \subseteq S(b(\mathscr{A})) \subseteq b(\mathscr{B}) \subseteq o(\mathscr{B}).$$

Thus S is order-bounded. □

In particular the bounded operators on an order-unit space are order-bounded. Converse results can be obtained under rather more general conditions.

PROPOSITION 1.7.2. *If $\mathscr{A}_+$ is generating and $\mathscr{B}_+$ is normal then every order-bounded linear operator $S; \mathscr{A} \mapsto \mathscr{B}$ is bounded.*

Proof. If S is not bounded, there exist $a_n \in \mathscr{A}_1$ with $\|Sa_n\| \geqslant 4^n$. Since $\mathscr{A}_+$ is $\alpha_\vee$-generating for some α, one has $a_n = a'_n - a''_n$ with $a'_n, a''_n \in \mathscr{A}_+ \cap \mathscr{A}_\alpha$. Replacing a_n by $-a_n$ if necessary it may be assumed that $\|Sa'_n\| > 2^{2n-1}$. Let $a = \Sigma 2^{-n} a'_n$. If S is order-bounded S maps $[0, a]$ into an interval $[b_1, b_2]$ so that $b_1 \leqslant 2^{-n} Sa'_n \leqslant b_2$ for all n. But this contradicts normality of $\mathscr{B}_+$. □

In particular, by choosing $\mathscr{B} = \mathbb{R}$, one deduces that every order-bounded linear functional over $\mathscr{A}$ is bounded when $\mathscr{A}_+$ is generating. Thus positive linear functionals over $\mathscr{A}$ are bounded.

Next we discuss properties of the ordered space $(\mathscr{L}, \mathscr{L}_+, \|\cdot\|)$. Throughout we assume $\mathscr{A}_+ \neq \mathscr{A}, \mathscr{B}_+ \neq \{0\}$.

THEOREM 1.7.3. *The cone* $\mathscr{L}_+$ *is normal if, and only if,* $\mathscr{A}_+$ *is generating and* $\mathscr{B}_+$ *is normal.*

Proof. Suppose $\mathscr{L}_+$ is $\alpha_\vee$-normal. For $\omega \in \mathscr{A}^*$ and $b \in \mathscr{B}$ let $\omega \otimes b$ be the rank-one operator with action

$$(\omega \otimes b)a = \omega(a)b.$$

If $\omega \in \mathscr{A}^*_+$ and $b_1 \leqslant b \leqslant b_2$ in $\mathscr{B}$ then $\omega \otimes b_1 \leqslant \omega \otimes b \leqslant \omega \otimes b_2$ and hence

$$\begin{aligned}\|\omega\|^* \|b\| &= \|\omega \otimes b\| \\ &\leqslant \alpha(\|\omega \otimes b_1\| \vee \|\omega \otimes b_2\|) = \alpha \|\omega\|^*(\|b_1\| \vee \|b_2\|),\end{aligned}$$

i.e., $\mathscr{B}_+$ is $\alpha_\vee$-normal. But if $b \in \mathscr{B}_+$ and $\omega_1 \leqslant \omega \leqslant \omega_2$ in $\mathscr{A}^*$ then $\omega_1 \otimes b \leqslant \omega \otimes b \leqslant \omega_2 \otimes b$ and a similar calculation gives

$$\|\omega\|^* \|b\| \leqslant \alpha \|b\| (\|\omega_1\|^* \vee \|\omega_2\|^*).$$

Thus $\mathscr{A}^*_+$ is $\alpha_\vee$-normal and $\mathscr{A}_+$ is approximately α_+-generating, by Theorem 1.1.4.

Conversely suppose $\mathscr{A}_+$ is α_+-generating and $\mathscr{B}_+$ is $\beta_\vee$-normal. Consider $S_1 \leqslant S \leqslant S_2$ in $\mathscr{L}$ and $a \in \mathscr{A}$. Then $a = a_1 - a_2$ where $a_1, a_2 \in \mathscr{A}_+$ and $\|a_1\| + \|a_2\| \leqslant \alpha \|a\|$. Now $S_1 a_j \leqslant S a_j \leqslant S_2 a_j$ so

$$\begin{aligned}\|S a_j\| &\leqslant \beta(\|S_1 a_j\| \vee \|S_2 a_j\|) \\ &\leqslant \beta \|a_j\| (\|S_1\| \vee \|S_2\|),\end{aligned}$$

and consequently

$$\begin{aligned}\|Sa\| &\leqslant \|Sa_1\| + \|Sa_2\| \\ &\leqslant \alpha\beta \|a\| (\|S_1\| \vee \|S_2\|).\end{aligned}$$

Thus $\mathscr{L}_+$ is $(\alpha\beta)_\vee$-normal. □

If, for example, one chooses $\mathscr{A} = \mathscr{B}$, then the theorem states that the cone $\mathscr{L}_+(\mathscr{B})$ of positive bounded linear operators is normal, if and only if, $\mathscr{B}_+$ is normal and generating. There is also a simple criterion for absolute monotonicity of the operator norm.

THEOREM 1.7.4. *The operator norm on* $\mathscr{L}$ *is absolutely monotone if, and only if,* $\mathscr{A}_+$ *is approximately absolutely dominating and the norm on* $\mathscr{B}$ *is absolutely monotone.*

Proof. The proof of necessity is almost identical to the proof in Theorem 1.7.3 except that one now uses $\omega_2 = -\omega_1, b_2 = -b_1$ and Theorem 1.3.1 replaces Theorem 1.1.4.

Conversely suppose that $\mathscr{A}_+$ is approximately absolutely dominating and the norm on $\mathscr{B}$ is absolutely monotone. Consider $S, T \in \mathscr{L}$ with $\pm S \leqslant T$. For $a \in \mathscr{A}$ and $\alpha > 1$ there exist $a_1, a_2 \in \mathscr{A}_+$ with $a = a_1 - a_2$ and $\| a_1 + a_2 \| \leqslant \alpha \| a \|$. Therefore

$$- Ta_1 - Ta_2 \leqslant Sa_1 - Sa_2 \leqslant Ta_1 + Ta_2.$$

Hence

$$\| Sa \| \leqslant \| T(a_1 + a_2) \| \leqslant \alpha \| T \| \| a \|$$

and $\| S \| \leqslant \alpha \| T \|$ for all $\alpha > 1$. Thus the operator norm is absolutely monotone. □

COROLLARY 1.7.5. *The operator norm on $\mathscr{L}(\mathscr{B}, \mathscr{B})$ is absolutely monotone if, and only if, the norm on $\mathscr{B}$ is a Riesz norm.*

This follows by setting $\mathscr{A} = \mathscr{B}$ in Theorem 1.7.4 and using the definition of a Riesz norm. The situation concerning monotonicity of the operator norm is not so clear.

PROPOSITION 1.7.6. *If the operator norm on $\mathscr{L}(\mathscr{A}, \mathscr{B})$ is monotone then $\mathscr{A}_+$ is approximately dominating and the norm on $\mathscr{B}$ is monotone.*

Proof. The proof is almost identical to the first part of the proof of Theorem 1.7.3 except that now $\omega_1 = 0$, $b_1 = 0$, and Theorem 1.2.2 replaces Theorem 1.1.4. □

In particular if the operator norm on $\mathscr{L}(\mathscr{B}, \mathscr{B})$ is monotone, then both the norm $\| \cdot \|$ on $\mathscr{B}$ and the dual norm $\| \cdot \|^*$ on $\mathscr{B}^*$ are monotone. But the converse, and hence the converse of Proposition 1.7.6, is false.

EXAMPLE 1.7.7. Let $\mathscr{B} = \mathbb{R}^2$, $\mathscr{B}_+ = \mathbb{R}^2_+$, and

$$\| (a_1, a_2) \| = | a_1 | \vee | a_2 | \vee | a_1 - a_2/2 |.$$

Then $\mathscr{B}_+$ is dominating and the norm is monotone. Define S, and T, in $\mathscr{L}(\mathscr{B}, \mathscr{B})$ by $S(a_1, a_2) = (a_2, a_1)$, and $T(a_1, a_2) = (a_2, a_1 + a_2/6)$. Then $0 \leqslant S \leqslant T$, $\| S \| = 5/4$, and $\| T \| = 7/6$, so the operator norm is not monotone.

More generally if the norm on $\mathscr{B}$ is monotone and the unit sphere in the first quadrant is symmetric about the line $a_1 = a_2$ then the operator norm is monotone if, and only if, the unit sphere in the second quadrant is symmetric about the line $a_1 = -a_2$. □

Monotonicity of the operator norm is related to a seemingly different property, positive attainment. To introduce this concept it is convenient to define

$$\| S \|_+ = \sup \{ \| Sa \| ; a \in \mathscr{A}_1 \cap \mathscr{A}_+ \}.$$

If $\mathscr{A}_+$ is weakly generating then $\| \cdot \|_+$ defines a norm on $\mathscr{L}(\mathscr{A}, \mathscr{B})$ with $\| S \|_+ \leqslant \| S \|$. Moreover if $\mathscr{A}_+$ is α_+-generating, then $\| S \| \leqslant \alpha \| S \|_+$ and the two norms $\| \cdot \|$, $\| \cdot \|_+$, are equivalent. If $\| S \|_+ = \| S \|$ for all $S \in \mathscr{L}_+$, the operator norm will be said to be *positively attained*.

PROPOSITION 1.7.8. *If either of the following conditions is satisfied,*

1. *$\mathscr{A}_+$ is approximately dominating and $\mathscr{B}_+$ is $1_\vee$-normal,*

2. *$\mathscr{A}_+$ is approximately absolutely dominating and the norm on $\mathscr{B}$ is absolutely monotone,*

then the operator norm on $\mathscr{L}(\mathscr{A}, \mathscr{B})$ is positively attained. Conversely, if the operator norm is positively attained, then $\mathscr{A}_+$ is approximately dominating.

Proof. We prove statement 1. The proof of statement 2 is identical except that $a_1 = a_2$.

Take $S \in \mathscr{L}_+$, $a \in \mathscr{A}$, and $\alpha > 1$. There exist $a_1, a_2 \in \mathscr{A}_+$ with $-a_2 \leqslant a \leqslant a_1$ and $\|a_i\| \leqslant \alpha \|a\|$. Then $-Sa_2 \leqslant Sa \leqslant Sa_1$ and

$$\|Sa\| \leqslant \|Sa_1\| \vee \|Sa_2\| \leqslant \alpha \|a\| \|S\|_+ .$$

Consequently $\|S\| \leqslant \|S\|_+$ and the norm is positively attained.

Conversely if the operator norm is positively attained, then for $\omega \in \mathscr{A}^*_+$ and $b \in \mathscr{B}_+$

$$\|\omega\|^* \|b\| = \|\omega \otimes b\| = \|\omega \otimes b\|_+ = N^*(\omega) \|b\|$$

by Proposition 1.6.2. So $\|\omega\|^* = N^*(\omega)$ for all $\omega \in \mathscr{A}^*_+$ and $\mathscr{A}_+$ is approximately dominating by Theorems 1.2.2 and 1.6.3. □

If the operator norm is absolutely monotone, then it is positively attained by Theorem 1.7.4 and Proposition 1.7.8. Conversely if the operator norm is positively attained, and the norm on $\mathscr{B}$ is monotone, then $0 \leqslant S \leqslant T$ implies $\|S\| = \|S\|_+ \leqslant \leqslant \|T\|_+ = \|T\|$ so the operator norm is monotone. Moreover if $\mathscr{B} = \mathbb{R}$, then $\mathscr{L} = \mathscr{A}^*$ and the operator norm is positively attained if, and only if, it is monotone, by Theorem 1.2.3.

It seems reasonable to expect that if the operator norm is monotone then it is positively attained. But this has only been established in two special cases,

THEOREM 1.7.9. *Suppose that the operator norm on $\mathscr{L}(\mathscr{A}, \mathscr{B})$ is monotone and*

either int $\mathscr{B}_+ \neq \emptyset$

or $\mathscr{A}_+$ *has a base.*

It follows that the operator norm is positively attained.

Proof. The proofs of the two statements are very similar and are based on three lemmas which are wholly or partly independent of the assumptions of the theorem.

LEMMA 1.7.10. *Let $S \in \mathscr{L}_+$, $a \in \mathscr{A}$, and let N be the canonical half-norm on $\mathscr{A}$. Then*

$$N(a)(\|S\| + \|S\|_+) \geqslant \|Sa\| - \|S\|_+ \|a\|.$$

Proof. If $a \leqslant a'$ then

$$\begin{aligned}\|Sa\| &\leqslant \|Sa'\| + \|S(a' - a)\| \\ &\leqslant (\|S\| + \|S\|_+)\|a'\| + \|S\|_+ \|a\|.\end{aligned}$$

But $N(a) = \inf\{\|a'\| ; a' \geqslant a\}$. □

LEMMA 1.7.11. *Suppose that* $\mathscr{A}_+$ *is approximately dominating and the norm on* $\mathscr{B}$ *is monotone. If* $S\in\mathscr{L}_+$, *then*

$$\|S\|_+ = \|S^*\|_+.$$

Proof. First note that the norm on $\mathscr{A}^*$ is monotone by Theorem 1.2.2. Now

$$\begin{aligned}\|S\|_+ &= \sup\{\|Sa\|;\, a\in\mathscr{A}_1\cap\mathscr{A}_+\}\\ &= \sup\{\omega(Sa);\, a\in\mathscr{A}_1\cap\mathscr{A}_+,\, \omega\in\mathscr{B}^*_1\cap\mathscr{B}^*_+\}\\ &= \sup\{(S^*\omega)(a);\, a\in\mathscr{A}_1\cap\mathscr{A}_+,\, \omega\in\mathscr{B}^*_1\cap\mathscr{B}^*_+\}\\ &= \sup\{\|S^*\omega\|^*;\, \omega\in\mathscr{B}^*_1\cap\mathscr{B}^*_+\} = \|S^*\|_+\end{aligned}$$

where the second and fourth equalities use Theorem 1.2.3. □

LEMMA 1.7.12. *Suppose that the operator norm is monotone,* $S\in\mathscr{L}_+$, $\|S\|=1$, *and* $\|S\|_+ = 1-2\delta$ *where* $\delta>0$. *Then for* $\varepsilon>0$ *and* $\theta<\varepsilon\delta^2(1-\delta)^{-2}$ *there exist* $\omega\in\mathscr{A}^*_+$ *and* $b\in\mathscr{B}_+$ *such that* $\|\omega\|^*\|b\|<\varepsilon$ *and* $\|S+\omega\otimes b\|>1+\theta$.

Proof. For $0<\delta'<\delta$ there exists $a\in\mathscr{A}_1$ with $\|Sa\|>1-\delta'$. Therefore $N(a)\geqslant$ $\geqslant(\delta-\delta'/2)(1-\delta)^{-1}$, by Lemma 1.7.10. Hence there exists an $\omega\in\mathscr{B}^*_1\cap\mathscr{B}^*_+$ with $\omega(a)\geqslant(\delta-\delta')(1-\delta)^{-1}$, by Proposition 1.6.2.

Next there exists an $\eta\in\mathscr{B}^*_1$ with $\eta(Sa)=\|Sa\|>1-\delta'$ so $\|S^*\eta\|^*>1-\delta'$. Therefore, applying a similar argument to S^* and η, instead of S and a, one deduces that there exists a $b\in\mathscr{B}_\varepsilon\cap\mathscr{B}_+$ with $\eta(b)\geqslant\varepsilon(\delta-\delta')(1-\delta)^{-1}$. Now

$$\begin{aligned}\|S+\omega\otimes b\| &\geqslant \eta((S+\omega\otimes b)(a))\\ &> 1-\delta'+\omega(a)\eta(b)\\ &> 1-\delta'+\varepsilon(\delta-\delta')^2(1-\delta)^{-2} > 1+\theta,\end{aligned}$$

if δ' is sufficiently small. □

Now we return to the proof of Theorem 1.7.9. First assume $\operatorname{int}\mathscr{B}_+\neq\emptyset$ and let $u\in\operatorname{int}\mathscr{B}_+$ with $\|u\|=1$. Therefore there is a $\beta>0$ such that $b\leqslant u$ for all $b\in\mathscr{B}_\beta$. Now suppose the operator norm is not positively attained. Thus there exists an $S\in\mathscr{L}_+$ with $\|S\|=1$ and $\|S\|_+=1-2\delta<1$. Then by Lemma 1.7.12 there exist $\theta>0$, $\omega\in\mathscr{A}^*_+\cap\mathscr{A}^*_1$, and $b\in\mathscr{B}_+\cap\mathscr{B}_{\beta\delta}$, with $\|S+\omega\otimes b\|=1+\theta$. Moreover $0\leqslant$ $\leqslant b\leqslant\delta u$. Now let $S_1=S+\delta\omega\otimes u$ so

$$\begin{aligned}\|S_1\| &\geqslant \|S+\omega\otimes b\| = 1+\theta,\\ \|S_1\|_+ &\leqslant \|S\|_+ + \delta = 1-\delta.\end{aligned}$$

Next choose $a\in\mathscr{A}_1$ such that $\|S_1a\|>1+\theta-\delta'$ where $0<\delta'<\delta$. Replacing a by $-a$, if necessary, one may assume $\omega(a)\geqslant 0$. Now there exists $\eta\in\mathscr{B}^*_1$ such that $\eta(S_1a)>$ $>1+\theta-\delta'$ and hence

$$\begin{aligned}\delta\omega(a) &\geqslant \delta\omega(a)\eta(u) =\\ &= \eta((S_1-S)a) > \theta-\delta'.\end{aligned}$$

But applying Lemma 1.7.10 to S_1^* and Proposition 1.7.6 and Lemma 1.7.11 to S_1 one has

$$N^*(-\eta) \geqslant (\delta + \theta - \delta')(2 + \theta - \delta')^{-1} > (\delta - \delta')/2.$$

By Proposition 1.6.2 there exists $b' \in \mathscr{B}_+ \cap \mathscr{B}_\beta$ with $\eta(b') < -\beta(\delta - \delta')/2$. Let $S_2 = S_1 - \delta\omega \otimes b' = S + \delta\omega \otimes (u - b')$ so $0 \leqslant S \leqslant S_2 \leqslant S_1$. But

$$\begin{aligned}\|S_2\| &\geqslant \eta(S_2 a) = \eta(S_1 a) - \delta\omega(a)\eta(b') \\ &> 1 + \theta - \delta' + (\theta - \delta')\beta(\delta - \delta')/2 \\ &> 1 + \theta = \|S_1\|\end{aligned}$$

for δ' sufficiently small. This contradicts the monotonicity of the operator norm and hence proves the first statement of the theorem.

Now suppose $\mathscr{A}_+$ has a base. Therefore $\mathscr{A}_+^*$ has an interior point ρ with $\|\rho\|^* = 1$, by Proposition 1.4.3. Hence there is a $\beta > 0$ such that $\omega \leqslant \rho$ for all $\omega \in \mathscr{A}_\beta^*$. Again suppose the operator norm is not positively attained, and hence there is an $S \in \mathscr{L}_+$ with $\|S\| = 1$ and $\|S\|_+ = 1 - 2\delta < 1$. There also exist $\theta > 0$, $\omega \in \mathscr{A}_+^* \cap \mathscr{A}_{\beta\delta}^*$, and $b \in \mathscr{B}_+ \cap \mathscr{B}_1$, with $\|S + \omega \otimes b\| = 1 + \theta$. Then $0 \leqslant \omega \leqslant \delta\rho$. Let $S_1 = S + \delta\rho \otimes b$ so $\|S_1\| > 1 + \theta$, and $\|S_1\|_+ \leqslant 1 - \delta$. Next choose $a \in \mathscr{A}_1$ such that $\|S_1 a\| > > 1 + \theta - \delta'$ where $0 < \delta' < \delta$. There exists an $\eta \in \mathscr{B}_1^*$ such that $\eta(S_1 a) > 1 + \theta - \delta'$, and we may assume that $\eta(b) \geqslant 0$. Now

$$\begin{aligned}\delta\eta(b) &\geqslant \delta\rho(a)\eta(b) \\ &= \eta((S_1 - S)a) > \theta - \delta'.\end{aligned}$$

But applying Lemma 1.7.10 to S_1 one obtains $N(-a) > (\delta - \delta')/2$. By Proposition 1.6.2 there exists $\omega' \in \mathscr{A}_+^* \cap \mathscr{A}_\beta^*$ with $\omega'(a) < -\beta(\delta - \delta')/2$. Let $S_2 = S_1 - \delta\omega' \otimes b = = S + \delta(\rho - \omega') \otimes b$ so $0 \leqslant S \leqslant S_2 \leqslant S_1$. But

$$\begin{aligned}\|S_2\| &\geqslant \eta(S_2 a) = \eta(S_1 a) - \delta\omega'(a)\eta(b) \\ &> 1 + \theta - \delta' + \beta(\delta - \delta')(\theta - \delta')/2 \\ &> \|S_1\|,\end{aligned}$$

if δ' is sufficiently small. Again this contradicts the monotonicity of the operator norm and hence proves the second statement of Theorem 1.7.9. □

2. Positive Semigroups

2.0. INTRODUCTION

In this part we review the basic theory of positive one-parameter semigroups, i.e., semigroups of bounded linear operators which act on an ordered Banach space and respect the order. The first objective is an infinitesimal characterization of such semigroups in terms of generators. Subsequently we discuss stricter notions of positivity, irreducibility criteria, and spectral properties.

If $(\mathscr{B}, \|\cdot\|)$ is a Banach space then a family $S = \{S_t\}_{t\geqslant 0}$ of bounded linear operators on $\mathscr{B}$ is defined to be a C_0-*semigroup* if it satisfies

1. $S_s S_t = S_{s+t}, \quad s, t \geqslant 0,$

2. $S_0 = I,$

3. $\lim_{t\to 0+} \|S_t a - a\| = 0, \quad a \in \mathscr{B}.$

Moreover if $\mathscr{B}$ is ordered by the positive cone $\mathscr{B}_+$ then S is defined to be *positive* if it also has the property

4. $S_t \mathscr{B}_+ \subseteq \mathscr{B}_+, \quad t > 0.$

Similarly if $(\mathscr{B}^*, \|\cdot\|^*)$ is a dual Banach space then a family $T = \{T_t\}_{t\geqslant 0}$ of bounded linear operators on $\mathscr{B}^*$ is defined to be a C_0^*-*semigroup* if it satisfies properties 1 and 2 above together with the weak*-continuity conditions

3*. a. $t \geqslant 0 \mapsto (T_t\omega)(a)$ is continuous for all $\omega \in \mathscr{B}^*$ and $a \in \mathscr{B}$,
b. $\omega \in \mathscr{B}^* \mapsto (T_t\omega)(a)$ is weak*-continuous for all $t \geqslant 0$ and $a \in \mathscr{B}$.

Again if $\mathscr{B}^*$ is ordered by the cone $\mathscr{B}^*_+$ then T is defined to be *positive* if one also has the property

4*. $T_t \mathscr{B}^*_+ \subseteq \mathscr{B}^*_+, \quad t > 0.$

These two types of semigroup are related by duality.

If $S = \{S_t\}_{t\geqslant 0}$ is a C_0-semigroup on $\mathscr{B}$ then the adjoint operators, $S^* = \{S_t^*\}_{t\geqslant 0}$ on $\mathscr{B}^*$, form a C_0^*-semigroup. Conversely if $T = \{T_t\}_{t\geqslant 0}$ is a C_0^*-semigroup on $\mathscr{B}^*$ then there exists a C_0-semigroup $T^* = \{T_t^*\}_{t\geqslant 0}$ on $\mathscr{B}$ which is adjoint to T, i.e., $(T_t^*)^* = T_t$ for all $t > 0$. This duality also extends to the generators of the semigroups.

The generator of a C_0-semigroup S on $\mathscr{B}$ is defined as the linear operator H whose domain $D(H)$ consists of those $a \in \mathscr{B}$ for which there exists a $b \in \mathscr{B}$ such that

$$\lim_{t\to 0+} \|(I - S_t)a/t - b\| = 0$$

and the action of H is then defined by $Ha = b$. The generator of a C_0^*-semigroup is defined similarly but with a weak*-derivative. Explicitly the generator K of the C_0^*-semigroup T on $\mathscr{B}^*$ has a domain $D(K)$ consisting of those $\omega \in \mathscr{B}^*$ for which there is an $\eta \in \mathscr{B}^*$ such that

$$\lim_{t\to 0+} ((I - T_t)\omega)(a)/t = \eta(a)$$

for all $a \in \mathscr{B}$ and then $K\omega = \eta$. For example, if S is a C_0-semigroup with generator H then the C_0^*-semigroup S^* has generator H^*.

The basic structural theorems of semigroup theory characterize those operators which generate C_0- and C_0^*-semigroups. The following statement incorporates both versions. Note that $R(X)$ denotes the range of X.

THE FELLER–MIYADERA–PHILLIPS THEOREM. *Let $\mathscr{B}$ be a Banach space (with a predual $\mathscr{B}_*$). The following conditions are equivalent:*

1. (1*). *H generates a C_0- (C_0^*-) semigroup,*
2. (2*). *H is a norm- (weak*-) densely defined norm- (weak*-) closed linear operator with*

$$R(I + \beta H) = \mathscr{B}$$

and

$$\|(I + \alpha H)^n a\| \geq (1 - \alpha\gamma)^n \|a\|/M$$

for all $0 < \alpha \leq \beta$, $a \in D(H^n)$ and $n \geq 1$, for some $M \geq 1$, $\gamma \in \mathbb{R}$, $\beta > 0$, with $\beta\gamma < 1$.

Moreover if these conditions are satisfied $\|S_t\| \leq M \exp\{\gamma t\}$.

The simpler and earlier version of this theorem for contraction semigroups, i.e., semigroups with $\|S_t\| \leq 1$ is as follows:

THE HILLE–YOSIDA THEOREM. *Let $\mathscr{B}$ be a Banach space (with a predual $\mathscr{B}_*$). The following conditions are equivalent:*

1. (1*). *H generates a C_0- (C_0^*-) contraction semigroup,*

   (2*). *H is a norm- (weak*-) densely defined norm- (weak*-) closed linear operator with*

$$R(I + \beta H) = \mathscr{B}$$

and

$$\|(I + \alpha H)a\| \geq \|a\|$$

for all $0 < \alpha \leq \beta$ and all $a \in D(H)$, for some $\beta > 0$.

In both these theorems the criterion for a generator ensures that $R(I + \alpha H) = \mathscr{B}$ for all small $\alpha > 0$ and the resolvents $(I + \alpha H)^{-1}$ exist with suitable bounds, e.g., in the Feller–Miyadera–Phillips theorem $\|(I + \alpha H)^{-n}\| \leq M(1 - \alpha\gamma)^{-n}$. Moreover, in all cases the semigroup S is constructible from the resolvents by a limit

$$S_t = \lim_{n \to \infty} (I + tH/n)^{-n}$$

in the strong, or pointwise weak*, topology.

In Sections 2.2 and 2.3 we derive various versions of these theorems for positive semigroups on suitable ordered Banach spaces, e.g., spaces with a Riesz norm or spaces with int $\mathscr{B}_+ \neq \emptyset$. In all these variants inequalities of the type $\|(I + \alpha H)a\| \geq \|a\|$

are replaced by analogous inequalities with respect to a half-norm. As a preliminary we examine these bounds in Section 2.1.

2.1. DISSIPATIVE OPERATORS

Throughout this section H denotes a norm-densely defined linear operator on a real Banach space $\mathscr{B}$ with domain $D(H)$. We are principally interested in operators which satisfy a dissipativity condition with respect to a half-norm p. There are a variety of equivalent definitions of this dissipativity, some expressed in terms of the sub-differentials of p, i.e., the sets of tangent functionals to p.

The *subdifferential* $\mathrm{d}p(a)$ of the half-norm p at the point $a \in \mathscr{B}$ is defined by

$$\mathrm{d}p(a) = \{\omega;\, \omega \in \mathscr{B}^*,\, \omega \leqslant p,\, \omega(a) = p(a)\}.$$

It follows from the Hahn–Banach theorem, Theorem A1 of the appendix, that $\mathrm{d}p(a)$ is non-empty and for $b \in \mathscr{B}$, $\lambda \in \mathbb{R}$, there exists $\omega \in \mathrm{d}p(a)$ with $\omega(b) = \lambda$ if, and only if,

$$(p(a) - p(a - tb))/t \leqslant \lambda \leqslant (p(a + tb) - p(a))/t \qquad (*)$$

for all $t > 0$. In fact since $t \mapsto p(a + tb)$ is convex one need only consider this relation in the limit $t \to 0+$.

Although $\mathrm{d}p(a)$ may contain more than one functional it cannot be large except for a 'few' elements $a \in \mathscr{B}$. Indeed Mazur's theorem [41] shows that if $\mathscr{B}$ is separable then there is a norm-dense set of $a \in \mathscr{B}$ at which $\mathrm{d}p(a)$ consists of a single functional. This fact makes the equivalence of Conditions 2 and 3 in the following theorem a little less surprising.

THEOREM 2.1.1. *Let H; $D(H) \mapsto \mathscr{B}$ be a norm densely-defined linear operator on the real Banach space $\mathscr{B}$ and p a half-norm on $\mathscr{B}$.*

The following conditions are equivalent:

1. $p((I + \alpha H)a) \geqslant p(a)$

for all (small) $\alpha > 0$, and all $a \in D(H)$,

2. $\omega(Ha) \geqslant 0$

for some $\omega \in \mathrm{d}p(a)$, and for all $a \in D(H)$,

3. $\omega(Ha) \geqslant 0$

for all $\omega \in \mathrm{d}p(a)$, and for all $a \in D(H)$.

Proof. $1 \Leftrightarrow 2$. This equivalence follows from the above relation (*) with the choice $b = Ha$.

$3 \Rightarrow 2$. This is obvious.

$1 \Rightarrow 3$. For $a, b \in D(H)$ and $t > 0$

$$\begin{aligned} p(a - tHa) &\leqslant p(a - tb) + tp(b - Ha) \\ &\leqslant p((1 + tH)(a - tb)) + tp(b - Ha) \\ &\leqslant p(a) + tp(Ha - b) + tp(b - Ha) + t^2 p(-Hb). \end{aligned}$$

Hence

$$\lim_{t \to 0_+} (p(a) - p(a - tHa))/t \geqslant -p(Ha - b) - p(b - Ha).$$

Since $D(H)$ is norm-dense and p is continuous the right-hand side may be made arbitrarily small. Therefore

$$\lim_{t \to 0+} (p(a) - p(a - tHa))/t \geqslant 0.$$

Now Condition 3 follows from (*), with $b = Ha$. □

A norm-densely defined linear operator H on the real Banach space $\mathscr{B}$ is defined to be *p-dissipative* if the equivalent conditions of Theorem 2.1.1 are satisfied. The term dissipative arises in the context of C_0-contraction semigroups; dissipativity is an infinitesimal form of contractivity.

Let S be a C_0-semigroup with generator H and assume that S is p-contractive in the sense

$$p(S_t a) \leqslant p(a)$$

for all $t \geqslant 0$ and $a \in \mathscr{B}$. Then if $\omega \in \mathrm{d}p(a)$ one has

$$\omega(S_t b) \leqslant p(S_t b) \leqslant p(b)$$

for all $b \in \mathscr{B}$ and

$$\omega(S_t a) \leqslant \omega(a)$$

because $p(a) = \omega(a)$. Thus for $a \in D(H)$

$$\omega(Ha) = \lim_{t \to 0_+} \omega((I - S_t)a)/t \geqslant 0.$$

i.e., H is p-dissipative. Conversely if H is p-dissipative, then by Condition 1 of Theorem 2.1.1, and the algorithm for S given in the introduction,

$$\begin{aligned} p(S_t a) &= \lim_{n \to \infty} p((I + tH/n)^{-n} a) \\ &\leqslant p(a) \end{aligned}$$

for all $a \in \mathscr{B}$, i.e., S is p-contractive. In particular a C_0-semigroup is a contraction semigroup if, and only if, its generator is norm-dissipative. This is in fact part of the statement of the Hille–Yosida theorem because norm-dissipativity corresponds to the bounds

$$\|(I + \alpha H)a\| \geqslant \|a\|$$

for all $a \in D(H)$ and all small $\alpha > 0$. Norm-dissipative operators are usually referred to simply as *dissipative* operators.

Another special example of dissipativity is for the canonical half-norm N on a Banach lattice $\mathscr{B}$. Then $N(a) = \|a_+\|$ by Example 1.6.4 and $\omega \leqslant N$ is equivalent to $\omega \in \mathscr{B}^*_+ \cap \mathscr{B}^*_1$ by Lemma 1.6.1. Therefore

$$\mathrm{d}N(a) = \{\omega; \omega \in \mathscr{B}^*_+ \cap \mathscr{B}^*_1, \omega(a) = \|a_+\|\}.$$

Thus H is N-dissipative if, and only if, for each $a \in D(H)$ there exists an $\omega \in \mathscr{B}^*_+ \cap \mathscr{B}^*_1$ with $\omega(a) = \|a_+\|$ and $\omega(Ha) \geqslant 0$. Such operators were called *dispersive* by Phillips [62].

It is an interesting question whether dispersive operators H on a Banach lattice are automatically dissipative. More generally it is of interest whether N-dissipative operators on a space with a Riesz norm are automatically dissipative. A short argument shows that this is the case if $I + \alpha H$ has a norm-dense range for all small $\alpha > 0$, but in this case much more can be said as will be seen in the next section. Furthermore, if $\mathscr{B}$ has the property that $N(\pm a) \leqslant N(\pm b)$ implies $\|a\| \leqslant \|b\|$, then N-dissipative operators are dissipative. Standard Banach lattices such as $C(X)$ and $L^p(X; \mathrm{d}\mu)$ have this property, as does the self-adjoint part of a C^*-algebra, or, more generally, an order-unit space.

Next we examine some simple properties of p-dissipative operators. In the sequel we encounter operators H for which there is a $\gamma \in \mathbb{R}$ such that $H + \gamma I$ is p-dissipative. In this context the following observation is useful.

PROPOSITION 2.1.2. *The following conditions are equivalent:*

1. *$H + \gamma I$ is p-dissipative,*
2. $$p((I + \alpha H)a) \geqslant (1 - \alpha\gamma)p(a)$$

   for all $\alpha > 0$, with $1 - \alpha\gamma > 0$, and all $a \in D(H)$.

Proof. The proof is an immediate consequence of the identity

$$p((I + \alpha H)a) = (1 - \alpha\gamma)p((I + \alpha(1 - \alpha\gamma)^{-1}(H + \gamma I))a)$$

which is valid for all α such that $1 - \alpha\gamma > 0$, and all $a \in D(H)$. □

The next result shows that p-dissipative operators are well-behaved.

THEOREM 2.1.3. *Let H be a p-dissipative operator, where p is a proper half-norm on $\mathscr{B}$. Then H is norm-closable and its closure $\bar{H}$ is p-dissipative.*

Proof. Suppose that $a_n \in D(H)$, $\|a_n\| \to 0$, $\|Ha_n - b\| \to 0$ and $b' \in D(H)$. For $t > 0$

$$\begin{aligned} p(a_n - tb) &\leqslant p(a_n - tb') + tp(b' - b) \\ &\leqslant p((I + tH)(a_n - tb')) + tp(b' - b) \\ &\leqslant p(a_n) + tp(b - b') + tp(b' - b) + tp(Ha_n - b) + t^2 p(-Hb'). \end{aligned}$$

It now follows from the continuity of p that in the limit $n \to \infty$

$$tp(-b) \leqslant tp(b - b') + tp(b' - b) + t^2 p(-Hb').$$

Dividing by t and taking the limit $t \to 0$ then gives

$$p(-b) \leqslant p(b - b') + p(b' - b).$$

But since $D(H)$ is norm-dense and p is continuous the right-hand side may be made arbitrarily small by suitable choice of b'. Therefore $p(-b) = 0$ and similarly $p(b) = 0$. Since p is proper it follows that $b = 0$ and H is closable.

The final statement of the theorem follows from Condition 1 of Theorem 2.1.1. □

Theorem 2.1.3 has one simple but practical implication for norm-closed operators. It is only necessary to verify the p-dissipativity conditions on a core of the operator H. This simplifies, for example, the discussion of partial differential operators for which one can usually choose a core of smooth functions.

There is a stronger version of Theorem 2.1.3. Suppose there exist $\lambda \in \mathbb{R}$ and $\varepsilon > 0$ such that for each $a \in D(H)$ with $p(a) = 1$ there is some $\omega \in \mathrm{d}p(a)$ with $|\omega(Ha) - \lambda| > \varepsilon$. Then H is norm-closable. A proof of this fact, in the case $p(\cdot) = \|\cdot\|$, may be found in [7].

We conclude this general discussion with some observations on order-unit spaces. Let $(\mathscr{B}, \mathscr{B}_+, \|\cdot\|)$ be an ordered Banach space for which $\mathscr{B}_+$ is proper and $\operatorname{int} \mathscr{B}_+ \neq \emptyset$. Let H be a norm-densely defined linear operator on $\mathscr{B}$. Hence there must exist a $u \in D(H) \cap \operatorname{int} \mathscr{B}_+$. Let N_u and $\hat{N}_u$ be the associated half-norms introduced in Section 1.6. Thus

$$N_u(a) = \inf\{\lambda; \lambda \in \mathbb{R}_+, a \leqslant \lambda u\},$$
$$\hat{N}_u(a) = \inf\{\lambda; \lambda \in \mathbb{R}, a \leqslant \lambda u\}.$$

PROPOSITION 2.1.4. *For each $u \in D(H) \cap \operatorname{int} \mathscr{B}_+$ the following conditions are equivalent:*

1. *if $a \in D(H)$ and $(I + \alpha H)a \in \mathscr{B}_+$ then $a \in \mathscr{B}_+$, for all small $\alpha > 0$,*
2. *$H + \gamma I$ is N_u-dissipative for some (or for all) $\gamma \geqslant \hat{N}_u(-Hu)$,*
3. *if $a \in D(H) \cap \mathscr{B}_+$, $\omega \in \mathscr{B}_+^*$, and $\omega(a) = 0$ then $\omega(Ha) \leqslant 0$.*

Proof. 1 ⇒ 2. Let $\gamma \geqslant \hat{N}_u(-Hu)$ then $(I + \alpha H)u \geqslant (1 - \alpha\gamma)u$. But

$$\begin{aligned} N_u((I + \alpha H)a) &= \inf\{\lambda; \lambda \geqslant 0, (I + \alpha H)a \leqslant \lambda u\} \\ &\geqslant \inf\{\lambda; \lambda \geqslant 0, (I + \alpha H)a \leqslant \lambda(1 - \alpha\gamma)^{-1}(I + \alpha H)u\} \\ &\geqslant \inf\{\lambda; \lambda \geqslant 0, a \leqslant \lambda(1 - \alpha\gamma)^{-1} u\} \\ &= (1 - \alpha\gamma)N_u(a) \end{aligned}$$

and $H + \gamma I$ is N_u-dissipative by Proposition 2.1.2.

2 ⇒ 1. If $(I + \alpha H)a \in \mathscr{B}_+$ then

$$0 = N_u(-(I + \alpha H)a) \geqslant (1 - \alpha\gamma)N_u(-a),$$

by Proposition 2.1.2. Thus $a \in \mathscr{B}_+$ provided $1 - \alpha\gamma > 0$.

2 ⇒ 3. It is easily checked that

$$\mathrm{d}N_u(a) = \{\omega; \omega \in \mathscr{B}_+^*, \omega(u) \leqslant 1, \omega(a) = N_u(a)\}.$$

Hence if $\omega \neq 0$ satisfies the hypotheses of Condition 3, then $\omega/\omega(u) \in \mathrm{d}N_u(-a)$. Therefore $\omega((H+\gamma I)(-a)) \geqslant 0$ by N_u-dissipativity and consequently $\omega(Ha) \leqslant 0$.

3⇒2. Let $\gamma \geqslant \hat{N}_u(-Hu)$, so $Hu \geqslant -\gamma u$. If $a \in B$ and $\omega \in \mathrm{d}N_u(a)\backslash\{0\}$ then $b = uN_u(a)/\omega(u) - a \in \mathscr{B}_+$, $\omega \in \mathscr{B}^*_+$, and $\omega(b) = 0$. Hence

$$\begin{aligned} 0 &\geqslant \omega(Hb) \geqslant -\gamma N_u(a) - \omega(Ha) \\ &= -\omega((H+\gamma I)a) \end{aligned}$$

Thus $H + \gamma I$ is N_u-dissipative. □

Condition 1 of Proposition 2.1.4 is a statement of positivity of $(I+\alpha H)^{-1}$, whenever this inverse operator exists, whilst Condition 3 states that H has 'negative off-diagonal elements'. For example if $\mathscr{B} = \mathbb{R}^n$, $\mathscr{B}_+ = \mathbb{R}^n_+$, and $H = (H_{ij})$ is an $n \times n$-matrix, then Condition 3 is equivalent to $H_{ij} \leqslant 0$ for $i \neq j$. Hence we refer to Condition 3 as the *negative off-diagonal property*.

Note that if N is the canonical half-norm and H is N-dissipative, then it has the negative off-diagonal property. This follows because

$$\mathrm{d}N(a) = \{\omega;\ \omega \in \mathscr{B}^*_+ \cap \mathscr{B}^*_1,\ \omega(a) = N(a)\}$$

by Lemma 1.6.1. Hence if $a \in D(H) \cap \mathscr{B}_+$, $\omega \in \mathscr{B}^*_+ \cap \mathscr{B}^*_1$, and $\omega(a) = 0$ then $\omega \in \mathrm{d}N(-a)$ and $\omega(-Ha) \geqslant 0$. A similar conclusion is valid for $\hat{N}$-dissipative operators, where $\hat{N}$ is the generalization of N given in Section 1.6. Of course these statements are empty if $D(H) \cap \mathscr{B}_+ = \emptyset$ but the assumption int $\mathscr{B}_+ \neq \emptyset$ is sufficient to ensure that $D(H)$ contains positive elements.

Next we consider N_u- and $\hat{N}_u$-dissipative operators in more detail. Again $\|\cdot\|_u$ denotes the order-unit norm associated with N_u, or $\hat{N}_u$.

COROLLARY 2.1.5. *For each $u \in D(H) \cap \operatorname{int} \mathscr{B}_+$ the following conditions are equivalent:*

1. *H is N_u-dissipative,*
2. *H is $\|\cdot\|_u$-dissipative and has the negative off-diagonal property,*
3. *$Hu \geqslant 0$ and H has the negative off-diagonal property.*

Moreover the following are equivalent:

$\hat{1}$. *H is $\hat{N}_u$-dissipative,*
$\hat{2}$. *H is $\|\cdot\|_u$-dissipative and $Hu \leqslant 0$,*
$\hat{3}$. *$Hu = 0$ and H has the negative off-diagonal property.*

Proof. 1⇔3. This is established in Proposition 2.1.4.

1⇒2. This follows from Proposition 2.1.4 and the fact that $\|a\|_u = N_u(a) \vee N_u(-a)$.

2⇒3. If $\omega \in \mathscr{B}^*_+\backslash\{0\}$ then $\omega/\omega(u)$ is in the subdifferential of $\|\cdot\|_u$ at u. Therefore if H is $\|\cdot\|_u$-dissipative $\omega(Hu) \geqslant 0$. Hence $Hu \geqslant 0$.

The proof of the equivalence of $\hat{1}$, $\hat{2}$, and $\hat{3}$ is similar. □

Note that in Corollary 2.1.5 the implication 1⇒2 does not depend upon the assumption that $u \in D(H)$. Similarly 2⇒1 if $u \in R(I+\alpha H)$, the range of $I + \alpha H$, for all small $\alpha > 0$.

EXAMPLE 2.1.6 (Matrices). Let $\mathscr{B} = \mathbb{R}^n$, $\mathscr{B}_+ = \mathbb{R}^n_+$, and $H = (H_{ij})$ a real $n \times n$ matrix. If $u = (u_i) \in \operatorname{int} \mathscr{B}_+$ then $u_i > 0$ for $i = 1, \ldots, n$ and N_u-dissipativity is equivalent to $(Hu)_i \geqslant 0$ for $i = 1, \ldots, n$ and $H_{ij} \leqslant 0$ for $i \neq j$. A particularly interesting case is $u = (1, 1, \ldots, 1)$ and then N_u-dissipativity is equivalent to

$$H_{ij} \leqslant 0,\ j \neq i, \quad \text{and} \quad \sum_{j=1}^{n} H_{ij} \geqslant 0,\ i = 1, \ldots, n;$$

$\hat{N}_u$-dissipativity is equivalent to

$$H_{ij} \leqslant 0,\ j \neq i, \quad \text{and} \quad \sum_{j=1}^{n} H_{ij} = 0,\ i = 1, \ldots, n;$$

and $\|\cdot\|_u$-dissipativity is equivalent to

$$H_{ii} - \sum_{j \neq i} |H_{ij}| \geqslant 0, \quad i = 1, \ldots, n.$$

Moreover in this case $\|\cdot\|_u$ coincides with the l^∞-norm $\|a\|_\infty = \max|a_i|$. If, alternatively, $\mathscr{B}$ is equipped with the l^1-norm $\|a\|_1 = \Sigma|a_i|$, then it follows that $\|\cdot\|_1$-dissipativity is equivalent to

$$H_{ii} - \sum_{j \neq i} |H_{ji}| \geqslant 0, \quad i = 1, \ldots, n.$$

The corresponding l^p-conditions, for $1 < p < \infty$, do not have any simple expression in terms of the matrix elements, but $\|\cdot\|_p$-dissipativity is implied by $\|\cdot\|_\infty$-dissipativity together with $\|\cdot\|_1$-dissipativity. □

EXAMPLE 2.1.7 (Hilbert space). If $\mathscr{B}$ is a (real) Hilbert space, the subdifferential of the norm at each $a \in \mathscr{B}$ with $\|a\| = 1$ consists of the unique element a. Thus H is (norm-) dissipative if, and only if, H is positive definite, i.e.,

$$(a, Ha) \geqslant 0, \quad a \in D(H).$$

But this is equivalent to positivity of the spectrum of H. Thus in this special context dissipativity is a spectral condition. □

Many elliptic differential operators on a wide variety of function spaces satisfy dissipativity conditions. As a simple illustration we consider N-dissipativity of the Laplacian with classical boundary conditions on L^2-spaces.

EXAMPLE 2.1.8 (The Laplacian). Let $\mathscr{B} = L^2(\Lambda; d^\nu x)$ where Λ is a bounded open subset of $\mathbb{R}^\nu$ with a piecewise differentiable boundary $\partial\Lambda$ and let $\mathscr{B}_+$ be the pointwise positive functions. Define H_σ by first specifying $D(H_\sigma)$ to consist of the functions f which are infinitely often differentiable in the interior of Λ and satisfy

$$\partial f/\partial n + \sigma f = 0$$

on $\partial\Lambda$ where $\partial/\partial n$ denotes the outward normal derivative and $\sigma \in \mathbb{R}$, then specify the action of H_σ by

$$(H_\sigma f)(x) = -\sum_{i=1}^{\nu} \frac{\partial^2 f}{\partial x_i^2}(x) = -(\nabla^2 f)(x)$$

for $f \in D(H_\sigma)$. Since $\mathscr{B}$ is a Banach lattice the canonical half-norm is given by $N(f) = \| f_+ \|$ where f_+ denotes the positive part of f, and $\mathrm{d}N(f)$ consists of the unique element $\omega = f_+/\| f_+ \|$ if $f_+ \neq 0$, and $0 \in \mathrm{d}N(f)$ if $f_+ = 0$. Therefore

$$\omega(H_\sigma f) = -(f_+, \nabla^2 f)/\| f_+ \|$$
$$= \left(\int_\Lambda \mathrm{d}^\nu x |\nabla f_+(x)|^2 + \sigma \int_{\partial\Lambda} \mathrm{d}s |f_+|^2 \right) \Big/ \| f_+ \| \geqslant 0.$$

for all $f \in D(H_\sigma)$ and all $\sigma \geqslant 0$. Thus H_σ is N-dissipative, and hence norm-dissipative, for all $\sigma \geqslant 0$. □

Dissipativity of bounded operators on C^*-algebras can be rephrased in an algebraic manner which is seemingly quite different to the original definition. Since discussion of this point is facilitated by the use of semigroup theory we postpone the details until the end of the next section.

2.2. C_0-SEMIGROUPS

In this section we examine versions of the Feller–Miyadera–Phillips theorem and the Hille–Yosida theorem for positive C_0-semigroups, and in the subsequent section we consider the analogous problem for C_0^*-semigroups. Many of the results, and their proofs, are very similar in both cases, but there are substantial differences for semigroups on order-unit spaces.

First, note that if $a \in \mathscr{B}_+$ and

$$a_\varepsilon = \varepsilon^{-1} \int_0^\varepsilon \mathrm{d}t\, S_t a,$$

then $a \in D(H) \cap \mathscr{B}_+$. Moreover $a_\varepsilon \to a$ in norm as $\varepsilon \to 0$. Therefore $D(H) \cap \mathscr{B}_+$ is norm-dense in $\mathscr{B}_+$.

Second, remark that

$$(I + \alpha H)^{-1} = \int_0^\infty \mathrm{d}t\, \mathrm{e}^{-t} S_{\alpha t}$$

and hence the resolvents $(I + \alpha H)^{-1}$ are positive. Conversely, if the resolvents are positive, then the formulae

$$S_t a = \lim_{n \to \infty} \left(I + \frac{t}{n} H \right)^{-n} a$$

show that S is positive.

The next result is a version of the Feller–Miyadera–Phillips theorem in which the canonical half-norm partially replaces the norm.

THEOREM 2.2.1. *Let $(\mathscr{B}, \mathscr{B}_+, \|\cdot\|)$ be an ordered Banach space for which the norm is monotone and the operator norm on $\mathscr{L}(\mathscr{B}, \mathscr{B})$ is positively attained. Furthermore let N be the canonical half-norm on $\mathscr{B}$, and M, β, γ, real numbers with $M \geqslant 1, \beta > 0, \beta\gamma < 1$.*

The following conditions are equivalent:

1. *H generates a positive C_0-semigroup S with*

$$\|S_t\| \leqslant M\,\mathrm{e}^{\gamma t}, \quad t > 0,$$

2. *H is a (norm-closed) norm-densely defined linear operator satisfying the range condition*

$$R(I + \beta H) = \mathscr{B}$$

and the dissipative conditions

$$N((I + \alpha H)^n a) \geqslant (1 - \alpha\gamma)^n N(a)/M$$

for all $a \in D(H^n)$, all $n \geqslant 1$, and all $0 < \alpha \leqslant \beta$.

Proof. $1 \Rightarrow 2$. First, the range condition $R(I + \beta H) = \mathscr{B}$, the norm-density, and norm-closedness, follow from the Feller–Miyadera–Phillips theorem. Second, since S is positive

$$\begin{aligned} N(S_t a) &= \inf\{\|b\|\,;\, b \geqslant S_t a\} \\ &\leqslant \inf\{\|S_t c\|\,;\, c \geqslant a\} \\ &\leqslant M\,\mathrm{e}^{\gamma t} \inf\{\|c\|\,;\, c \geqslant a\} = M\,\mathrm{e}^{\gamma t} N(a). \end{aligned}$$

Thus by Laplace transformation, and sublinearity of N,

$$\begin{aligned} N((I + \alpha H)^{-n} a) &\leqslant \int_0^\infty \mathrm{d}t \frac{t^{n-1}}{(n-1)!} \mathrm{e}^{-t} N(S_{\alpha t} a) \\ &\leqslant M(1 - \alpha\gamma)^{-n} N(a), \end{aligned}$$

and the dissipative conditions follow by rearrangement.

$2 \Rightarrow 1$. If $(I + \beta H)a = 0$, then

$$0 = N(\pm(I + \beta H)a) \geqslant (1 - \beta\gamma)N(\pm a)/M \geqslant 0.$$

Thus $N(\pm a) = 0$ and $a = 0$ because the norm is monotone on $\mathscr{B}$ and hence $\mathscr{B}_+$ is proper. Therefore $(I + \beta H)$ is invertible.

Next if $a \in \mathscr{B}_+$, then

$$0 = N(-a) \geqslant (1 - \beta\gamma)^n N(-(I + \beta H)^{-n} a)/M$$

and so $(I + \beta H)^{-n} a \in \mathscr{B}_+$, i.e., the operators $(I + \beta H)^{-n}$ are positive. Now since $a \in \mathscr{B}_+$, and the norm is monotone on $\mathscr{B}_+$,

$$\begin{aligned} \|(I + \beta H)^{-n} a\| &= N((I + \beta H)^{-n} a) \\ &\leqslant M(1 - \beta\gamma)^{-n} N(a) = M(1 - \beta\gamma)^{-n} \|a\|, \end{aligned}$$

by Theorem 1.6.3. But then

$$\|(I+\beta H)^{-n}\| \leqslant M(1-\beta\gamma)^{-n},$$

because the norm on $\mathscr{L}(\mathscr{B}, \mathscr{B})$ is positively attained.

Next a standard perturbation argument shows that $R(I+\alpha H) = \mathscr{B}$, and the resolvents $(I+\alpha H)^{-n}$ exist, are positive, and satisfy

$$\|(I+\alpha H)^{-n}\| \leqslant M(1-\alpha\gamma)^{-n}$$

whenever $\alpha > 0$ and $0 < \alpha \leqslant \beta$. Therefore H generates a C_0-semigroup S satisfying the bounds

$$\|S_t\| \leqslant M\,\mathrm{e}^{\gamma t},$$

by the Feller–Miyadera–Phillips theorem, and S is positive, because the resolvents $(I+\alpha H)^{-1}$ are positive. □

Note that neither of the assumptions on $(\mathscr{B}, \mathscr{B}_+, \|\cdot\|)$ are used in the proof of $1 \Rightarrow 2$ in the theorem.

There is an immediate corollary for contraction semigroups, i.e., an analogue of the Hille–Yosida theorem.

COROLLARY 2.2.2. *Let $(\mathscr{B}, \mathscr{B}_+, \|\cdot\|)$ be an ordered Banach space for which the norm is monotone and the operator norm on $\mathscr{L}(\mathscr{B}, \mathscr{B})$ is positively attained. Then the following conditions are equivalent:*

1. *H generates a positive C_0-semigroup of contractions,*
2. *H is a (norm-closed) norm-densely defined, N-dissipative operator, and $R(I+\alpha H) = \mathscr{B}$ for some $\alpha > 0$.*

This result follows immediately from Theorem 2.2.1 with $M = 1$ and $\gamma = 0$ when it is observed from Theorem 2.1.1 that N-dissipativity is equivalent to

$$N((I+\alpha H)a) \geqslant N(a), \quad a \in D(H),$$

and hence by iteration to

$$N((I+\alpha H)^n a) \geqslant N(a), \quad a \in D(H^n),\ n \geqslant 1.$$

In Theorem 2.2.1 and Corollary 2.2.2 the assumption of positive attainment is a somewhat implicit condition on $\mathscr{B}$. One can, however, combine these results with Proposition 1.7.8 to obtain the following statement.

COROLLARY 2.2.3. *If $(\mathscr{B}, \mathscr{B}_+, \|\cdot\|)$ is an ordered Banach space for which*

either $\|\cdot\|$ *is a Riesz norm,*
or $\mathscr{B}_+$ *is $1_\vee$-normal and approximately dominating,*

then the two conditions of Theorem 2.2.1 (resp. Corollary 2.2.2) are equivalent.

The first option of this corollary, the Riesz norm, covers Banach lattices, C^*-

algebras, and their duals. The second option is equivalent to the norm on $\mathscr{B}$ being an order-norm and the dual-norm being monotone.

Under weaker conditions on $\mathscr{B}$ it is possible to obtain the following weaker version of Theorem 2.2.1.

THEOREM 2.2.4. *Let $(\mathscr{B}, \mathscr{B}_+, \|\cdot\|)$ be an ordered Banach space for which $\mathscr{B}_+$ is approximately dominating and $\|\cdot\|$ is monotone. Furthermore, let N be the canonical half-norm on $\mathscr{B}$, and M, β, γ, real numbers with $M \geqslant 1, \beta > 0, \beta\gamma < 1$.*

The following conditions are equivalent:

1. *H generates a positive C_0-semigroup with*

$$\|S_t\|_+ \leqslant M\,\mathrm{e}^{\gamma t}, \quad t > 0,$$

2. *H is a norm-densely defined linear operator satisfying the range condition*

$$R(I + \beta H) = \mathscr{B}$$

and the dissipative conditions

$$N((I + \alpha H)^n a) \geqslant (1 - \alpha\gamma)^n N(a)/M$$

for all $a \in D(H^n)$, all $n \geqslant 1$, and all $0 < \alpha \leqslant \beta$.

Proof. $1 \Rightarrow 2$. Since $\mathscr{B}_+$ is approximately dominating

$$N(a) = \inf\{\|b\|; b \geqslant a, b \geqslant 0\},$$

The proof is now identical to that given in Theorem 2.2.1.

$2 \Rightarrow 1$. Let $\|\cdot\|_N$ be the order-norm,

$$\|a\|_N = N(a) \vee N(-a).$$

Since $\mathscr{B}_+$ is normal $\|\cdot\|_N$ and $\|\cdot\|$ are equivalent. But

$$\|(I + \alpha H)^n a\|_N \geqslant (1 - \alpha\gamma)^n \|a\|_N/M.$$

Therefore H generates a C_0-semigroup S with

$$\|S_t a\|_N \leqslant M\,\mathrm{e}^{\gamma t}\|a\|_N,$$

by the Feller–Miyadera–Phillips theorem. Moreover S is positive, as in Theorem 2.2.1. But since $\|\cdot\|$ is monotone $\|a\|_N = \|a\|$ for all $a \in \mathscr{B}_+$ by Theorem 1.6.3. Therefore

$$\|S_t a\| = \|S_t a\|_N \leqslant M\,\mathrm{e}^{\gamma t}\|a\|_N = M\,\mathrm{e}^{\gamma t}\|a\|$$

for $a \in \mathscr{B}_+$. Thus $\|S_t\|_+ \leqslant M \exp\{\gamma t\}$. □

Theorems 2.2.1 and 2.2.4 are not independent. In view of Proposition 1.7.8, Theorem 2.2.1 follows from 2.2.4. Conversely Theorem 2.2.4 can be deduced by applying 2.2.1 to $(\mathscr{B}, \mathscr{B}_+, \|\cdot\|_N)$.

Note also that in Theorem 2.2.4 the assumption of approximate domination is only used in the proof $1 \Rightarrow 2$ and norm-monotonicity is only used in the proof of $2 \Rightarrow 1$. Note also that if it is only assumed that $\mathscr{B}_+$ is normal then, because $\|\cdot\|_N$ and

$\|\cdot\|$ are equivalent, the above argument shows that Condition 2 of Theorem 2.2.4 implies that H generates a positive C_0-semigroup S for which $\|S_t\| \leqslant M' \exp\{\gamma t\}$ for some $M' \geqslant M$.

Again there is an analogous result for contraction semigroups.

COROLLARY 2.2.5. *Let $(\mathscr{B}, \mathscr{B}_+, \|\cdot\|)$ be an ordered Banach space for which $\mathscr{B}_+$ is approximately dominating and $\|\cdot\|$ is monotone. Then the following conditions are equivalent:*

1. *H generates a positive C_0-semigroup S satisfying*

$$\|S_t\|_+ \leqslant 1,$$

2. *H is a norm-densely defined, N-dissipative operator, and $R(I + \alpha H) = \mathscr{B}$ for some $\alpha > 0$.*

The following two-dimensional example shows the necessity of norm-monotonicity for the implication $2 \Rightarrow 1$ and domination for $1 \Rightarrow 2$.

EXAMPLE 2.2.6. Let $\mathscr{B} = \mathbb{R}^2$, $\mathscr{B}_+ = \mathbb{R}^2_+$, and $H(a_1, a_2) = (a_1, 0)$. Then $\exp\{-tH\}$ is positive. Moreover, if $\|(a_1, a_2)\| = |a_1| \vee |a_1 - 2a_2|$, then $N((a_1, a_2)) = a_1 \vee a_2 \vee 0$ and H is N-dissipative with $R(I + \alpha H) = \mathscr{B}$. Nevertheless $\|e^{-tH}\| = \|e^{-tH}\|_+ = 2 - e^{-t}$. In this case $\mathscr{B}_+$ is dominating but $\|\cdot\|$ is not monotone.

If $\|(a_1, a_2)\| = |a_1| \vee |a_1 + 2a_2|$, then $N((a_1, a_2)) = a_1 \vee a_2 \vee (a_1 + 2a_2) \vee 0$ and $\|e^{-tH}\|_+ = 1$. But H is not N-dissipative and $\|e^{-tH}\| = 2 - e^{-t}$. In this case $\|\cdot\|$ is monotone but $\mathscr{B}_+$ is not dominating. □

If $\mathscr{B}_+$ is normal and has interior points the characterization of generators can be considerably simplified. The following result shows that in this setting H generates a positive C_0-semigroup if, and only if, the resolvents $(I + \alpha H)^{-1}$ exist as positive operators for all small $\alpha > 0$.

THEOREM 2.2.7. *Let $(\mathscr{B}, \mathscr{B}_+, \|\cdot\|)$ be an ordered Banach space with $\mathscr{B}_+$ normal and $\operatorname{int} \mathscr{B}_+ \neq \emptyset$.*

The following conditions are equivalent:

1. *H generates a positive C_0-semigroup,*
2. *H is a norm-densely defined linear operator satisfying the range condition*

$$R(I + \beta H) = \mathscr{B}$$

and the positivity condition, $a \in D(H)$ and $(I + \alpha H)a \in \mathscr{B}_+$ imply $a \in \mathscr{B}_+$, for all $0 < \alpha \leqslant \beta$, for some $\beta > 0$.

REMARK. Proposition 2.1.4 establishes that the positivity requirement of Condition 2 is equivalent to a dissipativity condition, or the negative off-diagonal property discussed in Section 2.1.

Proof. $1 \Rightarrow 2$. If H generates a positive C_0-semigroup, then the resolvents $(I + \alpha H)^{-1}$ exist as positive bounded operators and so Condition 2 is satisfied.

$2 \Rightarrow 1$. Choose $u \in D(H) \cap \operatorname{int} \mathscr{B}_+$ then by Proposition 2.1.4 there is a $\gamma > 0$ such that $H + \gamma I$ is N_u-dissipative. Thus

$$N_u((I + \alpha H)a) \geqslant (1 - \alpha\gamma)N_u(a)$$

for $\alpha > 0$, $\alpha\gamma < 1$, and all $a \in D(H)$. Hence

$$\|(I + \alpha H)a\|_u \geqslant (1 - \alpha\gamma)\|a\|_u,$$

under the same restrictions, where $\|\cdot\|_u$ denotes the order-unit norm. By iteration

$$\|(I + \alpha H)^n a\|_u \geqslant (1 - \alpha\gamma)^n \|a\|_u$$

for $\alpha > 0$, $\alpha\gamma < 1$, and all $a \in D(H^n)$, $n \geqslant 1$. It then follows from the Feller–Miyadera–Phillips theorem that H generates a C_0-semigroup on $(\mathscr{B}, \|\cdot\|_u)$ satisfying $\|S_t\|_u \leqslant \exp\{\gamma t\}$, $t > 0$. But $\|\cdot\|_u$ is equivalent to $\|\cdot\|$, because $\mathscr{B}_+$ is normal, and hence S is a C_0-semigroup on $(\mathscr{B}, \|\cdot\|)$ with $\|S_t\| \leqslant M \exp\{\gamma t\}$ for some $M \geqslant 1$. Finally, positivity of the semigroup follows from positivity of the resolvents $(I + \alpha H)^{-1}$. □

The next corollary gives a variant of this result which stresses the N_u-dissipativity.

COROLLARY 2.2.8. *Assume $\mathscr{B}_+$ is normal and int $\mathscr{B}_+ \neq \emptyset$. Let N_u and $\hat{N}_u$ denote the half-norms associated with a $u \in$ int $\mathscr{B}_+$. Then the following conditions are equivalent:*

1. *H generates a positive C_0-semigroup S with $S_t u \leqslant u$ (respectively $S_t u = u$) for all $t > 0$,*
2. *H is norm-densely defined, $R(I + \alpha H) = \mathscr{B}$ for some $\alpha > 0$, and H is N_u-dissipative (respectively $\hat{N}_u$-dissipative),*

and these conditions imply $\|S_t\| \leqslant M$ for some $M \geqslant 1$ and all $t \geqslant 0$.

Proof. 1 ⇒ 2. If H generates a positive C_0-semigroup with $S_t u \leqslant u$ then $(I + \alpha H)^{-1}$ is positive for all $\alpha > 0$ and

$$(I + \alpha H)^{-1} a \leqslant N_u(a)(I + \alpha H)^{-1} u \leqslant N_u(a) u.$$

Therefore $N_u((I + \alpha H)^{-1} a) \leqslant N_u(a)$ and H is N_u-dissipative.

2 ⇒ 1. If H is N_u-dissipative then H is $\|\cdot\|_u$-dissipative and the range condition ensures that H generates a positive C_0-semigroup of contractions on $(\mathscr{B}, \mathscr{B}_+, \|\cdot\|_u)$. Moreover $(I + \alpha H)^{-1} u \leqslant u$ by N_u-dissipativity. Hence

$$S_t u = \lim_{n \to \infty} \left(I + \frac{t}{n} H \right)^{-n} u \leqslant u.$$

The bound on $\|S_t\|$ follows because $\|\cdot\|_u$ and $\|\cdot\|$ are equivalent, by normality of $\mathscr{B}_+$.

The proof of the $\hat{N}_u$-dissipative case is similar. □

It is interesting to note that positive C_0-semigroups automatically satisfy a stronger positivity condition, if int $\mathscr{B}_+ \neq \emptyset$.

PROPOSITION 2.2.9. *Let S be a positive C_0-semigroup on an ordered Banach space*

$(\mathscr{B}, \mathscr{B}_+, \|\cdot\|)$ with generator H. Then

$$S_t(\text{int}\,\mathscr{B}_+) \subseteq \text{int}\,\mathscr{B}_+, \qquad (I+\alpha H)^{-1}(\text{int}\,\mathscr{B}_+) \subseteq \text{int}\,\mathscr{B}_+,$$
$$(I+\alpha H)^{-1}(\text{qu.int}\,\mathscr{B}_+) \subseteq \text{qu.int}\,\mathscr{B}_+,$$

for all $t \geqslant 0$ and all (small) $\alpha \geqslant 0$.

Proof. We first show $S_t(\text{int}\,\mathscr{B}_+) \subseteq \text{int}\,\mathscr{B}_+$. If $\text{int}\,\mathscr{B}_+ = \emptyset$, there is nothing to show. Otherwise choose $u \in \text{int}\,\mathscr{B}_+$ then by continuity there is an $\varepsilon > 0$ such that $S_t u \in \text{int}\,\mathscr{B}_+$ for $0 \leqslant t \leqslant \varepsilon$. But if $v \in \text{int}\,\mathscr{B}_+$ then $\lambda u \leqslant v$ for some $\lambda > 0$ and $\lambda S_t u \leqslant S_t v$. Thus $S_t v \in \lambda\,\text{int}\,\mathscr{B}_+ + \mathscr{B}_+ \subseteq \text{int}\,\mathscr{B}_+$ whenever $0 \leqslant t \leqslant \varepsilon$. Finally, the semigroup property $S_t = (S_{t/n})^n$ applied for $n > t/\varepsilon$ shows that $S_t(\text{int}\,\mathscr{B}_+) \subseteq \text{int}\,\mathscr{B}_+$ for all $t \geqslant 0$.

Since $\text{int}\,\mathscr{B}_+ = \emptyset$ or $\text{int}\,\mathscr{B}_+ = \text{qu.int}\,\mathscr{B}_+$ it now suffices to prove that

$$(I+\alpha H)^{-1}(\text{qu.int}\,\mathscr{B}_+) \subseteq \text{qu.int}\,\mathscr{B}_+.$$

If $a \in \text{qu.int}\,\mathscr{B}_+$ and $\omega \in \mathscr{B}^*_+\backslash\{0\}$, then $\omega(a) > 0$ and $\omega(S_t a) \geqslant 0$ for all $t > 0$. Since $t \mapsto \omega(S_t a)$ is continuous it follows that

$$\omega((I+\alpha H)^{-1}a) = \int_0^\infty dt\, e^{-t}\omega(S_{\alpha t}a) > 0$$

and hence $(I+\alpha H)^{-1}a \in \text{qu.int}\,\mathscr{B}_+$. □

It is not, however, necessarily true that S maps quasi-interior points into quasi-interior points.

EXAMPLE 2.2.10. Let $\mathscr{B} = \{f\,;\, f \in C[0,\infty\rangle, f(0) = \lim_{x\to\infty} f(x) = 0\}$ equipped with the supremum norm and let $\mathscr{B}_+$ be the set of pointwise positive functions in $\mathscr{B}$. Then

$$\text{qu.int}\,\mathscr{B}_+ = \{f\,;\, f(x) > 0 \quad \text{for} \quad x > 0\}.$$

Define the positive C_0-semigroup S of right translations by

$$\begin{aligned}(S_t f)(x) &= 0 \quad \text{if } 0 \leqslant x \leqslant t\\ &= f(x-t) \quad \text{if } t \leqslant x\end{aligned}$$

then $S_t\mathscr{B}_+ \cap \text{qu.int}\,\mathscr{B}_+ = \emptyset$ for $t > 0$. □

Theorem 2.2.7 establishes that if $\mathscr{B}_+$ is normal with non-empty interior, then the generator property for H is equivalent to existence and positivity of the resolvents $(I+\alpha H)^{-1}$ for all small $\alpha > 0$. The next example shows that the condition $\text{int}\,\mathscr{B}_+ \neq \emptyset$ cannot be omitted, nor does it suffice that $\text{int}\,\mathscr{B}^*_+ \neq \emptyset$.

EXAMPLE 2.2.11. Let $\mathscr{B} = L^1(\mathbb{R})$, and let $\mathscr{B}_+$ be the cone of pointwise positive functions in $\mathscr{B}$. Furthermore let

$$\begin{aligned}v(x) &= 2^{-n} \quad \text{if } n - 2^{-n} \leqslant x \leqslant n,\ n \geqslant 2\\ &= 1 \quad \text{otherwise.}\end{aligned}$$

For $\alpha > 0$, $t \in \mathbb{R}$, one has

$$\int_t^\infty dx \frac{e^{-x/\alpha}}{v(x)} \leqslant \int_t^\infty dx\, e^{-x/\alpha} + \sum_{n \geqslant t} \int_{n-2^{-n}}^{n} dx\, 2^n e^{-x/\alpha}$$
$$\leqslant \alpha\, e^{-t/\alpha} + \sum_{n \geqslant t} e^{(1-n)/\alpha}$$
$$\leqslant \left(\alpha + \frac{e^{2/\alpha}}{e^{1/\alpha} - 1}\right) e^{-t/\alpha}.$$

Now define H by $D(H) = \{f \in \mathscr{B}: fv$ is absolutely continuous, and $(fv)'/v \in \mathscr{B}\}$, $Hf = (fv)'/v$. Suppose that $(1 + \alpha H)f = 0$. Then $\alpha(fv)' + fv = 0$, so $f(x) = c\, e^{-x/\alpha}/v(x)$ a.e. This contradicts the integrability of f, unless $f = 0$.

For $g \in \mathscr{B}$, let

$$k_\alpha(x) = \frac{1}{\alpha v(x)} \int_{-\infty}^{x} dt\, g(t) v(t)\, e^{-(x-t)/\alpha}.$$

Then

$$\int_{-\infty}^{\infty} dx\, |k_\alpha(x)| \leqslant \frac{1}{\alpha} \int_{-\infty}^{\infty} dt \int_t^\infty dx \frac{v(t)\, e^{-(x-t)/\alpha} |g(t)|}{v(x)}$$
$$\leqslant \left(1 + \frac{e^{2/\alpha}}{\alpha(e^{1/\alpha} - 1)}\right) \|g\|.$$

Thus $k_\alpha \in \mathscr{B}$. Furthermore $k_\alpha \in D(H)$ and $(I + \alpha H)k_\alpha = g$. This shows that $(I + \alpha H)$ is invertible and $(I + \alpha H)^{-1}$ is positive.

If g is a C^1-function of compact support, $f = g/v$, and $f_t(x) = g(x - t)/v(x)$, then $f_0 = f, f_t \in D(H)$ and $d/dt(f_t) = -Hf_t$. Hence $D(H)$ is norm-dense in $\mathscr{B}$; and if H generates a C_0-semigroup S, then $S_t f = f_t$ [17], Theorem 1.7. But if g_n is a C^1-function with

$$\begin{aligned} g_n(x) &= 1 \quad \text{if } n - \tfrac{1}{2} - 2^{-n} \leqslant x \leqslant n - \tfrac{1}{2} \\ &= 0 \quad \text{if } x \leqslant n - \tfrac{1}{2} - 2^{1-n} \text{ or if } x \geqslant n - \tfrac{1}{2} + 2^{-n}, \end{aligned}$$

and $f_n = g_n/v$, then $\|f_n\| \leqslant 3.2^{-n}$ but $\|S_{1/2} f_n\| \geqslant 1$ $(n \geqslant 2)$. This contradicts the fact that $S_{1/2}$ is bounded. □

A similar example may be constructed with $\mathscr{B} = \{f \in C(\mathbb{R}): \lim_{x \to \pm\infty} f(x) = 0\}$ [8]. So it does not seem possible to extend Theorem 2.2.6 to a larger class of ordered Banach spaces. Instead one may impose further conditions on the generator, and therefore on the semigroup. One example is the lower bounds

$$\|(I + \alpha H)^{-1} a\| \geqslant c_\alpha \|a\|, \quad a \in \mathscr{B}_+,$$

for some $c_\alpha > 0$. The resolvent identity

$$(\beta - \alpha)(I + \alpha H)^{-1}(I + \beta H)^{-1} = \beta(I + \beta H)^{-1} - \alpha(I + \alpha H)^{-1}$$

shows that this condition is independent of α.

PROPOSITION 2.2.12. *Let H be the generator of a positive C_0-semigroup on an* *ordered Banach space* $(\mathscr{B}, \mathscr{B}_+, \|\cdot\|)$ *satisfying* $\|S_t\| \leqslant M \exp\{\gamma t\}, t \geqslant 0$. *Let* $s > 0$, $\alpha > 0, \alpha\gamma < 1$ *and consider the following conditions:*

1. *there is a* $c_\alpha > 0$ *such that*

$$\|(I + \alpha H)^{-1}a\| \geqslant c_\alpha \|a\|, \quad a \in \mathscr{B}_+,$$

2. *there is a* $\lambda_s > 0$ *such that*

$$\|S_s a\| \geqslant \lambda_s \|a\|, \quad a \in \mathscr{B}_+.$$

Then $1 \Rightarrow 2$ *and if* $\mathscr{B}_+$ *has a base* $2 \Rightarrow 1$.

Proof. Let

$$\begin{aligned}\varphi(t) &= \inf\{\|S_t a\|; a \in \mathscr{B}_+, \|a\| = 1\}\\ &= \sup\{\lambda \geqslant 0; \|S_t a\| \geqslant \lambda \|a\| \quad \text{for all } a \in \mathscr{B}_+\}.\end{aligned}$$

Then φ is upper semi-continuous and the semigroup property $S_s S_t = S_{s+t}$ shows that

$$\varphi(s)\varphi(t) \leqslant \varphi(s + t) \leqslant \|S_s\| \varphi(t).$$

Hence Condition 2 is independent of s.

$1 \Rightarrow 2$. Suppose Condition 2 is false. Then by the above discussion there exists, for any $s > 0, \varepsilon > 0$, an $a \in \mathscr{B}_+$ with $\|a\| = 1$ and $\|S_s a\| \leqslant \varepsilon$. Now

$$\begin{aligned}&\|S_t a\| \leqslant M\,\mathrm{e}^{\gamma t} \qquad \text{if } 0 \leqslant t \leqslant s,\\ &\|S_t a\| \leqslant \varepsilon M\,\mathrm{e}^{\gamma(t-s)} \quad \text{if } s \leqslant t.\end{aligned}$$

Hence

$$\begin{aligned}\|(I + \alpha H)^{-1}a\| &= \left\|\int_0^\infty \mathrm{d}t\, \mathrm{e}^{-t} S_{\alpha t} a\right\|\\ &\leqslant M \int_0^{s/\alpha} \mathrm{d}t\, \mathrm{e}^{-(1-\alpha\gamma)t} + \varepsilon M \int_{s/\alpha}^\infty \mathrm{d}t\, \mathrm{e}^{-\gamma s}\, \mathrm{e}^{-(1-\alpha\gamma)t}\\ &\leqslant Ms/\alpha + \varepsilon M\, \mathrm{e}^{-s/\alpha}/(1 - \alpha\gamma).\end{aligned}$$

But since ε and s are arbitrarily small, this shows that Condition 1 is false.

$2 \Rightarrow 1$. Suppose $\mathscr{B}_+$ has a base, so that there is a $\beta > 0$ such that

$$\beta \left\|\sum_{i=1}^n a_i\right\| \geqslant \sum_{i=1}^n \|a_i\|$$

for all $n \geqslant 1$ and all $a_i \in \mathscr{B}_+$. Hence for $a \in \mathscr{B}_+$ one can use a Riemann approximation to establish that

$$\begin{aligned}\|(I + \alpha H)^{-1}a\| &= \left\|\int_0^\infty \mathrm{d}t\, \mathrm{e}^{-t} S_{\alpha t} a\right\|\\ &\geqslant \beta^{-1} \int_0^\infty \mathrm{d}t\, \mathrm{e}^{-t} \varphi(\alpha t) \|a\|.\end{aligned}$$

But Condition 2 implies that $\varphi(s) > 0$ for all s and hence Condition 1 is valid. □

In general Condition 2 of Proposition 2.1.12 does not imply Condition 1. Nor does the existence of a base ensure that the conditions are satisfied.

EXAMPLE 2.2.13. Let $\mathscr{B} = C(\mathbb{T})$, the continuous functions on the circle, ordered by the pointwise positive functions $\mathscr{B}_+$, and with the supremum norm. Define the positive C_0-semigroup of rotations S on $\mathscr{B}$ by

$$(S_t f)(s) = f(s - t).$$

Then S satisfies Condition 2 of Proposition 2.2.12 with $\lambda_s = 1$, but

$$((I + \alpha H)^{-1} f)(s) = \int_0^\infty dt\, e^{-t} f(s - \alpha t)$$

and Condition 1 is not satisfied. □

EXAMPLE 2.2.14. Let $\mathscr{B} = L^1(\mathbb{R}_+ ; dx)$ with $\mathscr{B}_+$ the pointwise positive functions in $\mathscr{B}$. Define a positive semigroup S on $\mathscr{B}$ by

$$(S_t f)(x) = f(x + t).$$

Although $\mathscr{B}$ has a base the conditions of Proposition 2.2.12 are not valid for S. □

Despite this last example there are many semigroups on base spaces which satisfy the conditions of Proposition 2.2.12. These semigroups can be characterized in various ways. Typically their adjoints must map the interior of the dual cone into itself. This is clearly not the case in Example 2.2.14 because the adjoint semigroup on $L^\infty(\mathbb{R}_+ ; dx)$ has the action

$$\begin{aligned}(S_t f)(x) &= f(x - t) \quad & t < x,\\ &= 0 & 0 \leqslant x \leqslant t.\end{aligned}$$

We will return to a more detailed discussion of this point in the next section. But we next demonstrate that the bounds

$$\|(I + \alpha H)^{-1} a\| \geqslant c_\alpha \|a\|, \quad a \in \mathscr{B}_+,$$

provide sufficient extra information to ensure that H is a generator.

THEOREM 2.2.15. *Let $(\mathscr{B}, \mathscr{B}_+, \|\cdot\|)$ be an ordered Banach space for which $\mathscr{B}_+$ is normal and generating. Let H be a norm-densely defined linear operator satisfying*

1. $R(I + \beta H) = \mathscr{B}$,
2. *if* $0 < \alpha \leqslant \beta$ *and* $(I + \alpha H)a \in \mathscr{B}_+$, *then* $a \in \mathscr{B}_+$,
3. $\|(I + \beta H)^{-1} a\| \geqslant c\|a\|, \quad a \in \mathscr{B}_+$,

for some $\beta > 0$ *and* $c > 0$.

Then H generates a positive C_0-semigroup satisfying bounds

$$\|S_t\| \leqslant M\, e^{t/\beta}, \qquad \|S_t a\| \geqslant \lambda_t \|a\|, \quad a \in \mathscr{B}_+,$$

for some $M \geqslant 1$ and $\lambda_t > 0$.

Proof. The resolvent $(I+\beta H)^{-1}$ exists and is positive, hence bounded, by Proposition 1.7.2. Let

$$\|a\|' = \inf\{\|(I+\beta H)^{-1}a_1\| + \|(I+\beta H)^{-1}a_2\|; a = a_1 - a_2, a_1, a_2 \in \mathscr{B}_+\}.$$

Therefore $\|a\|' \geqslant c\|a_1\| + c\|a_2\| \geqslant c\|a\|$ by Condition 3 and the triangle inequality. But if $\mathscr{B}_+$ is λ_+-generating, there is a decomposition $a = a_1 - a_2$ with $\|a_1\| + \|a_2\| \leqslant \leqslant \lambda\|a\|$. Hence $\|a\|' \leqslant \lambda\|(I+\beta H)^{-1}\|\,\|a\|$. Thus $\|\cdot\|'$ and $\|\cdot\|$ are equivalent norms.

Next, by perturbation theory $(I+\alpha H)^{-1}$ must exist for α in an open neighbourhood of β and by Condition 2 $(I+\alpha H)^{-1}$ must be positive. Suppose $0 < \alpha < \beta$ then the resolvent identity gives

$$\begin{aligned}(\beta-\alpha)(I+\beta H)^{-1}(I+\alpha H)^{-1} &= \beta(I+\beta H)^{-1} - \alpha(I+\alpha H)^{-1}\\ &\leqslant \beta(I+\beta H)^{-1}\end{aligned}$$

and hence by iteration

$$(\beta-\alpha)^n(I+\beta H)^{-1}(I+\alpha H)^{-n} \leqslant \beta^n(I+\beta H)^{-1}.$$

Now if $(I+\alpha H)^n a = a_1 - a_2$ with $a_1, a_2 \in \mathscr{B}_+$, and if $\|\cdot\|$ is μ-monotone, then

$$\begin{aligned}(\beta-\alpha)^n\|a\|' &\leqslant (\beta-\alpha)^n\{\|(I+\beta H)^{-1}(I+\alpha H)^{-n}a_1\| +\\ &\quad + \|(I+\beta H)^{-1}(I+\alpha H)^{-n}a_2\|\}\\ &\leqslant \mu\beta^n\{\|(I+\beta H)^{-1}a_1\| + \|(I+\beta H)^{-1}a_2\|\}.\end{aligned}$$

Hence

$$\mu^{-1}(1-\alpha\beta^{-1})^n\|a\|' \leqslant \|(I+\alpha H)^n a\|'.$$

It now follows by a standard perturbation argument and the Feller–Miyadera–Phillips theorem that H generates a C_0-semigroup S on $\mathscr{B}$ with the bounds

$$\|S_t\|' \leqslant \mu\, e^{t/\beta} \quad \text{and hence} \quad \|S_t\| \leqslant M e^{t/\beta}$$

for some $M \geqslant 1$. Finally, positivity of S follows from positivity of the resolvents and the lower bounds $\|S_t a\| \geqslant \lambda_t\|a\|$, $a \in \mathscr{B}_+$, follow from Condition 3 and Proposition 2.2.12. □

We conclude this section with two examples. The first is a comment on dissipativity in an algebraic setting.

EXAMPLE 2.2.16 (Dissipations on C^*-algebras). Let $\mathscr{B}$ be the self-adjoint part of a C^*-algebra with an identity $\mathbb{1}$, ordered by the positive elements $\mathscr{B}_+$. Let H be a bounded linear operator on $\mathscr{B}$ with the negative off-diagonal property and define H' by

$$H'(a) = H(a) - \{aH(\mathbb{1}) + H(\mathbb{1})a\}/2.$$

If $a \in \mathscr{B}_+$, $\omega \in \mathscr{B}^*_+$, and $\omega(a) = 0$, then $\omega(ab) = 0$ for all $b \in \mathscr{B}$ by the Cauchy–Schwarz inequality. Hence, with $b = H(\mathbb{1})$, one concludes that H' has the negative off-diagonal property and therefore generates a positive C_0-semigroup S on $\mathscr{B}$. Moreover $H'(\mathbb{1}) = 0$ and so $S_t(\mathbb{1}) = \mathbb{1}$ for all $t \geq 0$. But an inequality of Kadison [44] then gives

$$S_t(a^2) \geq S_t(a)^2$$

for all $a \in \mathscr{B}$ and $t \geq 0$. Differentiating at $t = 0$ one obtains

$$H'(a^2) \leq aH'(a) + H'(a)a$$

or, equivalently,

$$H(a^2) + aH(\mathbb{1})a \leq aH(a) + H(a)a, \quad a \in \mathscr{B}. \qquad (*)$$

Conversely, suppose H satisfies $(*)$ and let $a \in \mathscr{B}_+$, $\omega \in \mathscr{B}^*_+$, with $\omega(a) = 0$. It then follows from the Cauchy–Schwarz inequality that $\omega(a^{1/2}b) = 0$ for all $b \in \mathscr{B}$ and hence applying $(*)$, with a replaced by $a^{1/2}$, one finds $\omega(H(a)) \leq 0$. Thus H has the negative off-diagonal property.

In conclusion, the negative off-diagonal property for H is equivalent to the algebraic dissipation condition $(*)$. □

The final example partially motivates the study of C_0^*-semigroups.

EXAMPLE 2.2.17. Let $\mathscr{B} = L^\infty(X; d\mu)$ and $\mathscr{B}_+$ the cone of pointwise positive functions in $\mathscr{B}$. It can be shown that any positive C_0-semigroup on $\mathscr{B}$ is automatically uniformly continuous [50]. But $\mathscr{B}$ is the dual of $\mathscr{B}_* = L^1(X; d\mu)$ so it is natural to study C_0^*-semigroups on $\mathscr{B}$, e.g., translations are a C_0^*-semigroup on $L^\infty(\mathbb{R}; dx)$. □

2.3. C_0^*-SEMIGROUPS

In this section we discuss positive C_0^*-semigroups and their generators. Throughout the section $(\mathscr{B}, \mathscr{B}_+, \|\cdot\|)$ is an ordered Banach space, which is the dual of another ordered Banach space $(\mathscr{B}_*, \mathscr{B}_{*+}, \|\cdot\|_*)$, and H denotes a weak*-densely defined, weak*-closed, linear operator on $\mathscr{B}$. Moreover H^* denotes the norm-closed, norm-densely defined operator on $\mathscr{B}_*$ which is adjoint to H. If H generates a C_0^*-semigroup S, then positivity of S is again equivalent to positivity of the resolvents $(I + \alpha H)^{-1}$, and if S is positive, then $D(H) \cap \mathscr{B}_+$ is weak*-dense in $\mathscr{B}_+$.

The first result is a C_0^*-version of Theorem 2.2.1.

THEOREM 2.3.1. *Suppose that $\|\cdot\|$ is monotone and the operator norm on $\mathscr{L}(\mathscr{B}, \mathscr{B})$ is positively attained. Let N be the canonical half-norm on $\mathscr{B}$ and M, β, and γ, real numbers with $M \geq 1$, $\beta > 0$, and $\beta\gamma < 1$.*

The following conditions are equivalent:

1. *H generates a C_0^*-semigroup S satisfying*

$$\|S_t\| \leq M\, e^{\gamma t}, \quad t \geq 0,$$

2. $$R(I + \beta H) = \mathscr{B}$$

and

$$N((I+\alpha H)^n a) \geqslant (1-\alpha\gamma)^n N(a)/M$$

for all $a \in D(H^n)$, *all* $n \geqslant 1$, *and* $0 < \alpha \leqslant \beta$.

In particular these conditions are equivalent if

either $\|\cdot\|$ *is a Riesz norm*

or $\mathscr{B}_+$ *is* 1_+*-generating and* $\|\cdot\|$ *is monotone.*

The proof of equivalence of the two conditions is identical to the proof of Theorem 2.2.1, but one uses the C_0^*-version of the Feller–Miyadera–Phillips theorem. In fact it is sufficient that $\|\cdot\|$ is monotone and the operator norm is positively attained for positive weak*-continuous operators. This happens if, and only if, $\mathscr{B}_+$ is dominating and the operator norm on $\mathscr{L}(\mathscr{B}_*, \mathscr{B}_*)$ is positively attained. In this case Theorem 2.3.1 could also be deduced from the C_0-version, Theorem 2.2.1, applied to the adjoint H_* of H on $\mathscr{B}_*$. Moreover, the foregoing conditions are ensured if $\|\cdot\|$ is a Riesz norm or $\mathscr{B}_+$ is 1_+-generating and $\|\cdot\|$ is monotone. These statements all follow from Proposition 1.7.8, Lemma 1.7.11, and the earlier duality results.

There is also a C_0^*-version of Theorem 2.2.4.

THEOREM 2.3.2. *Suppose* $\mathscr{B}_+$ *is dominating and* $\|\cdot\|$ *is monotone. Then Conditions* 1 *and* 2 *of Theorem* 2.3.1 *are equivalent, but with the modified bound* $\|S_t\|_+ \leqslant M\exp\{\gamma t\}$ *in Condition* 1.

The proof is very similar to that of Theorem 2.2.4. Alternatively one may deduce the result by applying Theorem 2.2.4 to the adjoint H^* of H on $\mathscr{B}_*$ and then using the identity $\|S_t\|_+ = \|S_t^*\|_+$ of Lemma 1.7.11.

So far the results in this section have been in almost exact parallel with Section 2.2. But now we discuss the special case $\operatorname{int}\mathscr{B}_+ \neq \emptyset$ and a number of differences occur.

Let $\mathscr{B} = L^\infty(\mathbb{R}_+; dx)$ with $\mathscr{B}_+$ the pointwise positive functions in $\mathscr{B}$ and consider the C_0^*-semigroup S defined by

$$\begin{aligned}(S_t f)(x) &= f(x-t), \quad t \leqslant x\\ &= 0, \quad 0 \leqslant x < t.\end{aligned}$$

(This is the adjoint of the semigroup considered in Example 2.2.14.) Let H denote the generator of S. Although $\operatorname{int}\mathscr{B}_+ \neq \emptyset$ and $D(H)\cap\mathscr{B}_+$ is weak*-dense in $\mathscr{B}_+$ one has $D(H)\cap\operatorname{int}\mathscr{B}_+ = \emptyset$.

The C_0^*-analogue of Theorem 2.2.7 also fails in general. The adjoint of the operator H of Example 2.2.11 is a counterexample.

Nevertheless it is possible to characterize positive C_0^*-semigroups whose generators satisfy $D(H)\cap\operatorname{int}\mathscr{B}_+ \neq \emptyset$ and then to characterize the generators of such semigroups.

PROPOSITION 2.3.3. *Suppose that* $\operatorname{int}\mathscr{B}_+ \neq \emptyset$. *Let* S *be a positive* C_0^**-semigroup on* $\mathscr{B}$ *satisfying bounds* $\|S_t\| \leqslant M\exp\{\gamma t\}$ *for all* $t \geqslant 0$. *Let* s *and* α *be strictly positive numbers, with* $\alpha\gamma < 1$, *and let* $u \in \operatorname{int}\mathscr{B}_+$.

The following conditions are equivalent:

1. $D(H) \cap \operatorname{int} \mathscr{B}_+ \neq \emptyset$,
2. $(I+\alpha H)^{-1} u \in \operatorname{int} \mathscr{B}_+$,
3. $S_s u \in \operatorname{int} \mathscr{B}_+$,
4. *there is a* $c_\alpha > 0$ *such that*

$$\|(I+\alpha H^*)^{-1}\omega\|_* \geqslant c_\alpha \|\omega\|_*, \quad \omega \in \mathscr{B}_{*+},$$

5. *there is a* $\lambda_s > 0$ *such that*

$$\|S_s^* \omega\|_* \geqslant \lambda_s \|\omega\|_*, \quad \omega \in \mathscr{B}_{*+}.$$

Proof. $1 \Rightarrow 2$. Let $v \in D(H) \cap \operatorname{int} \mathscr{B}_+$. Since $u \in \operatorname{int} \mathscr{B}_+$ there exists $\lambda > 0$ such that $(I+\alpha H)v \leqslant \lambda u$. Hence $v \leqslant \lambda (I+\alpha H)^{-1} u$ and so $(I+\alpha H)^{-1} u \in \operatorname{int} \mathscr{B}_+$.

$2 \Rightarrow 1$. Clearly $(I+\alpha H)^{-1} u \in D(H) \cap \operatorname{int} \mathscr{B}_+$.

$2 \Leftrightarrow 4$. Let $K = \{\omega; \omega \in \mathscr{B}_{*+}, \omega(u) = 1\}$. Then K is a base of $\mathscr{B}_{*+}$ and

$$\operatorname{int} \mathscr{B}_+ = \{a; a \in \mathscr{B}_+, \inf_{\omega \in K} \omega(a) > 0\}$$

by Proposition 1.4.2. Hence Condition 2 is equivalent to

$$\inf_{\omega \in K} \omega((I+\alpha H)^{-1} u) > 0.$$

Since there are constants $\alpha > 0$, $\beta > 0$, such that $\alpha \|\omega\|_* \leqslant \omega(u) \leqslant \beta \|\omega\|_*$, for all $\omega \in \mathscr{B}_{*+}$, this is equivalent to Condition 4.

$3 \Leftrightarrow 5$. The argument is similar to the above.

$4 \Leftrightarrow 5$. Since $\mathscr{B}_{*+}$ has a base, this follows from Proposition 2.2.12. □

Now one can deduce a version of Theorem 2.2.7 for a restricted class of C_0^*-semigroups.

THEOREM 2.3.4. *Suppose* $\mathscr{B}_+$ *is normal and* $\operatorname{int} \mathscr{B}_+ \neq \emptyset$.
The following conditions are equivalent:

1. *H generates a positive* C_0^**-semigroup S such that*

$$S_t(\operatorname{int} \mathscr{B}_+) \subseteq \operatorname{int} \mathscr{B}_+, \quad t \geqslant 0,$$

2. *a.* $D(H) \cap \operatorname{int} \mathscr{B}_+ \neq \emptyset$,

   b. $R(I+\beta H) = \mathscr{B}$,

   c. if $0 < \alpha \leqslant \beta$ *and* $(I+\alpha H)a \in \mathscr{B}_+$ *then* $a \in \mathscr{B}_+$,

   for some $\beta > 0$.

Proof. $1 \Rightarrow 2$. This follows from Proposition 2.3.3 and general semigroup theory.

$2 \Rightarrow 1$. Choose $u \in D(H) \cap \operatorname{int} \mathscr{B}_+$ and let $\gamma = \hat{N}_u(-Hu)$. It follows from the argument used to prove $1 \Rightarrow 2$ in Proposition 2.1.4 that

$$N_u((I+\alpha H)a) \geqslant (1-\alpha\gamma) N_u(a)$$

for all $a \in D(H)$ and hence

$$\| (I + \alpha H)^n a \|_u \geqslant (1 - \alpha\gamma)^n \| a \|_u$$

for all $\alpha > 0$, $\alpha\gamma < 1$, and all $a \in D(H^n)$. Thus H generates a C_0^*-semigroup S satisfying $\| S_t \|_u \leqslant \exp\{\gamma t\}$ by the C_0^*-version of the Feller–Miyadera–Phillips theorem. Positivity of S follows from Condition 2c and $S_t(\text{int}\,\mathscr{B}_+) \subseteq \text{int}\,\mathscr{B}_+$ by Proposition 2.3.3. □

Theorem 2.3.4 can also be proved by applying Theorem 2.2.15 to the adjoint H^* of H on $\mathscr{B}_*$ and taking note of the equivalence $1 \Leftrightarrow 4$ in Proposition 2.3.3.

One can also derive a C_0^*-version of Theorem 2.2.15. The proof is almost identical once it is observed that the implication $1 \Rightarrow 2$ in Proposition 2.2.12 is valid for C_0^*-semigroups.

THEOREM 2.3.5. *Suppose that $\mathscr{B}_+$ is normal and generating and there exist $\beta > 0$ and $c > 0$ such that*

1. $R(I + \beta H) = \mathscr{B}$,
2. *if* $0 < \alpha \leqslant \beta$ *and* $(I + \alpha H)a \in \mathscr{B}_+$ *then* $a \in \mathscr{B}_+$
3. $\| (I + \beta H)^{-1} a \| \geqslant c \| a \|, \quad a \in \mathscr{B}_+$.

Then H generates a positive C_0^*-semigroup S satisfying $\| S_t \| \leqslant M \exp\{t/\beta\}$ for some $M \geqslant 1$ and

$$\| S_t a \| \geqslant \lambda_t \| a \|$$

for all $a \in \mathscr{B}_+$ and some $\lambda_t > 0$.

For ordered spaces $(\mathscr{B}, \mathscr{B}_+, \|\cdot\|)$ with a predual $(\mathscr{B}_*, \mathscr{B}_{*+}, \|\cdot\|_*)$ there is a weaker notion of interior point which differs from the quasi-interior point introduced in Section 1.4. A point $a \in \mathscr{B}_+$ is said to be an *N-interior point* of $\mathscr{B}_+$ if $\omega(a) > 0$ for all $\omega \in \mathscr{B}_{*+} \backslash \{0\}$. The set of N-interior points is denoted by $N.\text{int}\,\mathscr{B}_+$ and one has in general

$$\text{int}\,\mathscr{B}_+ \subseteq \text{qu.int}\,\mathscr{B}_+ \subseteq N.\text{int}\,\mathscr{B}_+ .$$

If H is the generator of a positive C_0^*-semigroup S then the argument used at the end of the proof of Proposition 2.2.9 shows that $(I + \alpha H)^{-1}(N.\text{int}\,\mathscr{B}_+) \subseteq N.\text{int}\,\mathscr{B}_+$. Moreover if $a \in N.\text{int}\,\mathscr{B}_+$, $\omega \in \mathscr{B}_{*+}$, $t > 0$, and $\omega(S_t a) = 0$, then $S_t^* \omega = 0$. Hence if $\text{int}\,\mathscr{B}_+ \neq \emptyset$ and $S_t\,\text{int}\,\mathscr{B}_+ \subseteq \text{int}\,\mathscr{B}_+$, then $S_t(N.\text{int}\,\mathscr{B}_+) \subseteq N.\text{int}\,\mathscr{B}_+$. But this last inclusion is possible even if S does not map the interior of $\mathscr{B}_+$ into itself.

EXAMPLE 2.3.6. Let $\mathscr{B} = L^\infty(\mathbb{R}_+; dx)$ and $\mathscr{B}_+$ the pointwise positive functions in $\mathscr{B}$. Then

$$\text{int}\,\mathscr{B}_+ = \text{qu.int}\,\mathscr{B}_+ = \{f\,;\, f \in \mathscr{B},\, f(x) \geqslant K \text{ a.e.} \quad \text{for some } K > 0\}$$
$$N.\text{int}\,\mathscr{B}_+ = \{f\,;\, f \in \mathscr{B},\, f(x) > 0 \quad \text{a.e.}\}.$$

Now define a positive C_0^*-semigroup S by

$$S_t f(x) = f(x-t) \quad t \leqslant x$$
$$= 0 \qquad 0 \leqslant x < t.$$

This is the adjoint of the C_0-semigroup of Example 2.2.14 and

$$((I+\alpha H)^{-1} f)(x) = \int_0^{x/\alpha} dt\, e^{-t} f(x - \alpha t).$$

So $S_t(\text{int}\,\mathscr{B}_+) \nsubseteq N.\text{int}\,\mathscr{B}_+$ and $(I+\alpha H)^{-1}(\text{int}\,\mathscr{B}_+) \nsubseteq \text{int}\,\mathscr{B}_+$.

Alternatively define the C_0^*-semigroup T on $\mathscr{B}$ by

$$(T_t f)(x) = e^{-t/x} f(x).$$

Then $S_t(N.\text{int}\,\mathscr{B}_+) \subseteq N.\text{int}\,\mathscr{B}_+$ but $S_t(\text{int}\,\mathscr{B}_+) \nsubseteq \text{int}\,\mathscr{B}_+$. In fact $S_t(\mathscr{B}_+) \cap \text{int}\,\mathscr{B}_+ = \emptyset$ for all $t > 0$. □

Von Neumann algebras provide one setting in which the theory of C_0^*-semigroups is of relevance. A von Neumann algebra can be characterized either abstractly as a C^*-algebra with a predual, or concretely as a weakly closed C^*-algebra $\mathscr{M}$ acting on a Hilbert space $\mathscr{H}$. In this second framework there is an interplay between C_0-semigroups on $\mathscr{H}$ and C_0^*-semigroups on $\mathscr{M}$.

EXAMPLE 2.3.7 (von Neumann algebras). Let $\mathscr{B}$ be the self-adjoint part of a von Neumann algebra on a complex Hilbert space $\mathscr{H}$, ordered by the positive elements $\mathscr{B}_+$. Let U be a C_0-semigroup on $\mathscr{H}$ with generator K and suppose $U_t^* \mathscr{B} U_t \subseteq \mathscr{B}$ for all $t \geqslant 0$. Then one can define a positive C_0^*-semigroup S on $\mathscr{B}$ by

$$S_t a = U_t^* a U_t.$$

If H denotes the generator of S, one can establish that $a \in D(H)$ if, and only if, $aD(K) \subseteq D(K^*)$ and $\varphi \mapsto aK\varphi + K^* a\varphi$ is a bounded operator from $D(K)$ into $\mathscr{H}$; then Ha is the continuous extension of this operator to $\mathscr{H}$ ([11] Proposition 3.2.55). □

2.4. SEMIGROUP TYPE AND SPECTRAL BONDS

In this section we examine a number of relations between the spectrum $\sigma(S_t)$ of a semigroup S and the spectrum $\sigma(H)$ of its generator H. We consider both C_0- and C_0^*-semigroups, and in the latter case we implicitly assume that the space has a predual. In the context of positive semigroups we further assume that the space and its predual have dual orderings.

If H is bounded then $S_t = \exp\{-tH\}$ and the spectral mapping theorem gives

$$\sigma(S_t) = \{e^{-tz};\, z \in \sigma(H)\}.$$

In particular the spectral radius $\rho(S_t)$ of S_t is $\exp\{-tm_H\}$ where

$$m_H = \inf\{\text{Re}\, z;\, z \in \sigma(H)\}.$$

If H is unbounded the spectral mapping theorem is no longer valid but one can conclude that

$$\sigma(S_t) \supseteq \{e^{-tz}; z \in \sigma(H)\}$$

and hence

$$\rho(S_t) \geqslant \exp\{-tm_H\}$$

(see, for example, [17] Theorem 2.16). Furthermore

$$\rho(S_t) = \exp\{t\gamma_S\}$$

where

$$\begin{aligned}\gamma_S &= \lim_{t\to\infty} t^{-1}\log\|S_t\| \\ &= \inf\{\gamma; \|S_t\| \leqslant M e^{\gamma t}, t \geqslant 0, \text{for some } M\}\end{aligned}$$

so that $-\infty \leqslant \gamma_S < \infty$ [17], Theorem 1.22. Thus $\gamma_S \geqslant -m_H$.

The numbers γ_S and m_H are known as the *type* of S and the *spectral bound* of H, respectively. In the absence of positivity γ_S may be strictly larger than $-m_H$ even in such special cases as a Hilbert space [86] or a C^*-algebra [37]. In the sequel we examine conditions which ensure that $\gamma_S = -m_H$ for a positive semigroup. Clearly $\gamma_S = \gamma_{S^*}$ and $m_H = m_{H^*}$ if S is a C_0-semigroup so it makes no difference whether we consider C_0- or C_0^*-semigroups.

Note that if $\lambda > \gamma_S$, then the resolvent $(\lambda I + H)^{-1}$ is given by the Laplace transform

$$(\lambda I + H)^{-1} = \int_0^\infty dt\, e^{-\lambda t} S_t,$$

where the integral is norm-convergent. Furthermore

$$\int_0^\infty dt\, e^{-\lambda t}\|S_t a\| \leqslant \int_0^\infty dt\, e^{-\lambda t} M e^{\gamma t}\|a\| < \infty,$$

where $\gamma_S < \gamma < \lambda$. The following lemma gives a converse.

LEMMA 2.4.1. *Let S be a C_0- or C_0^*-semigroup and λ a real number.*

1. *If for each $a \in \mathscr{B}$*

$$\int_0^\infty dt\, e^{-\lambda t}\|S_t a\| < \infty$$

then $\lambda \geqslant \gamma_S$.

2. *If for each $a \in \mathscr{B}$ and $\omega \in \mathscr{B}^*$ (or $\omega \in \mathscr{B}_*$ if S is a C_0^*-semigroup)*

$$\int_0^\infty dt\, e^{-\lambda t}|\omega(S_t a)| < \infty$$

then $\lambda > -m_H$. Moreover if $\lambda' > \lambda$ then

$$(\lambda' I + H)^{-1} = \int_0^\infty \mathrm{d}t\, \mathrm{e}^{-\lambda' t} S_t$$

where the integral is norm-convergent.

Proof. 1. Suppose that $\lambda < \gamma_S$. Then $\|S_n\| \exp\{-\lambda n\} \to \infty$ and, by the uniform boundedness theorem, there exists an $a \in \mathscr{B}$ and a sequence of integers $n_i \to \infty$ such that $\|S_{n_i} a\| \exp\{-\lambda n_i\} \geqslant 1$. Thus, if

$$K = \sup\{\mathrm{e}^{-\lambda t}\|S_t\|; 0 \leqslant t \leqslant 1\},$$

then

$$\mathrm{e}^{-\lambda t}\|S_t a\| \geqslant K^{-1}$$

for $t \in [n_i - 1, n_i]$ and

$$\int_0^\infty \mathrm{d}t\, \mathrm{e}^{-\lambda t}\|S_t a\| = \infty.$$

2. In order to use spectral arguments it is appropriate to introduce a complexification of $\mathscr{B}$ and to extend operators by linearity. There is a large class of equivalent norms that can be used for this complexification, but this ambiguity does not affect the following spectral argument, and in special cases, such as L^p-spaces or C^*-algebras, there is usually a natural choice of norm. We will make no notational distinction between the real and complex structures.

Suppose $t \mapsto \omega(S_t a)\exp\{-\lambda t\}$ is integrable for all $a \in \mathscr{B}$ and $\omega \in \mathscr{B}^*$ (or $\omega \in \mathscr{B}_*$ in the C_0^*-case). Define L_t by

$$L_t(z) = \int_0^t \mathrm{d}s\, \mathrm{e}^{-zs} S_s$$

for complex z. Then two applications of the uniform boundedness theorem show that $\{L_t(\lambda); t \geqslant 0\}$ is uniformly bounded. Moreover if $\operatorname{Re} z > \lambda$, the formula

$$L_t(z) = \mathrm{e}^{-(z-\lambda)t} L_t(\lambda) + (z - \lambda)\int_0^t \mathrm{d}s\, \mathrm{e}^{-(z-\lambda)s} L_s(\lambda)$$

shows that $L_t(z)$ converges in norm, as $t \to \infty$, to a limit $L(z)$ which is an analytic function of z. But if $\operatorname{Re} z > \gamma_S$, then $L(z) = (zI + H)^{-1}$ and hence, by the uniqueness of analytic functions $-z \notin \sigma(H)$ and $L(z) = (zI + H)^{-1}$ for $\operatorname{Re} z > \lambda$. Finally the estimate

$$|\omega((zI + H)^{-1} a)| = \left|\int_0^\infty \mathrm{d}t\, \mathrm{e}^{-zt}\omega(S_t a)\right| \leqslant \int_0^\infty \mathrm{d}t\, \mathrm{e}^{-\lambda t}|\omega(S_t a)|$$

for $\operatorname{Re} z > \lambda$ and two applications of the uniform boundedness theorem show that $\{(zI + H)^{-1}; \operatorname{Re} z > \lambda\}$ is uniformly bounded so $m_H > -\lambda$ by perturbation theory. □

Part 1 of Lemma 2.4.1 can be improved to show that $\lambda > \gamma_S$.

Next consider a real $n \times n$-matrix H acting on $\mathbb{R}^n$, ordered by $\mathbb{R}^n_+$. Then $\exp\{-tH\}$ is a positive semigroup if, and only if, H has negative off-diagonal matrix elements. In this case $\lambda I - H$ has positive matrix elements for sufficiently large positive λ and it follows from the Perron–Frobenius theorem that

$$n_H = \sup\{\mathrm{Re}\, z; z \in \sigma(\lambda I - H)\}$$

is an eigenvalue of $\lambda I - H$ with an eigenvector which may be chosen to be positive. Therefore m_H is an eigenvalue of H, with positive eigenvector. The following result is a partial infinite-dimensional analogue of these facts. A further extension is given in Theorem 2.4.6.

THEOREM 2.4.2. *Let S be a positive C_0- or C_0^*-semigroup on the ordered Banach space $(\mathscr{B}, \mathscr{B}_+, \|\cdot\|)$ and suppose $\mathscr{B}_+$ is normal and generating. Let H denote the generator of S. It follows that*

1. *if $\sigma(H) \neq \emptyset$ then $m_H \in \sigma(H)$ and*

$$0 = \inf\{\|(m_H I - H)a\|; a \in D(H) \cap \mathscr{B}_+, \|a\| = 1\}$$

2. *if $\lambda > -m_H$ then*

$$(\lambda I + H)^{-1} = \int_0^\infty dt\, e^{-\lambda t} S_t$$

where the integral is norm-convergent, and in particular, $(\lambda I + H)^{-1}$ is positive.

Proof. Suppose $m_H \notin \sigma(H)$, so there exists a $\lambda < -m_H$ such that $\sigma(H)$ does not intersect the real interval $\langle -\infty, -\lambda]$. For $a \in \mathscr{B}_+$ and $\omega \in \mathscr{B}^*_+$ (or $\omega \in \mathscr{B}_{*+}$ in the C_0^*-case) the Laplace transform of the positive function $t \mapsto \omega(S_t a)$ exists for $\mathrm{Re}\, z > \gamma_S$ and has an analytic extension to a region containing $[\lambda, \infty\rangle$ given by $z \mapsto \omega((zI + H)^{-1}a)$. By the Pringsheim–Landau theorem [83] the Laplace transform exists for $\mathrm{Re}\, z \geqslant \lambda$. In particular $t \mapsto \omega(S_t a) \exp\{-\lambda t\}$ is integrable over $\mathbb{R}_+$ for $a \in \mathscr{B}_+$ and $\omega \in \mathscr{B}^*_+$, and hence for all $a \in \mathscr{B}$ and $\omega \in \mathscr{B}_*$ because $\mathscr{B}_+$ and $\mathscr{B}^*_+$ are generating. It then follows from Lemma 2.5.1 that $-m_H < \lambda$ which is a contradiction.

The above argument shows that $t \mapsto \omega(S_t a) \exp\{-\lambda t\}$ is integrable over $\mathbb{R}_+$ for $\lambda > -m_H$ and hence the second statement of the theorem follows from Lemma 2.4.1.

Finally, suppose $\lambda_n < m_H$ but $\lambda_n \to m_H$ as $n \to \infty$. Therefore $\|(\lambda_n I - H)^{-1}\| \to \infty$ and one can choose $a_n \in \mathscr{B}$ such that $\|a_n\| \to 0$ but $\|(\lambda_n I - H)^{-1} a_n\| = 1$. Moreover since $\mathscr{B}_+$ is generating one can assume $a_n \in \mathscr{B}_+$. Therefore setting $b_n = (\lambda_n I - H)^{-1} a_n$ one has $b_n \in \mathscr{B}_+$ by Statement 2 of the theorem and $\|b_n\| = 1$ But

$$\|(m_H I - H)b_n\| \leqslant (m_H - \lambda_n) + \|a_n\| \to 0.$$

□

COROLLARY 2.4.3. *If H generates a positive C_0-, or C_0^*-, group on an ordered Banach space for which $\mathscr{B}_+$ is normal and generating then $\sigma(H) \neq \emptyset$.*

Proof. Suppose $\sigma(H) = \emptyset$. Since $\pm H$ generate positive C_0- or C_0^*-semigroups, it follows from Theorem 2.4.2 that $(\pm H)^{-1}$ are both positive. But this is impossible since $\mathscr{B}_+$ is proper and $\mathscr{B}_+ \neq \{0\}$.

□ If S is of type $-\infty$ then $\sigma(H) = \emptyset$. This is, for example, the case if $\mathscr{B} = L^1(0, 1)$ and

$$\begin{aligned}(S_t f)(x) &= f(x-t) \quad \text{if } x - t \in [0,1] \\ &= 0 \quad \text{otherwise.}\end{aligned}$$

But $\sigma(H)$ can be empty even for positive C_0-semigroups of finite type. This will be demonstrated in Example 2.4.8.

THEOREM 2.4.4. *Let S be a positive C_0- or C_0^*-semigroup on the ordered Banach space $(\mathscr{B}, \mathscr{B}_+, \|\cdot\|)$ and suppose $\mathscr{B}_+$ is generating and has a base. Let H denote the generator of S. It follows that $m_H = -\gamma_S$.*

Proof. If $\lambda > -m_H$ then

$$(\lambda I + H)^{-1} a = \int_0^\infty \mathrm{d}t\, \mathrm{e}^{-\lambda t} S_t a,$$

where the integral is norm-convergent, by Theorem 2.4.2. Now $\|\cdot\|$ is α-additive for some $\alpha \geq 1$, by Proposition 1.4.2, so by the use of Riemann approximants one deduces that

$$\begin{aligned}\int_0^\infty \mathrm{d}t\, \mathrm{e}^{-\lambda t} \|S_t a\| &\leq \alpha \left\| \int_0^\infty \mathrm{d}t\, \mathrm{e}^{-\lambda t} S_t a \right\| \\ &= \alpha \|(\lambda I + H)^{-1} a\| < +\infty,\end{aligned}$$

at least for a in the set

$$\mathscr{S} = \{a; a \in \mathscr{B}_+, \lim_{t \to 0} \|S_t a - a\| = 0\}.$$

If S is a C_0-semigroup, $\mathscr{S} = \mathscr{B}_+$ which spans $\mathscr{B}$ and $\gamma_S \leq \lambda$ by Lemma 2.4.1. Hence $\gamma_S \leq -m_H$ and the proof is complete since the reverse inequality is always true.

If S is a C_0^*-semigroup then $a \in \mathscr{B}_+$ is the weak*-limit of the bounded sequence $a_n \in \mathscr{S}$ where

$$a_n = n \int_0^{1/n} \mathrm{d}t\, S_t a.$$

Therefore

$$\begin{aligned}\int_0^\infty \mathrm{d}t\, \mathrm{e}^{-\lambda t} \|S_t a\| &\leq \liminf_{n \to \infty} \int_0^\infty \mathrm{d}t\, \mathrm{e}^{-\lambda t} \|S_t a_n\| \\ &\leq \alpha \|(\lambda I + H)^{-1}\| \liminf \|a_n\| < +\infty.\end{aligned}$$

Again the result follows from Lemma 2.4.1. □

COROLLARY 2.4.5. *If H generates a positive C_0- or C_0^*-semigroup on an ordered Banach space for which $\mathscr{B}_+$ is normal and α-directed, then $m_H = -\gamma_S$.*

Proof. Since $\mathscr{B}_+^*$ (or $\mathscr{B}_{*+}$ in the C_0^*-case) is generating and has a base, by Proposition 1.4.2 and the subsequent remarks, this result follows by applying Theorem 2.4.4 to S^*. □

It follows from Theorem 2.4.4 and Corollary 2.4.5 that $m_H = -\gamma_S$ for positive semigroups on L^1- or L^∞-spaces. A completely different argument [57] establishes that this is also the case on L^2-spaces, but the result appears to be unknown for general L^p-spaces.

Theorem 2.4.4 established that if $\mathscr{B}$ has a predual $\mathscr{B}_*$ and $\mathscr{B}_+$ is generating with a base K then $m_H = -\gamma_S$ for all positive C_0^*-semigroups. But if K is weak*-compact, which is equivalent to $\operatorname{int}\mathscr{B}_{*+} \neq \emptyset$ by Proposition 1.4.2, then m_H is in fact an eigenvalue of H.

THEOREM 2.4.6. *Let S be a positive C_0^*-semigroup on an ordered Banach space $(\mathscr{B}, \mathscr{B}_+, \|\cdot\|)$ with an ordered predual. Suppose $\mathscr{B}_+$ is generating and has a weak*-compact base K. Then there exists an $a \in K$ such that*

$$S_t a = e^{\gamma_S t} a, \; t \geqslant 0.$$

Proof. Let H denote the generator of S. By Theorems 2.4.2 and 2.4.4 one has $-\gamma_S \in \sigma(H)$ and

$$\inf\{\|(\gamma_S I + H)a\|; a \in D(H) \cap \mathscr{B}_+, \|a\| = 1\} = 0.$$

Next note that for each $a \in \mathscr{B}_+$ there is a $\lambda_K(a)$ such that $a \in \lambda_K(a)K$ and $\alpha\|a\| \leqslant \leqslant \lambda_K(a) \leqslant \beta\|a\|$ for some $0 < \alpha \leqslant \beta$. Therefore one can choose a net $a_\alpha \in D(H) \cap K$ such that $\|(\gamma_S I + H)a_\alpha\| \to 0$ and, by weak*-compactness of K, one may also assume a_α is weak*-convergent to a limit $a \in K$. Since H is weak*-closed it follows that $a \in D(H) \cap K$ and $Ha + \gamma_S a = 0$. Thus

$$S_t a = e^{\gamma_S t} a, \quad t \geqslant 0.$$ □

COROLLARY 2.4.7. *Let S be a positive C_0-semigroup on an ordered Banach space for which $\mathscr{B}_+$ is normal and has an interior point u, and let H denote the generator of S. The following conditions are equivalent:*

1. *For each $\omega \in \mathscr{B}_+^* \backslash \{0\}$ there exists a $t > 0$ such that $\omega(S_t u) < \omega(u)$,*
2. *For each $\omega \in \mathscr{B}^*$ and $a \in \mathscr{B}$, $\omega(S_t a) \to 0$ as $t \to \infty$,*
3. $\|S_t\| \to 0$ *as* $t \to \infty$,
4. $\gamma_S < 0$,
5. *There exists a $v \in D(H) \cap \operatorname{int}\mathscr{B}_+$ and a $\lambda > 0$ such that $Hv \geqslant \lambda v$.*

Proof. $1 \Rightarrow 4$. This follows from Theorem 2.4.6.

$4 \Rightarrow 3 \Rightarrow 2 \Rightarrow 1$. These implications are obvious.

$4 \Rightarrow 5$. Take $0 < \lambda < -\gamma_S$ then $v = (H - \lambda I)^{-1} u \in D(H) \cap \operatorname{int} \mathscr{B}_+$ by the proof of Proposition 2.2.9. But $Hv - \lambda v = u \geqslant 0$.

$5 \Rightarrow 3$. Since $Hv \geqslant \lambda v$ one has $(I + tH/n)^{-1} v \leqslant (1 + t\lambda/n)^{-1} v$ and hence, by iteration $S_t v \leqslant \exp\{-\lambda t\} v$. Therefore $\|S_t a\|_v \leqslant \exp\{-\lambda t\} \|a\|_v$ for all $a \in \mathscr{B}$ where $\|\cdot\|_v$ denotes the order-unit norm defined by v. But $\|\cdot\|_v$ and $\|\cdot\|$ are equivalent, because $\mathscr{B}_+$ is normal, so $\|S_t\| \leqslant M \exp\{-\lambda t\}$ for some $M \geqslant 1$. □

Note that the equivalence of Conditions 4 and 5 in Corollary 2.4.7 establishes that if S is a positive C_0-semigroup with generator H on a space for which $\mathscr{B}_+$ is normal and $\operatorname{int} \mathscr{B}_+ \neq \emptyset$, then

$$m_H = -\gamma_S = \sup\{\lambda ; \lambda \in \mathbb{R}, Hu \geqslant \lambda u \text{ for some } u \in D(H) \cap \operatorname{int} \mathscr{B}_+\}.$$

We conclude the section with a number of illustrative examples.

EXAMPLE 2.4.8. Let

$$\mathscr{B} = \{f ; f \in C(\mathbb{R}_+), \lim_{x \to \infty} f(x) = 0, \int_0^\infty dx\, e^{x^2} |f(x)| < +\infty\}$$

and let $\mathscr{B}_+$ be the pointwise positive functions in $\mathscr{B}$. Moreover define the norm by

$$\|f\| = \|f\|_1 + \|f\|_\infty$$

where

$$\|f\|_1 = \int_0^\infty dx\, e^{x^2} |f(x)|,$$

$$\|f\|_\infty = \sup_{x \in \mathbb{R}_+} |f(x)|.$$

Then $\mathscr{B}$ is a Banach lattice and there is a positive C_0-semigroup of lattice homomorphisms on $\mathscr{B}$ defined by

$$(S_t f)(x) = f(x + t).$$

Moreover $\|S_t\| = 1$ and $\gamma_S = 0$.

Now for real λ

$$\left| \int_0^t ds\, e^{-\lambda s} (S_s f)(x) \right| \leqslant c_\lambda \int_0^t ds\, e^{(x+s)^2} |f(x + s)| \leqslant c_\lambda \|f\|_1,$$

where $c_\lambda = \sup\{\exp(-\lambda s - s^2); s \geqslant 0\}$, and

$$\left\| \int_0^t ds\, e^{-\lambda s} S_s f \right\|_1 \leqslant \int_0^\infty ds\, e^{-\lambda s} \int_s^{s+t} dy\, e^{(y-s)^2} |f(y)|$$

$$\leqslant \int_0^\infty ds\, e^{-\lambda s - s^2} \|f\|_1.$$

Thus the operators

$$L_t(\lambda) = \int_0^t ds\, e^{-\lambda s} S_s$$

are uniformly bounded, for all $t \geq 0$, and it follows as in part 2 of Lemma 2.4.1 that $\sigma(H) \cap \{z; \text{Re}\, z < -\lambda\} = \emptyset$. Since λ is arbitrary $\sigma(H) = \emptyset$.

This example can be modified in several ways. Taking

$$\mathscr{B} = L^p(\mathbb{R}_+ ; e^{px^2} dx) \cap L^q(\mathbb{R}_+ ; dx)$$

with $1 < p < q < \infty$ one can arrange that $\mathscr{B}$ is reflexive, $\gamma_S = 0$, but $\sigma(H) = \emptyset$. Alternatively taking

$$\mathscr{B} = \left\{ f ; f \in C(\mathbb{R}_+), \lim_{x \to \infty} f(x) = 0, \int_0^\infty dx\, e^x |f(x)| < \infty \right\}$$

one can arrange that $\gamma_S = 0$ but $m_H = 1$. □

EXAMPLE 2.4.9 (von Neumann Algebras). Consider Example 2.3.7. One has $\|S_t\| = \|S_t \mathbb{1}\| = \|U_t^* U_t\| = \|U_t\|^2$ where $\mathbb{1}$ is the identity in $\mathscr{B}$. Therefore $\gamma_S = 2\gamma_U$. Moreover by Corollary 2.4.5 one has $m_H = -\gamma_S$. More specifically the following conditions are equivalent:

1. $m_H > 0$,
2. $\gamma_U < 0$,
3. For each $b \in \mathscr{B}_+$ there exists an $a \in D(H) \cap \mathscr{B}_+$ such that $Ha = b$,
4. There exists an $a \in \mathscr{B}_+$ such that $Ha = \mathbb{1}$,
5. For all $\varphi \in \mathscr{H}$ one has

$$\int_0^\infty dt\, \|U_t \varphi\|^2 < \infty,$$

6. For all $\varphi, \psi \in \mathscr{H}$ and $a \in \mathscr{B}$ one has

$$\int_0^\infty dt\, |(\varphi, (S_t a)\psi)| < \infty.$$

The equivalence $1 \Leftrightarrow 2$ follows from the above discussion; $1 \Rightarrow 3$ by Theorem 2.4.2; $3 \Rightarrow 4$ is trivial; $4 \Rightarrow 5$ by the estimate

$$\begin{aligned}\int_0^t ds\, \|U_s \varphi\|^2 &= \int_0^t ds (\varphi, (S_s Ha)\varphi) \\ &= (\varphi, (a - S_t a)\varphi) \leq (\varphi, a\varphi);\end{aligned}$$

5 ⇒ 6 by the estimate

$$\left|\int_0^\infty \mathrm{d}t(\varphi, (S_t a)\varphi)\right| \leqslant \|a\| \int_0^\infty \mathrm{d}t(\varphi, (S_t \mathbb{1})\varphi)$$

$$= \|a\| \int_0^\infty \mathrm{d}t \, \|U_t\varphi\|^2$$

and polarization. Finally, $\mathscr{B}_{*+}$ consists of infinite runs of functionals $a \mapsto (\varphi, a\varphi)$, $\varphi \in \mathscr{H}$, and by a uniform boundedness argument Condition 6 implies

$$\int_0^\infty \mathrm{d}t \, |\omega(S_t a)| < \infty$$

for all $\omega \in \mathscr{B}_*$ so $m_H > 0$ by Lemma 2.4.1. □

2.5. IRREDUCIBILITY AND STRICT POSITIVITY

Theorems 2.4.2 and 2.4.6 give infinite-dimensional analogues of the simplest features of the Perron–Frobenius spectral theory of positive matrices and one could expect to derive other analogues by further extension of the matrix theory. The purpose of the present section is to pursue this point.

Recall that the matrix semigroup $S_t = \exp\{-tH\}$ acting on $\mathscr{B} = \mathbb{R}^n$, ordered by $\mathscr{B}_+ = \mathbb{R}^n_+$, is positive if, and only if, the matrix $A = \|H\| I - H$ is positive, i.e., has positive entries. Thus spectral properties of S can be derived from application of the Perron–Frobenius theory either directly to the S_t or to A. This theory states that the spectral radius of a positive matrix is an eigenvalue with a positive eigenvector, and strictly positive, or positive 'irreducible', matrices have even stronger spectral properties, e.g., the spectral radius is a simple eigenvalue with a strictly positive eigenvector. The first of these points was generalized in Theorem 2.4.2 and 2.4.6 and we now examine extensions of the second. The initial problem is to find a suitable irreducibility criterion. There are various equivalent definitions of irreducibility for a positive matrix and we begin our discussion by investigating the extent to which these definitions remain equivalent in infinite dimensions.

Let $(\mathscr{B}, \mathscr{B}_+, \|\cdot\|)$ be an ordered Banach space. A subcone $\mathscr{C}$ of $\mathscr{B}_+$ is said to be *hereditary* if $0 \leqslant a \leqslant b$ and $b \in \mathscr{C}$ implies $a \in \mathscr{C}$, i.e., $\mathscr{C}$ is an extremal subset of $\mathscr{B}_+$. A subset $\mathscr{S}$ of the positive bounded operators $\mathscr{L}(\mathscr{B}, \mathscr{B})_+$ is said to be *norm-irreducible* if there is no proper norm-closed $\mathscr{S}$-invariant hereditary subcone of $\mathscr{B}_+$; *norm-ergodic* if for each $a \in \mathscr{B}_+ \backslash \{0\}$ the smallest $\mathscr{S}$-invariant hereditary subcone of $\mathscr{B}_+$ containing a is norm-dense in $\mathscr{B}_+$; *strictly positive* if $T(\mathscr{B}_+ \backslash \{0\}) \subseteq \text{qu.int}\, \mathscr{B}_+$ for all $T \in \mathscr{S}$. Similarly a subset $\mathscr{S}^*$ of $\mathscr{L}(\mathscr{B}^*, \mathscr{B}^*)_+$ is said to be *weak*-irreducible* if there is no proper weak*-closed $\mathscr{S}^*$-invariant hereditary subcone of $\mathscr{B}^*_+$; *weak*-ergodic* if for each $\omega \in \mathscr{B}^*_+ \backslash \{0\}$ the smallest $\mathscr{S}^*$-invariant hereditary subcone of $\mathscr{B}^*_+$ containing ω is weak*-dense in $\mathscr{B}^*_+$; **-strictly positive* if $T(\mathscr{B}^*_+ \backslash \{0\}) \subseteq N.\text{int}\, \mathscr{B}^*_+$ for all $T \in \mathscr{S}^*$.

Note that if $\mathscr{S}$ is a semigroup, i.e., if $\mathscr{S}\mathscr{S} \subseteq \mathscr{S}$, then the smallest $\mathscr{S}$-invariant hereditary subcone of $\mathscr{B}_+$ containing a is given by

$$\left\{ b; b \in \mathscr{B}_+, b \leqslant \sum_{i=1}^{n} \lambda_i T_i a \text{ for some } \lambda_i \geqslant 0, T_i \in \mathscr{S} \right\}.$$

Moreover, if $\mathscr{S}$ is a general subset of $\mathscr{L}(\mathscr{B}, \mathscr{B})_+$ and $\mathscr{S}^* = \{T^*; T \in \mathscr{S}\}$, then strict positivity of $\mathscr{S}$ and *-strict positivity of $\mathscr{S}^*$ are each equivalent to the condition,

$$\text{if } \omega \in \mathscr{B}_+^* \backslash \{0\} \text{ and } a \in \mathscr{B}_+ \backslash \{0\} \text{ then } \omega(Ta) > 0 \text{ for all } T \in \mathscr{S}. \qquad (*)$$

Finally, if $\mathscr{S} = \{T\}$ contains the unique operator T, we describe T as norm-irreducible instead of $\{T\}$, etc.

The various notions of irreducibility, ergodicity, and strict positivity, are generally not equivalent, as we will subsequently see. But first we examine the interrelationships between these properties for continuous semigroups. For this purpose it is useful to introduce various notions of sharpness of the cone $\mathscr{B}_+$.

First, for $\mathscr{A} \subseteq \mathscr{B}_+$ define $\mathscr{A}^\perp$ by

$$\mathscr{A}^\perp = \{\omega; \omega \in \mathscr{B}_+^*, \omega(a) = 0 \text{ for all } a \in \mathscr{A}\}$$

and for $\mathscr{A} \subseteq \mathscr{B}_+^*$ define $\mathscr{A}^\perp$ by

$$\mathscr{A}^\perp = \{a; a \in \mathscr{B}_+, \omega(a) = 0 \text{ for all } \omega \in \mathscr{A}\}.$$

Now $\mathscr{B}_+(\mathscr{B}_+^*)$ is defined to satisfy the *positive bipolar property* (*positive *-bipolar property*) if

$$\mathscr{C}^{\perp\perp} = \bar{\mathscr{C}}$$

for each hereditary subcone $\mathscr{C}$ of $\mathscr{B}_+(\mathscr{B}_+^*)$ where the bar denotes norm- (weak*-) closure. Less stringently $\mathscr{B}_+(\mathscr{B}_+^*)$ is defined to be *sharp* (**-sharp*) if each hereditary subcone $\mathscr{C}$ of $\mathscr{B}_+(\mathscr{B}_+^*)$ for which $\mathscr{C}^\perp = \{0\}$ is norm- (weak*-) dense in $\mathscr{B}_+(\mathscr{B}_+^*)$. Finally $\mathscr{B}_+(\mathscr{B}_+^*)$ is defined to be *n-sharp* (*w*-sharp*) if there is no proper norm- (weak*-) closed hereditary subcone $\mathscr{C}$ of $\mathscr{B}_+(\mathscr{B}_+^*)$ for which $\mathscr{C}^\perp = \{0\}$.

If $\mathscr{B}_+(\mathscr{B}_+^*)$ satisfies the positive bipolar (*-bipolar) property, then it is sharp (*-sharp), and if it is sharp (*-sharp), then it is *n*-sharp (*w**-sharp). Conversely, if $\mathscr{B}_+(\mathscr{B}_+^*)$ is *n*-sharp (*w**-sharp) and the norm-closure (weak*-closure) of every hereditary cone in $\mathscr{B}_+(\mathscr{B}_+^*)$ is hereditary, then $\mathscr{B}_+(\mathscr{B}_+^*)$ is sharp (*-sharp).

After these preliminaries we return to the examination of irreducibility criteria for semigroups.

Strict positivity is generally stronger than irreducibility, but this latter property is usually equivalent to ergodicity, at least with some sharpness assumption.

THEOREM 2.5.1. *Let $(\mathscr{B}, \mathscr{B}_+, \|\cdot\|)$ be an ordered Banach space with $\mathscr{B}_+$ proper and weakly generating. Let S be a positive C_0-semigroup of type γ on $\mathscr{B}$ with generator H, and let $\alpha > 0$ satisfy $\alpha\gamma < 1$. Consider the following conditions:*

1. *S is norm-ergodic,*

1*. *S^* is weak*-ergodic,*

2. *S is norm-irreducible,*

2*. *S* is weak*-irreducible,*

3. $(I + \alpha H)^{-1}$ *is norm-irreducible,*

3*. $(I + \alpha H^*)^{-1}$ *is weak*-irreducible,*

4. $(I + \alpha H)^{-1}$ *is strictly positive,*

4*. $(I + \alpha H^*)^{-1}$ *is *-strictly positive,*

5. *for each* $a \in \mathscr{B}_+\backslash\{0\}$ *and* $\omega \in \mathscr{B}^*_+\backslash\{0\}$ *there is a* $t > 0$ *such that* $\omega(S_t a) > 0$.

It follows that $1 \Rightarrow 2 \Leftrightarrow 3 \Rightarrow 4 \Leftrightarrow 5$ *and* $1^* \Rightarrow 2^* \Leftrightarrow 3^* \Rightarrow 4^* \Leftrightarrow 5$. *Moreover, if* $\mathscr{B}_+$ *is n-sharp* $5 \Rightarrow 2$; *if* $\mathscr{B}_+$ *is sharp* $5 \Rightarrow 1$; *if* $\mathscr{B}^*_+$ *is w*-sharp* $5 \Rightarrow 2^*$; *and if* $\mathscr{B}^*_+$ *is *-sharp* $5 \Rightarrow 1^*$.

Proof. The proofs of $1 \Rightarrow 2 \Rightarrow 5$, $1^* \Rightarrow 2^* \Rightarrow 5$, and the converse statements under sharpness conditions are all elementary.

$4 \Leftrightarrow 5 \Leftrightarrow 4^*$. This follows from the formulae

$$\omega((I + \alpha H)^{-1}a) = ((I + \alpha H^*)^{-1}\omega)(a) = \int_0^\infty \mathrm{d}t\, \mathrm{e}^{-t}\omega(S_{\alpha t}a).$$

$3 \Rightarrow 2$. Any S-invariant norm-closed (hereditary) subcone is invariant under $(I + \alpha H)^{-1}$ because of the relation

$$(I + \alpha H)^{-1} = \int_0^\infty \mathrm{d}t\, \mathrm{e}^{-t}S_{\alpha t}.$$

$2 \Rightarrow 3$. Let $\mathscr{C}$ be a norm-closed hereditary subcone of $\mathscr{B}_+$ which is invariant under $(I + \alpha H)^{-1}$. If $\alpha(2 - \alpha\gamma)^{-1} < \beta \leqslant \alpha$ the relations

$$0 \leqslant (I + \beta H)^{-1} \leqslant (\alpha/\beta) \sum_{n \geqslant 0} \left(\frac{\alpha - \beta}{\beta}\right)^n (I + \alpha H)^{-n-1}$$

show that $\mathscr{C}$ is invariant under $(I + \beta H)^{-1}$. Iteration of the argument establishes that $\mathscr{C}$ is invariant under $(I + \beta H)^{-1}$ whenever $0 < \beta \leqslant \alpha$. Since

$$S_t a = \lim_{n \to \infty} \left(I + \frac{t}{n} H\right)^{-n} a$$

it follows that $\mathscr{C}$ is S-invariant.

$2^* \Leftrightarrow 3^*$. The proof is analogous to the above. □

Theorem 2.5.1 demonstrates a disparity between strict positivity of a C_0-semigroup S and strict positivity of the resolvents $(I + \alpha H)^{-1}$. If $\mathscr{B}_+$ is n-sharp strict positivity of $(I + \alpha H)^{-1}$ is equivalent to norm-irreducibility of S, but this is generally a weaker property than strict positivity of S, e.g., the semigroup S of rotations described in Example 2.2.13 is norm-irreducible but not strictly positive.

Although spaces exist for which the cone $\mathscr{B}_+$ is not sharp this property does occur in many important cases, as the subsequent examples show. In the first we use the following criterion: if $\mathscr{C}$ is a (norm-closed) hereditary subcone of $\mathscr{B}_+$ then $\mathscr{B}_+$ is (n-) sharp if, and only if, $\mathscr{C}$ is norm-dense in $\mathscr{B}_+$ whenever $\mathscr{C} - \mathscr{B}_+$ is norm-dense in

$\mathscr{B}$. This follows from the bipolar theorem, Theorem A3 of the appendix. A similar criterion is valid for *-sharpness.

EXAMPLE 2.5.2 (Order-unit spaces). Let $(\mathscr{B}, \mathscr{B}_+, \|\cdot\|)$ be an ordered Banach space with int $\mathscr{B}_+ \neq \emptyset$. Then $\mathscr{B}_+$ is sharp, but $\mathscr{B}_+$ does not necessarily have the positive bipolar property even for $\mathscr{B}$ finite-dimensional. To prove sharpness suppose $u \in \operatorname{int} \mathscr{B}_+$, so $\{a; \|u - a\| < \varepsilon\} \subset \operatorname{int} \mathscr{B}_+$ for some $\varepsilon > 0$. Moreover let $\mathscr{C}$ be a hereditary subcone of $\mathscr{B}_+$ such that $\mathscr{C} - \mathscr{B}_+$ is norm-dense in $\mathscr{B}$. Then for $a \in \mathscr{B}_+$ there exist $b \in \mathscr{B}_+$ and $c \in \mathscr{C}$ such that $\|a + u + b - c\| < \varepsilon$. Therefore $c - a - b \in \mathscr{B}_+$ so $0 \leqslant a \leqslant c - b \leqslant c \in \mathscr{C}$. Since $\mathscr{C}$ is hereditary, $a \in \mathscr{C}$. Thus $\mathscr{B}_+$ is sharp. If, however, $\mathscr{B}$ is three-dimensional and $\mathscr{B}_+$ is the cone with base

$$\{(x, y, z); z = 1, (x, y) \in (K_+ \cup K_- \cup K)\}$$

where $K_\pm = \{(x, y); x^2 + (y \pm 1)^2 \leqslant 1\}$, $K = \{(x, y), |x| \vee |y| \leqslant 1\}$, then $\mathscr{B}_+$ does not have the positive bipolar property, e.g., $\mathscr{C}^{\perp\perp} \neq \mathscr{C}$ for the closed hereditary cone $\mathscr{C} = \{(x, x, x); x \geqslant 0\}$. In fact $\mathscr{C}^{\perp\perp} = \{(x, y, x); |y| \leqslant x\}$. □

EXAMPLE 2.5.3 (Banach lattices). Let $(\mathscr{B}, \mathscr{B}_+, \|\cdot\|)$ be a Banach lattice then $\mathscr{B}_+$ has the positive bipolar property. To deduce this suppose $\mathscr{C}$ is an hereditary subcone of $\mathscr{B}_+$. Since $\mathscr{C} \subseteq \mathscr{C}^{\perp\perp}$, it suffices to prove that if $\omega \in \mathscr{B}^*$ and $\omega(\mathscr{C}) = 0$, then $\omega(\mathscr{C}^{\perp\perp}) = 0$. But this follows because $(\pm\omega \vee 0)(\mathscr{C}) = 0$, by the formula quoted in the proof of Theorem 1.5.4, and hence $\omega \in \mathscr{C}^\perp - \mathscr{C}^\perp$. It can also be shown that the dual cone $\mathscr{B}_+$ has the positive *-bipolar property.

EXAMPLE 2.5.4 (C^*-algebras). Let $(\mathscr{B}, \mathscr{B}_+, \|\cdot\|)$ be the self-adjoint part of a C^*-algebra $\mathfrak{A}$ ordered by the positive elements. The (norm-closed) hereditary subcones of $\mathscr{B}_+$ are the sets of the form $\mathscr{B}_+ \cap L$ where L is a (norm-closed) left ideal in $\mathfrak{A}$. Furthermore $\mathscr{B}_+$ has the positive bipolar property and $\mathscr{B}^*_+$ is w^*-sharp, but not necessarily *-sharp [23, 60]. If S is strongly positive in the sense that

$$\|S_t\| S_t(a^* a) \geqslant S_t(a)^* S_t(a)$$

for all $a \in \mathfrak{A}$, where S is extended linearly to $\mathfrak{A}$, then L is S-invariant if, and only if, $\mathscr{B}_+ \cap L$ is S-invariant. Hence each of the conditions of Theorem 2.5.1 is satisfied if, and only if, there are no proper norm-closed S-invariant left ideals in $\mathfrak{A}$. □

EXAMPLE 2.5.5 (von Neumann algebras). Let $(\mathscr{B}, \mathscr{B}_+, \|\cdot\|)$ be the self-adjoint part of a von Neumann algebra $\mathscr{M}$. The weak*-closed hereditary subcones of $\mathscr{B}_+$ are the sets of the form $\mathscr{B}_+ \cap L$ where L is a weak*-closed left ideal in $\mathscr{M}$, or, equivalently, the sets of the form $\{x; x \in \mathscr{B}_+, xp = 0\}$ where p is a projection in $\mathscr{B}$. Furthermore $\mathscr{B}_+$ has the positive *-bipolar property and $\mathscr{B}_{*+}$, the positive cone of the predual $\mathscr{B}_*$, has the positive bipolar property [23, 60]. Hence a positive C_0^* semigroup on $\mathscr{M}$ is weak*-irreducible, or weak*-ergodic, if, and only if, there is no non-trivial projection $p \in \mathscr{M}$ with $S_t(p)(1 - p) = 0$ for all t. □

After these positive examples we give two examples for which sharpness fails.

EXAMPLE 2.5.6. Let $\mathscr{A}$ be the space of l^1-sequences $a = \{a_n\}_{n \geq 1}$, equipped with the l^1-norm $\|a\| = \Sigma |a_n|$, and ordered by the cone $\mathscr{A}_+$ of positive sequences in $\mathscr{A}$. Moreover, let $\mathscr{E}$ be the one-dimensional subspace spanned by the sequence $\{(-2)^{-n}\}_{n \geq 1}$ and $\mathscr{B} = \mathscr{A}/\mathscr{E}$ the quotient space, equipped with the quotient norm, and ordered by the image of $\mathscr{E}_+$ under the quotient map. One can establish that $\mathscr{B}_+$ is generating with a base, but $\mathscr{B}_+$ is not sharp. □

EXAMPLE 2.5.7. Let $\mathscr{B}$ be the space of l^∞-sequences $a = \{a_n\}_{n \geq 1}$ such that $a_n \to 0$ as $n \to \infty$ with the l^∞-norm, $\|a\| = \sup |a_n|$. Then $\mathscr{B}^*$ is isomorphic to the space of l^1-sequences $\omega = \{\omega_n\}_{n \geq 1}$ with the norm $\|\omega\|^* = \Sigma |\omega_n|$. The duality is given by $\omega(a) = \Sigma \omega_n a_n$. Let

$$F = \{\omega \in \mathscr{B}^* ; \omega_1 = 1, \omega_n \geq 0 \text{ if } n \geq 3, \sum_{n \geq 3} n\omega_n \leq 1, \sum_{n \geq 2} \omega_n = 0\}$$

$$X = \{\omega \in \mathscr{B}^* ; \omega_1 = 1, \omega_n \geq 0 \text{ if } n \geq 2, \sum_{n \geq 2} \omega_n \leq 1\}$$

and let K be the convex hull of $F \cup X$. Define $\mathscr{B}^*_+$ as the cone with the base K. Then $\mathscr{B}^*_+$ is generating with a weak*-compact base, but $\mathscr{B}^*_+$ is not *-sharp. □

The theory of positive semigroups in finite dimensions is particularly simple because strict positivity and norm-irreducibility coincide. The next example shows that this simplification also occurs for norm-continuous semigroups on $C(X)$ i.e., on a commutative C^*-algebra.

EXAMPLE 2.5.8 (Continuous functions). If $\mathscr{B} = C(X)$, ordered by the pointwise positive functions $\mathscr{B}_+$, then the (norm-closed) hereditary subcones of $\mathscr{B}_+$ are of the form

$$\mathscr{C}_E = \{f ; f \geq 0, f(x) = 0 \text{ for } x \in E\}$$

for (closed) subsets E of X. Hence each of the nine conditions of Theorem 2.5.1 is satisfied if, and only if, for each proper closed E there exists an $f \in \mathscr{B}$ and $t > 0$ such that f is supported by E but $S_t f$ is not.

If S is uniformly continuous and positive, its generator H is bounded and $\|H\| I - H$ is positive, [17] Theorem 7.21. The expansion

$$S_t = e^{-t\|H\|} \sum_{n \geq 0} (\|H\| I - H)^n t^n / n!$$

shows that if S is norm-irreducible then $(\|H\| I - H)$ is norm-irreducible. Conversely, if $(\|H\| I - H)$ is norm-irreducible, then for each $\omega \in \mathscr{B}^*_+ \backslash \{0\}$ and $f \in \mathscr{B}_+ \backslash \{0\}$ one has $\omega((\|H\| I - H)^n f) > 0$ for some $n \geq 0$ so $\omega(S_t f) > 0$. Thus S is strictly positive. This establishes equivalence of the following conditions

1. S is norm-irreducible,
2. S is strictly positive,
3. $\|H\| I - H$ is norm-irreducible. □

The equivalence of strict positivity and norm-irreducibility extends to a large class of positive C_0-semigroups on the classical function spaces. The key features of this extension are the Riesz interpolation property for the space and an analyticity property of the semigroup.

A C_0-semigroup on the Banach space $\mathscr{B}$ is *holomorphic* if for some $\theta > 0$ there exists an extension of S to a holomorphic family $\{S_z ; z \in \mathbb{C}, |\arg z| < \theta\}$ of bounded linear operators on the complexification of $\mathscr{B}$ such that $S_{z_1} S_{z_2} = S_{z_1 + z_2}$ and

$$\lim_{\substack{z \to 0 \\ |\arg z| < \theta}} \|S_z a - a\| = 0$$

for all $a \in \mathscr{B}$. For example all uniformly continuous C_0-semigroups are holomorphic.

THEOREM 2.5.9. *Let $(\mathscr{B}, \mathscr{B}_+, \|\cdot\|)$ be an ordered Banach space with the Riesz interpolation property for which $\mathscr{B}_+$ is normal and generating. Furthermore, let S be a positive holomorphic semigroup on $\mathscr{B}$.*

The following conditions are equivalent:

1. *S is strictly positive,*
2. *S is norm-irreducible,*
3. *S is norm-ergodic,*
4. *for each $a \in \mathscr{B}_+ \setminus \{0\}$ and $\omega \in \mathscr{B}^*_+ \setminus \{0\}$ there is a $t > 0$ such that*

$$\omega(S_t a) > 0.$$

The proof relies upon two lemmas.

LEMMA 2.5.10. *Let $(\mathscr{B}, \mathscr{B}_+, \|\cdot\|)$ be an ordered Banach space with the Riesz interpolation property and suppose $\mathscr{B}_+$ is α-monotone and β-dominating. Suppose $a, a_n \in \mathscr{B}_+$ and $\Sigma \|a - a_n\| < +\infty$. Then there exists $b \in \mathscr{B}_+$ such that $b \leqslant a_n$, $n \geqslant 1$, and*

$$\|a - b\| \leqslant \alpha\beta \Sigma \|a - a_n\|.$$

Proof. There exist $c_n \in \mathscr{B}_+$ such that $c_n \geqslant a - a_n$ and $\|c_n\| \leqslant \beta \|a - a_n\|$. Now $a - c_1 \leqslant a_1$, $a - c_1 \leqslant a$, $0 \leqslant a_1$, $0 \leqslant a$, so by the Riesz interpolation property there exists b_1 such that $a - c_1 \leqslant b_1 \leqslant a_1$, $0 \leqslant b_1 \leqslant a$. Since $0 \leqslant a - b_1 \leqslant c_1$

$$\|a - b_1\| \leqslant \alpha \|c_1\| \leqslant \alpha\beta \|a - a_1\|.$$

Now $b_1 - c_2 \leqslant a_2$, $b_1 - c_2 \leqslant b_1$, $0 \leqslant a_2$, $0 \leqslant b_1$, and so there exists a b_2 such that $b_1 - c_2 \leqslant b_2 \leqslant b_1$, $0 \leqslant b_2 \leqslant a_2$. Since $0 \leqslant b_1 - b_2 \leqslant c_2$

$$\|b_1 - b_2\| \leqslant \alpha \|c_2\| \leqslant \alpha\beta \|a - a_2\|.$$

Iteration of this argument gives a sequence $b_n \in \mathscr{B}_+$ such that

$$b_n \leqslant b_{n-1} \leqslant a, \qquad b_n \leqslant a_n, \qquad \|b_n - b_{n-1}\| \leqslant \alpha\beta \|a - a_n\|.$$

Hence, as $n \to \infty$, b_n converges to a limit b with the required properties. □

LEMMA 2.5.11. *Let* $(\mathscr{B}, \mathscr{B}_+, \|\cdot\|)$ *be an ordered Banach space with the Riesz interpolation property and suppose* $\mathscr{B}_+$ *is normal and generating. Let S be a positive holomorphic semigroup on* $\mathscr{B}$. *If* $a \in \mathscr{B}_+$, $\omega \in \mathscr{B}^*_+$ *then*

$$\text{either} \quad \omega(S_t a) = 0 \quad \text{for all } t > 0$$
$$\text{or} \quad \omega(S_t a) > 0 \quad \text{for all } t > 0.$$

Proof. Suppose $\omega(S_{t_0} a) = 0$ for some $t_0 > 0$. Let t_n be a sequence of positive numbers decreasing to 0 such that $\Sigma \| S_{t_n} a - a \| < \infty$. By Lemma 2.5.10 there is a sequence $a_m \in \mathscr{B}_+$ such that $a_m \leqslant S_{t_m} a$ for $m \leqslant n$ and $\| a_m - a \| \to 0$ as $m \to \infty$. Therefore,

$$0 \leqslant \omega(S_{t_0 - t_n} a_m) \leqslant \omega(S_{t_0} a) = 0$$

for $n \geqslant m$. Thus the zeros of the analytic function $t \mapsto \omega(S_t a_m)$ have a limit point at $t = t_0$, so $\omega(S_t a_m) = 0$ for all t. Therefore $\omega(S_t a) = 0$ for all t, by continuity. □

Now the proof of Theorem 2.5.9 is immediate.

Proof of 2.5.9. $1 \Leftrightarrow 4$. This follows from Lemma 2.5.11.

$2 \Leftrightarrow 3 \Leftrightarrow 4$. Since $\mathscr{B}$ has the Riesz interpolation property, $\mathscr{B}_+$ has the positive bipolar property by the argument used in Example 2.5.3. The required equivalence then follows from Theorem 2.5.1. □

Under the conditions of Theorem 2.5.9 strict positivity of S is also equivalent to weak*-irreducibility of the adjoint semigroup. This follows from Theorem 2.5.1 because the Riesz interpolation property implies that the dual cone $\mathscr{B}^*_+$ is w^*-sharp.

If $(\mathscr{B}, \mathscr{B}_+, \|\cdot\|)$ does not have the Riesz interpolation property, the conclusion of Theorem 2.5.9 can fail, see Example 2.5.14 below.

EXAMPLE 2.5.12 (L^p-spaces) Let $\mathscr{B} = L^p(X; d\mu)$, where $1 \leqslant p < \infty$, ordered by the pointwise positive functions $\mathscr{B}_+$ in $\mathscr{B}$. The norm-closed hereditary subcones of $\mathscr{B}_+$ are of the form

$$\mathscr{C}_E = \{ f\,;\, f \in \mathscr{B}_+, f(x) = 0 \text{ for almost all } x \in E \}$$

for measurable subsets E of X. at least if $\mathscr{B}$ is separable. Each of the nine conditions of Theorem 2.5.1 is satisfied if, and only if, for each E there exists an $f \in \mathscr{B}_+$ and $t > 0$ such that f vanishes almost everywhere in E but $S_t f$ does not, even if $\mathscr{B}$ is non-separable.

Now suppose S is a positive holomorphic semigroup on $\mathscr{B}$ so that the nine conditions of Theorem 2.5.1 are equivalent to strict positivity of S, by Theorem 2.5.9. If $f \in \mathscr{B}$ and $\omega \in \mathscr{B}^* = L^q(X; d\mu)$, where $p^{-1} + q^{-1} = 1$, then $f = h|f|$, $\omega = k|\omega|$, for some $h, k \in L^\infty(X; d\mu)$ with $|h| = |k| = 1$ and

$$|\omega|(S_t|f|) = \omega(kS_t(hf)).$$

Hence, if S_t is strictly positive, there is no proper closed subspace of $\mathscr{B}$ which is invariant under both S and the action of multiplication by $L^\infty(X; d\mu)$.

Conversely, suppose S is not strictly positive, so $\omega(S_t f) = 0$ for some, and hence all, $t > 0$ and for some $f \in \mathscr{B}_+ \setminus \{0\}$, $\omega \in \mathscr{B}_+^* \setminus \{0\}$. But if $g \in L^\infty(X; d\mu)$, then $|gS_t f| \leqslant \leqslant \|g\|_\infty S_t f$ and hence $\omega(gS_t f) = 0$. Moreover, $|S_s(gS_t f)| \leqslant \|g\|_\infty S_{s+t} f$ so that $S_s(gS_t f) = hS_{s+t} f$ for some $h \in L^\infty(X; d\mu)$. Thus the closed linear span of $\{gS_t f; t > 0, g \in L^\infty(X; d\mu)\}$ is a proper subspace invariant under both S and the action of $L^\infty(X; d\mu)$.

Thus S is strictly positive if, and only if, there are no proper subspaces invariant under S and multiplication by $L^\infty(X; d\mu)$.

If $\mathscr{B} = L^\infty(X; d\mu)$ ordered by the cone of pointwise positive functions in $\mathscr{B}$, then any positive C_0-semigroup is uniformly continuous, Example 2.2.17, and therefore holomorphic. Thus the nine conditions of Theorem 2.5.1 are equivalent to strict positivity of S (see Example 2.5.8). Irreducibility criteria for C_0^*-semigroups are given by the above discussion of C_0-semigroups on the predual $L^1(X; d\mu)$ (see also Example 2.5.5). □

Finally we examine analogues of the second part of the Perron–Frobenius theory of positive matrices, the spectral properties of positive irreducible matrices. Frobenius' theorem states that if A is a positive irreducible $n \times n$ matrix, there is an mth root of unity θ, for some $m \in [1, n]$, such that

1. $\{z; z \in \sigma(A), |z| = \rho\} = \{\rho\theta^r; 1 \leqslant r \leqslant m\}$,
2. $\rho\theta^r$ is a simple eigenvalue of A, $1 \leqslant r \leqslant n$,
3. ρ is the unique eigenvalue of A with a positive eigenvector,
4. $\theta\sigma(A) = \sigma(A)$.

Here ρ denotes the spectral radius of A and this information is about the periphery of the spectrum. This suggests that one might be able to derive structural results about the periphery of the spectrum of positive semigroups.

Let S be a positive semigroup with generator H then the *peripheral (point) spectrum* is the set of points z in the (point) spectrum of H such that $\operatorname{Re} z = m_H$, Theorem 2.4.2 shows that m_H is in the peripheral spectrum if $\sigma(H) \neq \emptyset$. Little more is known about this spectrum except for Banach lattices, and to some extent C^*-algebras. The only results for general ordered Banach spaces concern the nature of the poles of the resolvent in the peripheral spectrum [35]. For Banach lattices it is possible to obtain fairly strong results even for reducible semigroups.

THEOREM 2.5.13. [34]. *Let H be the generator of a positive C_0-semigroup on a Banach lattice. Suppose that $\sigma(H) \neq \emptyset$.and*

$$\sup_{\lambda < m_H} (m_H - \lambda) \|(\lambda I - H)^{-1}\| < +\infty.$$

If $m_H + iy \in \sigma(H)$ then $m_H + iny \in \sigma(H)$, $n \in \mathbb{Z}$.

It is not known whether the growth condition on the resolvents can be omitted in this theorem, but it is automatically satisfied if $m_H = -\gamma_S$.

These conclusions fail, however, for non-commutative C^*-algebras.

EXAMPLE 2.5.14. Let $\mathscr{B}$ be the space of hermitian 2×2 matrices and $\mathscr{B}_+$ the cone of positive definite matrices in $\mathscr{B}$. Define a positive C_0-semigroup S on $\mathscr{B}$ by

$$S_t A = \begin{pmatrix} 1 & 0 \\ 0 & e^{it} \end{pmatrix} A \begin{pmatrix} 1 & 0 \\ 0 & e^{-it} \end{pmatrix}.$$

Then

$$HA = \begin{pmatrix} 0 & 0 \\ 0 & i \end{pmatrix} A - A \begin{pmatrix} 0 & 0 \\ 0 & i \end{pmatrix}.$$

and $\sigma(H) = \{0, i, -i\}$. □

For irreducible semigroups on Banach lattices and C^*-algebras one can often say something about the peripheral point spectrum.

THEOREM 2.5.15. *Let H be the generator of a norm-irreducible positive C_0-semigroup on a Banach lattice $\mathscr{B}$. Suppose that the peripheral point spectrum of H is nonempty and that m_H is an eigenvalue of H^* with a positive eigenvector. It follows that*

1. *m_H is an eigenvalue of H with an eigenvector in* qu.int $\mathscr{B}_+$,
2. *m_H is the only eigenvalue of H with a positive eigenvector,*
3. *if $m_H + iy_1$ and $m_H + iy_2$ are eigenvalues of H, then $m_H + i(y_1 - y_2)$ is also an eigenvalue,*
4. *if $m_H + iy$ is an eigenvalue of H, then $\sigma(H) + iy = \sigma(H)$.*

The condition that m_H is an eigenvalue of H^* with a positive eigenvector is automatically satisfied if H satisfies the other hypotheses of Theorem 2.5.15 and also satisfies the bound

$$\sup_{\lambda < m_H} (m_H - \lambda) \| (\lambda I - H)^{-1} \| < +\infty.$$

THEOREM 2.5.16 [36]. *Let H be the generator of a norm-irreducible C_0-semigroup S on the (self-adjoint) part of a C^*-algebra $\mathfrak{A}$ with identity $\mathbb{1}$. Suppose that*

$$S_t(\mathbb{1}) = \mathbb{1},$$
$$S_t(a^* a) \geqslant S_t(a)^* S_t(a), \quad a \in \mathfrak{A},\ t \geqslant 0.$$

Then statements 2, 3 and 4 of Theorem 2.5.15 are valid (with $m_H = 0$).

In comparing the last two theorems, note that under the hypotheses of Theorem

2.5.16, $\gamma_S = -m_H$ and m_H is an eigenvalue of H^* with a positive eigenvector. This follows from Theorem 2.4.4 and 2.4.6.

Appendix: Hahn–Banach Theorems

In this appendix, we state the various forms of the Hahn–Banach Theorem used in the preceding review.

Let E be a real vector space. A functional $p: E \to (-\infty, \infty]$ is *sublinear*, if

$$p(\lambda a) = \lambda p(a), \quad a \in E, \lambda \in \mathbb{R}_+,$$

$$p(a+b) \leqslant p(a) + p(b), \quad a, b \in E.$$

THEOREM A1. (Hahn–Banach Theorem). *Let $p: E \to \mathbb{R}$ be a finite-valued sublinear functional, and $a \in E$. There is a linear functional $\omega: E \to \mathbb{R}$ with $\omega(a) = p(a)$ and $\omega(b) \leqslant p(b)$ for all $b \in E$. Furthermore, for $c \in E$ and $\lambda \in \mathbb{R}$, there exists a linear functional $\omega: E \to \mathbb{R}$ with $\omega(a) = p(a)$, $\omega(c) = \lambda$, and $\omega(b) \leqslant p(b)$ for all $b \in E$ if, and only if,*

$$\frac{p(a) - p(a - tc)}{t} \leqslant \lambda \leqslant \frac{p(a + tc) - p(a)}{t}, \quad t \geqslant 0.$$

Now let p_j $(1 \leqslant j \leqslant n)$ be sublinear functionals on E, and ω be a linear functional on E with $\omega \leqslant p_j$. If $\Sigma_{i=1}^n a_i = 0$, then

$$\sum_{i=1}^n p_i(a_i) \geqslant \sum_{i=1}^n \omega(a_i) = \omega\left(\sum_{i=1}^n a_i\right) = 0.$$

The following is a converse to this.

THEOREM A2 (Multiple Hahn–Banach Theorem). *Let $p_i: E \to (-\infty, \infty]$ be sublinear functionals $(1 \leqslant i \leqslant n)$ with p_1 finite-valued, and suppose that*

$$a_i \in E, \quad \sum_{i=1}^n a_i = 0 \Rightarrow \sum_{i=1}^n p_i(a_i) \geqslant 0.$$

Then there is a linear functional $\omega: E \to \mathbb{R}$ with $\omega \leqslant p_i$, for $1 \leqslant i \leqslant n$.

Proof. If $\Sigma_{i=1}^n a_i = a$, then

$$\sum_{i=1}^n p_i(a_i) \geqslant p_1(a_1 - a) - p_1(-a) + \sum_{i=2}^n p_i(a_i) \geqslant -p_1(-a).$$

Thus, if

$$p(a) = \inf\left\{ \sum_{i=1}^n p_i(a_i);\ \sum_{i=1}^n a_i = a \right\},$$

then p is a finite-valued sublinear functional with $-p_1(-a) \leqslant p(a) \leqslant p_i(a)$, $1 \leqslant i \leqslant n$. By Theorem A1, there is a linear functional ω with $\omega \leqslant p \leqslant p_i$. □

The Hahn–Banach Separation Theorem appears in several forms, which are summarised in Theorem A3 below. If $\mathscr{B}$ is a Banach space, the theorem may be applied either with $E = \mathscr{B}$ in the norm topology, or with $E = \mathscr{B}^*$ in the weak*-topology so that $E^* = \mathscr{B}$.

Let E be a real locally convex space, and E^* be the dual of E. For a subset F of E, let F^0 be the polar of F in E^*, so that

$$F^0 = \{\omega \in E^*; \omega(a) \leqslant 1 \text{ for all } a \in F\}.$$

Let F^{00} be the bipolar of F, that is, the polar of F^0 in E, so

$$F^{00} = \{a \in E; \omega(a) \leqslant 1 \text{ for all } \omega \in F^0\}.$$

THEOREM A3 (Hahn–Banach Separation Theorem). *Let E be a real locally convex space.*

(i) *Let C be a convex subset of E with non-empty interior* int C. *For* $a \in E \backslash \operatorname{int} C$, *there exists* $\omega \in E^*$ *such that* $\omega(a) \leqslant \omega(c)$ *for all* $c \in C$.

(ii) *Let C be a closed convex subset of E not containing* 0. *There exists* $\omega \in \mathscr{B}^*$ *such that* $\omega(c) \geqslant 1$ *for all* $c \in C$.

(iii) *For any subset F of E,* F^{00} *is the closed convex hull of* $F \cup \{0\}$ *in E.*

Notes and Remarks

Part 1. §1.0. The abstract theory of ordered topological vector spaces originated in the 1930s in the work of Kantorovich, M. Krein, F. Riesz and their collaborators. Monographs on the subject include [42, 58, 61, 84], all of which emphasise topological, rather than metric, properties. A close approximation to our approach is adopted in [6], Chapter 2.

§1.1. A duality between normality and generation was first discovered by Grosberg and Krein [38], who showed 1 ⇔ 2 in Theorem 1.1.4. The reverse duality 1′ ⇔ 2′ (without regard to the constant α) was proved in [3]; the invariance of α was established in [27]. Extended introductions to the theory of C^*-algebras (Example 1.1.7) may be found in [11, 60, 81], and to compact convex sets (Example 1.1.8) in [2]. Example 1.1.9 appears in [6].

§1.2. Proposition 1.2.1 can be found in [13] – see also [58]. Theorems 1.2.2 and 1.2.3 were proved in [71].

§1.3. Theorem 1.3.1 and Proposition 1.3.2 are both mainly due to Davies [16]. There is a detailed discussion of absolute domination and Riesz norms in [84]. Proposition 1.3.2 in its full generality appears in [67].

§1.4. Order-unit spaces were introduced by Kadison [43] who subsequently completed the proof of the equivalence 1 ⇔ 3 in Theorem 1.4.1 [45] – see also [77]. Base-norm spaces were introduced by Edwards [21] and Ellis [27], and they showed 1 ⇔ 2 and 1′ ⇔ 2′ in Theorem 1.4.1. Further details may be found in [2] and [6]. Quasi-interior points in a topological vector lattice were introduced independently by Fullerton [32] and Schaefer [74] using a definition which is equivalent to ours in their context, but differs in general.

§1.5. A comprehensive account of the theory of Banach lattices may be found in [75]. [31] and [82] consider more general types of vector lattices.

Detailed information about AL- and AM-spaces may be found in [75, 18, 78 and 33]. Theorem 1.5.1 was discovered independently by Kakutani [47, 9] and (in part) by M. and S. Krein [51]. A similar result was obtained by Stone [80]. Theorem 1.5.2 is also due to Kakutani [46].

The decomposition and separation properties were introduced by F. Riesz [64, 65], who also proved the implication 1 ⇒ 2 of Theorem 1.5.4. The converse was obtained by Andô [3]. Choquet simplexes were originally introduced in [14]. A common definition of a Choquet simplex is a compact convex set K for which $A(K)^*$ is a lattice, and this definition together with Theorem 1.5.4 and Theorem 1.4.1 reduces Statement 1 of Theorem 1.5.5 to a tautology. But the Choquet–Meyer theorem [15] permits simplexes to be defined alternatively in terms of uniqueness of representing measures. Many other characterizations of simplexes may be found in [2, 6]. Part 1 of Theorem 1.5.5 was obtained independently by Edwards [22], Linderstrauss [53] and Semadeni [77] without reference to Theorem 1.5.4. Effros [24–26] studied in detail 'simplex spaces', those ordered Banach spaces whose duals are AL-spaces.

§1.6. Half-norms were explicitly introduced by Arendt *et al.* [4], but the canonical half-norm, defined as in Proposition 1.6.2, was used by Calvert [13] to prove a version of Theorem 2.2.4. Half-norms have been studied in detail by Robinson and Yamamuro [69–71, 85]. Proposition 1.6.2 and Theorem 1.6.3 originated in [71].

§1.7. An account of the order properties of continuous linear mappings between ordered topological vector spaces is given in [61]. They were first studied by Kantorovich, who showed that $\mathscr{L}(\mathscr{A}, \mathscr{B})$ is a lattice if $\mathscr{A}_+$ is normal, generating and has the Riesz separation property and $\mathscr{B}$ is an order-complete topological vector lattice [48]. For positive operators, Proposition 1.7.2 was stated by Nachbin [56] – see also [58, 75]; for order-bounded operators, it appeared in [10]. Theorem 1.7.3 was proved in an abstract form in [73], and in this form in [10]. Corollary 1.7.5 appears in [85]. The notion of positive attainment was introduced by Robinson, and Proposition 1.7.8 arises from [66–67]. Another sufficient condition (of L^p-type) for positive attainment has been given by Yamamuro [85].

Part 2. §2.0. There are numerous books on one-parameter semigroups of operators, for example [12, 17, 40]. The Hille–Yosida theorem was proved in 1948 and the Feller–Miyadera–Phillips theorem in 1952, and both are named after their originators who worked independently. Original references are given in the books mentioned above.

§2.1. In the case when $p = \|\cdot\|$, the equivalence $2 \Leftrightarrow 3$ of Theorem 2.1.1 was proved in [7]; in the more general form, it appeared in [4]. The history of the equivalence $1 \Leftrightarrow 2$ is obscure. Norm-dissipative operators were introduced by Lumer and Phillips [54], who proved Theorem 2.1.3 for $p = \|\cdot\|$. The general version of Theorem 2.1.3, as well as Proposition 2.1.4, were given in [4]. Dispersive operators on Banach lattices were introduced by Phillips [62] – see also [39, 72]. The negative off-diagonal property as a criterion for positivity of uniformly continuous semigroups was discussed by Evans and Hanche-Olsen [29]. A version of Corollary 2.1.5 occurs in [4].

§2.2. Theorem 2.2.1 originated in [66], where it was stated for Riesz norms (Corollary 2.2.3). Theorem 2.2.4 was proved in [67]. Theorem 2.2.7 was proved in [4], and was further discussed in [10], which also included Proposition 2.2.9 and Example 2.2.10. An example similar to 2.2.11, but on the Banach space $C_0(\mathbb{R})$, was given in [8]. The algebraic characterization of bounded positive generators on C^*-algebras given in Example 2.2.16 is due to Evans and Hanche-Olsen [29]. The absence of positive C_0-semigroups on $L^\infty(X; d\mu)$ (Example 2.2.17) was discovered by Kishimoto and Robinson [50].

§2.3. Theorems 2.3.1 (for Riesz norms) and 2.3.2 appeared in [64] and [65], respectively. Proposition 2.3.3, Theorem 2.3.4 and Example 2.3.6 were all given in [10]. The ideas underlying Example 2.3.7 are discussed in detail in [11].

§2.4. The Perron–Frobenius theory of matrices is summarized in [75], and is included in many books on matrices. Part 2 of Lemma 2.4.1 is due to Greiner *et al.* [35], who also proved Theorem 2.4.2 and Corollary 2.4.3. Derndinger [19] proved Theorem 2.4.4 for C_0-semigroups on AL-spaces and AM-spaces with unit, and the C_0-version of Corollary 2.4.5 appeared in [8]. Theorem 2.4.6 is a semigroup version of a celebrated result of Krein and Rutman [52]. This version and Corollary 2.4.7 (for C^*-algebras with identity), as well as Example 2.4.9, appeared in [37]. The phenomenon of Example 2.4.8 was exhibited in [35]. Further discussion of the equality $\gamma_S = -m_H$ may be found in [57].

§2.5. The general relationship between irreducibility, strict positivity and the positive bipolar property, was discussed in [10]. The latter property is related to Kadison's notion of 'Archimedean' order ideals [43] – see also [2, 70]. Irreducibility had previously been considered separately for Banach lattices [75] and for C^*-algebras [28]. Example 2.5.6 originated in [5]. Theorem 2.5.9 is the end of a chain of results through [79, 49, 55]. The criterion for strict positivity given in Example 2.5.12 developed from the case of self-adjoint semigroups on $L^2(X; d\mu)$ (see [49, 63] and the references cited therein) to its present form in [50]. Schaefer [76] has surveyed the theory of positive semigroups on Banach lattices up to 1980. Theorems 2.5.13 and 2.5.15 are analogues (due to Greiner) of older results for single operators [75]. Further spectral information may be found in [19, 20, 34, 35]. Theorem 2.5.16 is a semigroup version of a result in [36], which itself extended a finite-dimensional result in [20]. Another theorem of Perron–Frobenius type for operator algebras may be found in [1].

References

1. Albeverio, S. and Høegh-Krohn, R.: *Commun. Math. Phys.* **64** (1978), 83–94.
2. Alfsen, E. M.: *Compact Convex Sets and Boundary Integrals*, Springer-Verlag, Berlin, 1971.
3. Andô, T.: *Pacif. J. Math.* **12** (1962), 1163–1169.
4. Arendt, W., Chernoff, P. R. and Kato, T.: *J. Operator Theory* **8** (1982), 167–180.
5. Asimow, L.: *Pacif. J. Math.* **35** (1970), 11–21.
6. Asimow, L. and Ellis, A. J.: *Convexity Theory and Its Applications in Functional Analysis*, Academic Press, London, 1980.
7. Batty, C. J. K.: *J. Lond. Math. Soc.* **18** (1978), 527–533.
8. Batty, C. J. K. and Davies, E. B.: 'Positive semigroups and resolvents'. *J. Operator Theory* **10** (1983), 357–363.
9. Bohnenblust, F. and Kakutani, S.: *Ann. of Math.* **42** (1941), 1025–1028.
10. Bratteli, O., Digernes, T. and Robinson, D. W.: *J. Operator Theory* **9** (1983), 371–400.
11. Bratteli, O. and Robinson, D. W.: *Operator Algebras and Quantum Statistical Mechanics*, vol. I, Springer-Verlag, Berlin, 1979.
12. Butzer, P. L. and Berens, H.: *Semigroups of Operators and Approximation*, Springer-Verlag, Berlin, 1967.
13. Calvert, B. D.: *J. Math. Soc. Japan*, **23** (1971), 311–319.
14. Choquet, G.: 'Existence des représentations intégrales au moyen des points extrémaux dans les cônes convexes', *Sem. Bourbaki*, **139** (1956).
15. Choquet, G. and Meyer, P. A.: *Ann. Inst. Fourier (Grenoble)*, **13** (1963), 139–154.
16. Davies, E. B.: *Trans. Amer. Math. Soc.* **131** (1968), 544–555.
17. Davies, E. B.: *One-parameter Semigroups*, Academic Press, London, 1980.
18. Day, M. M.: *Normed Linear Spaces*, Springer-Verlag, Berlin, 1973.
19. Derndinger, R.: *Math. Z.* **172** (1980), 281–293.
20. Derndinger, R. and Nagel, R.: *Math. Ann.* **245** (1979), 159–177.
21. Edwards, D. A.: *Proc. Lond. Math. Soc.* **14** (1964), 399–414.
22. Edwards, D. A.: *C.R. Acad. Sci. Paris*, **261** (1965), 2798–2800.
23. Effros, E. G.: *Duke Math. J.* **30** (1963), 391–412.
24. Effros, E. G.: *Acta Math.* **117** (1967), 103–121.
25. Effros, E. G.: *J. Funct. Anal.* **1** (1967), 361–391.
26. Effros, E. G. and Gleit, A.: *Trans. Amer. Math. Soc.* **142** (1969), 355–379.
27. Ellis, A. J.: *J. Lond. Math. Soc.* **39** (1964), 730–744.
28. Enomoto, M. and Watatani, S.: *Math. Japon.* **24** (1979), 53–63.
29. Evans, D. E. and Hanche-Olsen, H.: *J. Funct. Anal.* **32** (1979), 207–212.
30. Evans, D. E. and Høegh-Krohn, R.: *J. Lond. Math. Soc.* **17** (1978), 345–355.
31. Fremlin, D. H.: *Topological Riesz Spaces and Measure Theory*, Cambridge University Press, Cambridge, 1974.
32. Fullerton, R. E.: *Quasi-interior Points of Cones in a Linear Space*, ASTIA Doc. No. AD-120406 (1957).
33. Goullet de Rugy, A.: *J. Math. Pures Appl.* **51** (1972), 331–373.
34. Greiner, G.: *Math. Z.* **177** (1981), 401–423.
35. Greiner, G., Voigt, J. and Wolff, M.: *J. Operator Theory*, **5** (1981), 245–256.
36. Groh, U.: *Math. Z.* **176** (1981), 311–318.
37. Groh, U. and Neubrander, F.: *Math. Ann.* **256** (1981), 509–516.
38. Grosberg, J. and Krein, M. G.: *C.R. Dokl. Acad. Sci. URSS*, **25** (1939), 723–726.
39. Hasegawa, M.: *J. Math. Soc. Japan*, **18** (1966), 290–302.
40. Hille, E. and Phillips, R. S.: *Amer. Math. Soc. Coll. Publ.* **31** (1957), Providence, R. I.
41. Holmes, R. B.: *Geometric Functional Analysis and Its Applications*, Springer-Verlag, Berlin, 1975.
42. Jameson, G. J. O.: 'Ordered linear spaces', *Lecture Notes in Math.*, vol. 141, Springer-Verlag, Berlin, 1970.
43. Kadison, R. V.: *Mem. Amer. Math. Soc.* **7** (1951).
44. Kadison, R. V.: *Ann. of Math.* **56** (1952), 494–503.
45. Kadison, R. V.: *Topology*, **3** (1965), 177–198.

46. Kakutani, S.: *Ann. of Math.* **42** (1941), 523–537.
47. Kakutani, S.: *Ann. of Math.* **42** (1941), 994–1024.
48. Kantorovich, L. V.: *Dokl. Akad. Nauk. SSSR*, **1** (1936), 271–276.
49. Kishimoto, A. and Robinson, D. W.: *Commun. Math. Phys.* **75** (1980), 85–101.
50. Kishimoto, A. and Robinson, D. W.: *J. Austral. Math. Soc.* (Series A), **31** (1981), 59–76.
51. Krein, M. G. and Krein, S.: *C.R. Dokl. Acad. Sci. URSS*, **27** (1940), 427–430.
52. Krein, M. G. and Rutman, M. A.: *Uspeki Mat. Nauk.* **3** (1948), 3–95; *Amer. Math. Soc. Transl.* **10** (1950), 199–325.
53. Lindenstrauss, J., *Mem. Amer. Math. Soc.* **48** (1964).
54. Lumer, G. and Phillips, R. S.: *Pacif. J. Math.* **11** (1961), 697–698.
55. Majewski, A. and Robinson, D. W.: 'Strictly positive and strongly positive semigroups', *J. Austral. Math. Soc.* (Series B). (To appear).
56. Nachbin, L.: *Proc. International Congress of Mathematicians 1950*, vol. I, Amer, Math. Soc., Providence, R. I., 1952, pp. 464–465.
57. Nagel, R.: 'Zur Characterisierung stabiler Operatorhalbgruppen', *Semesterbericht Funktionanalysis, Tübingen*, 1981/82, pp. 99–119.
58. Namioka, I.: *Mem. Amer. Math. Soc.* **24** (1957).
59. Ogasawara, T.: *J. Sci. Hiroshima Univ.* **18** (1955), 307–309.
60. Pedersen, G. K.: *C*-algebras and their Automorphism Groups*, Academic Press, London, 1979.
61. Peressini, A. L.: *Ordered Topological Vector Spaces*, Harper and Row, New York, 1967.
62. Phillips, R. S.: *Czech. Math. J.* **12** (1962), 294–313.
63. Reed, M. and Simon, B.: *Methods of Modern Mathematical Physics IV: Analysis of Operators*, Academic Press, New York, 1978.
64. Riesz, F.: *Atti. det Congresso Bologna*, **3** (1928), 143–148.
65. Riesz, F.: *Ann. of Math.* **41** (1940), 174–206.
66. Robinson, D. W.: 'Continuous semigroups on ordered Banach spaces', *J. Funct. Anal.* (To appear).
67. Robinson, D. W.: 'On Positive Semigroups', *Publ. RIMS, Kyoto. Univ.* (To appear).
68. Robinson, D. W. and Yamamuro, S.: *J. Austral. Math. Soc.*, Series A
69. Robinson, D. W. and Yamamuro, S.: 'The Jordan decomposition and half-norms', *Pac. J. Math.* **110** (1984), 345–353.
70. Robinson, D. W. and Yamamuro, S.: 'Hereditary cones, order ideals and half-norms', *Pac. J. Math.* **110** (1984), 335–343.
71. Robinson, D. W. and Yamamuro, S.: 'The canonical half-norms and monotonic norms', *Tôhoku Math. J.* **35** (1983), 375–386.
72. Sato, K.: *J. Math. Soc. Japan*, **20** (1968), 423–436.
73. Schaefer, H. H.: *Math. Ann.* **138** (1959), 259–286.
74. Schaefer, H. H.: *Math. Ann.* **141** (1960), 113–142.
75. Schaefer, H. H.: *Banach Lattices and Positive Operators*, Springer-Verlag, Berlin, 1974.
76. Schaefer, H. H.: *Jber. Dt. Math. Ver.* **82** (1980), 33–50.
77. Semadeni, Z.: *Bull. Acad. Sci. Pol.* **13** (1965), 141–146.
78. Semadeni, Z.: *Banach spaces of continuous functions*, Polish Scientific Publ. Warsaw. 1971.
79. Simon, B.: *J. Funct. Anal.* **12** (1973), 335–339.
80. Stone, M. H.: *Proc. Nat. Acad. Sci. USA*, **27** (1941), 83–87.
81. Takesaki, M.: *Theory of Operator Algebras* I, Springer-Verlag, Berlin, 1979.
82. Vulikh, B. C.: *Introduction to the Theory of Partially Ordered Spaces*, Wolters-Noordhoff, Groningen, 1967.
83. Widder, D. V.: *The Laplace Transform*, Princeton Univ. Press, Princeton, NJ, 1946.
84. Wong, Y. C. and Ng, K. F.: *Partially Ordered Topological Vector Spaces*, Clarendon Press, Oxford, 1973.
85. Yamamuro, S.: 'On linear operators on ordered Banach spaces', Preprint, 1982.
86. Zabczyk, J.: *Bull. Acad. Pol. Sci.* **23** (1975), 895–898.

Acta Applicandae Mathematicae **2**, 297–309. 0167–8019/84/0023–0297$01.95

Asymptotic Behavior of One-Parameter Semigroups of Positive Operators

W. KERSCHER and R. NAGEL
Mathematisches Institut, Auf der Morgenstelle 10, D-7400 Tübingen, F.R. Germany

(Received: 24 February 1984)

Abstract. In this paper we survey the Perron–Frobenius spectral theory for positive semigroups on Banach lattices and indicate its applications to stability theory of retarded differential equations and quasi-periodic flows.

AMS (MOS) subject classifications (1980). 47B55, 47D05, 47A10, 47A40, 47A50.

Key words. One-parameter semigroups, positive operators, Perron–Frobenius spectral theory, asymptotic stability, quasi-periodic flows.

0. Introduction

A strongly continuous one-parameter semigroup $\mathscr{T} = (T(t))_{t \geqslant 0}$ of bounded linear operators on a Banach space E can be thought of as the solution of the abstract Cauchy problem (ACP)

$$\frac{\mathrm{d}}{\mathrm{d}t} f(t) = Af(t), \quad f(0) = f_0 \in D(A),\ t \geqslant 0,$$

where the operator A with domain $D(A)$ is the 'generator' of $\mathscr{T}$ (see [8, 25]). Therefore the behavior of $T(t)$ as $t \to \infty$ is of great theoretical and practical importance and has been studied intensively. The classical theorem of Liapunov, for example, shows that (on finite-dimensional Banach spaces) the existence of $\lim_{t \to \infty} T(t)$ is determined by the location of the eigenvalues of the generator A (e.g., see [5]). Due to the existence of different operator topologies, of semigroups with unbounded generators and of spectral values not being eigenvalues, the situation in infinite-dimensional Banach spaces is much more complex and, in fact, very little can be said in general (see [6, 21]).

Since many semigroups appearing in applications (e.g., in probability theory, transport theory, population dynamics) are semigroups of *positive* operators on some naturally *ordered* Banach space, it is worthwhile studying the asymptotic behavior of these special semigroups. In particular the recently developed analogue of the Perron–Frobenius theory for generators of positive semigroups (see [10]) shows that the spectrum of these generators possesses interesting symmetries. In this paper we discuss how these symmetries influence and sometimes determine the asymptotic behavior of positive semigroups $\mathcal{T}$.

As an application, we discuss in Section 4 certain functional differential equations whose solutions form a positive semigroup.

1. The Framework

In the following $\mathcal{T} = (T(t))_{t \geqslant 0}$ always denotes a strongly continuous semigroup of bounded linear operators on some Banach space E. To each semigroup $\mathcal{T}$ is associated its 'generator' A being a closed operator with dense domain $D(A)$. Moreover, we denote the spectrum of A by $\sigma(A)$, the resolvent set by $\rho(A) := \mathbb{C} \backslash \sigma(A)$ and the resolvent by $R(\lambda, A) := (\lambda - A)^{-1}$ for $\lambda \in \rho(A)$. We recall that $R(\lambda, A)$ is obtained as the Laplace transform

$$R(\lambda, A) = \int_0^\infty e^{-\lambda s}\, T(s)\, ds$$

as soon as $\operatorname{Re} \lambda > \omega$, where

$$\omega = \omega(\mathcal{T}) := \inf\{w \in \mathbb{R} : \exists M_w \text{ such that } \|T(t)\| \leqslant M_w e^{wt},\ t \geqslant 0\}$$

is the *growth bound* (or *type*) of $\mathcal{T}$ (see [6], p. 38). In order to describe the location of $\sigma(A)$, we introduce the *spectral bound*

$$s(A) := \sup\{\operatorname{Re} \lambda : \lambda \in \sigma(A)\}$$

and observe that always $s(A) \leqslant \omega(\mathcal{T})$. We now restrict the class of spaces and operators to be considered.

The Spaces: By E we always denote a (real or complex) *Banach lattice* with positive cone E_+. We use the terminology and results from H. H. Schaefer's monograph [27] and mention only the most standard examples: $\mathbb{R}^n$ (resp. $\mathbb{C}^n$), $C_0(X)$ for some locally compact space X, $L^p(X, \Sigma, \mu)$ for $1 \leqslant p \leqslant \infty$ and some measure space (X, Σ, μ).

The Semigroups: By $\mathcal{T} = (T(t))_{t \geqslant 0}$ we always denote a strongly continuous semigroup of *positive operators* on E, i.e., $T(t)E_+ \subset E_+$ for every $t \geqslant 0$. The semigroup $\mathcal{T}$ is positive if and only if the resolvent operators $R(\lambda, A)$ are positive for $\lambda > \omega(\mathcal{T})$. More direct characterizations of generators of positive semigroups have been found in many cases (see [3] or [1]). Here we mention the standard examples:

(i) matrix semigroups $T(t) := \exp(tA)$ where $A = (a_{ij})_{n \times n}$ and $a_{ij} \geqslant 0$ for $i \neq j$.

(ii) translation semigroups $T(t)f := f \circ \varphi_t$ induced by some 'flow' $\varphi_t : X \to X$, $t \geqslant 0$, on some appropriate function space on X.

(iii) convolution semigroups $T(t)f := k_t * f$ on the function spaces $C_0(G)$, $L^p(G)$, for some locally compact Abelian group G (see [4], p. 87).

Finally we recall from [27], III.8.1 that the semigroup $\mathcal{T}$ is called *irreducible* if there exists no nontrivial $\mathcal{T}$-invariant closed ideal in E (see [3]). For example, the translation semigroup (ii) is irreducible in $C(X)$, resp. $L^p(X, \Sigma, \mu)$, $p < \infty$, if and only if the corresponding flow $(\varphi_t)_{t \geqslant 0}$ is minimal, resp. ergodic (see [27], p. 193).

2. The Perron–Frobenius Theory

Positive operators on Banach lattices admit an interesting spectral theory which originated with the classical results of O. Perron and G. Frobenius on the eigenvalues of positive matrices. These 'Perron–Frobenius theorems', as well as their infinite-dimensional generalizations, express a certain rotational symmetry of the spectral values with maximal modulus (see [27], Chap. V, for a modern and systematic presentation).

Let now $\mathcal{T} = (T(t))_{t \geqslant 0}$ be a strongly continuous semigroup of positive operators and denote its generator by A. Then the heuristic (!) formulas

$$T(t) = \exp(tA), \text{ resp. } A = \frac{1}{t} \log T(t)$$

suggest the validity of a 'Perron–Frobenius theory' for A. More precisely, one may expect some kind of translation symmetry for the spectral values of A having maximal real part, i.e., for $\sigma(A) \cap (s(A) + i\mathbb{R})$.

The first step toward such a theory was the following extension of the integral representation of the resolvent to all λ satisfying $\operatorname{Re} \lambda > s(A)$. Observe that it may happen that $s(A) < \omega(\mathcal{T})$, even for positive semigroups (see [13] or [3], Section 2.4).

LEMMA [10, 3]: *Let $\mathcal{T} = (T(t))_{t \geqslant 0}$ be a strongly continuous semigroup of positive operators on some Banach lattice E. Then for every $\lambda \in \mathbb{C}$ such that $\operatorname{Re} \lambda > s(A)$ the resolvent of the generator A is given as*

$$R(\lambda, A)f = \int_0^\infty e^{-\lambda s}\, T(s)f \,ds, \quad f \in E.$$

As a simple consequence of this lemma, one obtains

$$\| R(\lambda, A) \| \leqslant \| R(\operatorname{Re} \lambda, A) \| \quad \text{for } \operatorname{Re} \lambda > s(A)$$

and, therefore, the analogue of Perron's theorem (see [27], II.2.2).

PROPOSITION (e.g. [3]): *The spectral bound $s(A)$ of the generator A of a strongly continuous semigroup of positive operators is always a spectral value of A.*

In the sequel G. Greiner and R. Derndinger developed an analogue for generators of positive semigroups of the complete Perron–Frobenius theory (see [7, 10, 11]). In the following theorem we collect the most important properties.

THEOREM [10, 11]: *Let $\mathcal{T}$ be a bounded, strongly continuous semigroup of positive operators with generator A and assume $s(A) = 0$. Then the following statements hold:*

(0) *the peripheral spectrum $\sigma(A) \cap i\mathbb{R}$ is additively cyclic, i.e., if $\lambda \in \sigma(A) \cap i\mathbb{R}$ then $n\lambda \in \sigma(A)$ for every $n \in \mathbb{Z}$.*

If in addition $\mathcal{T}$ is irreducible and the set

$$\mathcal{K} := P\sigma(A) \cap i\mathbb{R}$$

of purely imaginary eigenvalues is nonempty, then

(i) *the spectral bound* 0 *is a simple eigenvalue having an eigenvector which is a quasi-interior point* $u \in E_+$,
(ii) *every eigenvalue* $\lambda \in \mathscr{K}$ *is simple and the corresponding eigenvectors* f *satisfy* $|f| = \alpha u$, $\alpha \in \mathbb{R}$,
(iii) $\mathscr{K}$ *is a subgroup of* $i\mathbb{R}$,
(iv) $\sigma(A) + \mathscr{K} = \sigma(A)$.
(v) *If* 0 *is a pole of the resolvent then each purely imaginary spectral value is a pole of first order and* $\sigma(A) \cap i\mathbb{R} = i\alpha\mathbb{Z}$ *for some* $\alpha \geqslant 0$.

The proof of (i)–(v) is based on the fact that A and $A + \lambda$, $\lambda \in \kappa$, are 'conjugated' in the sense that

$$A = M_{\bar{v}_\lambda}(A - \lambda)M_{v_\lambda}$$

for the 'multiplication operator' M_{v_λ} associated to the eigenfunction v_λ satisfying $Av_\lambda = \lambda v_\lambda$ (see [10], 1.14).

The most standard application of the above theorem is as follows: Let $\mathscr{T}$ be an eventually norm continuous semigroup (e.g., $\mathscr{T}$ is analytic or differentiable or eventually compact, see [25], Chap. 2). Then it follows that $\sigma(A) \cap (\beta + i\mathbb{R})$ is compact for every $\beta \in \mathbb{R}$ (Greiner and Neubrander, to appear). Therefore, if in addition $\mathscr{T}$ is bounded and positive, it follows from (0) that

$$\sigma(A) \cap i\mathbb{R} \subset \{0\}.$$

In particular, we can draw the following conclusion.

COROLLARY. *Let* $\mathscr{T}$ *be a strongly continuous semigroup of positive operators and assume that* $\mathscr{T}$ *is eventually compact. If the spectral bound* $s(A)$ *of the generator* A *is zero, then* 0 *is a* strictly dominant eigenvalue, *i.e., there exists* $\varepsilon > 0$ *such that* $\operatorname{Re} \lambda \leqslant -\varepsilon$ *for every* $0 \neq \lambda \in \sigma(A)$.

Proof. Eventually compact semigroups are eventually norm continuous ([25], p. 48) and, therefore, the spectral mapping theorem

$$\sigma(T(t))\backslash\{0\} = \exp(t\sigma(A)), \quad t > 0,$$

holds (see [6], p. 46). Therefore, $\sigma(A)$ consists of isolated eigenvalues and the assertion follows from the above considerations. □

3. Asymptotic Stability

Liapunov's stability theorem for systems of linear differential equations can be formulated in the following way:

> If the spectral bound $s(A)$ of the matrix $A = (a_{ij})_{n \times n}$ is less than zero, then $\lim_{t \to \infty} \|\exp(tA)\| = 0$, i.e., the semigroup generated by A is (*uniformly*) *stable*.

Unfortunately, the analogous result does not hold for generators of arbitrary semigroups on infinite-dimensional Banach spaces (see [13]). One either needs a stronger hypothesis on the semigroup or must be satisfied with weaker conclusions. It was Neubrander [24] who realized that for most purposes it is good enough to have 'stability' for the actual solutions of (ACP), i.e., for $t \mapsto T(t)f$, $f \in D(A)$ only. For positive semigroups he obtained the following stability result.

THEOREM A [24]: *Let $\mathcal{T}$ be a strongly continuous semigroup of positive operators. If $s(A) < 0$, then $\mathcal{T}$ is* strongly exponentially stable, *i.e., there exists $v > 0$ such that*

$$\| T(t)f \| \leqslant M_f \, \mathrm{e}^{-vt}$$

for every $f \in D(A)$, $t \geqslant 0$ and appropriate constants M_f.

Proof. Take $s(A) < \alpha < 0$ and $f \in D(A)$. Elementary semigroup theory implies

$$f - \mathrm{e}^{-\alpha t} T(t) f = \int_0^t \mathrm{e}^{-\alpha s} T(s)(\alpha - A) f \, \mathrm{d}s = \int_0^t \mathrm{e}^{-\alpha s} T(s) g \, \mathrm{d}s$$

for $g = (\alpha - A)f$. But $\lim_{t \to \infty} \int_0^t \mathrm{e}^{-\alpha s} T(s) g \, \mathrm{d}s$ exists by the Lemma in Section 2. and equals f. Hence, $\lim_{t \to \infty} \mathrm{e}^{-\alpha t} T(t) f = 0$ and the assertion follows. □

As pointed out above, uniform stability of $\mathcal{T}$ does not follow from $s(A) < 0$. But since the growth bound $\omega(\mathcal{T})$ of $\mathcal{T}$ is related to the spectral raldius of the operators $T(t)$ by

$$r(T(t)) = \mathrm{e}^{t\omega(\mathcal{T})}, \quad t > 0,$$

it suffices to require some form of a 'spectral mapping theorem', e.g.,

$$\sigma(T(t)) \setminus \{0\} = \overline{\exp(t \cdot \sigma(A))}.$$

We recall that this holds for eventually norm continuous semigroups on Banach spaces or for normal semigroups on Hilbert spaces (see [6], 2.19).

On the other hand, due to the special properties of positive operators, it has been shown in a series of papers that the analogue of Liapunov's theorem still holds for positive semigroups on particular Banach lattices.

THEOREM B [2, 7, 12]: *Let $\mathcal{T}$ be a strongly continuous semigroup of positive operators on $E = C_0(X)$, X locally compact, or on $E = L^p(X, \Sigma, \mu)$ for $p = 1$ or $p = 2$. If $s(A) < 0$ for the spectral bound $s(A)$ of the generator A, then $\lim_{t \to \infty} \| T(t) \| = 0$.*

For a more complete discussion of this circle of ideas we refer to [23] or [3], Section 2.4 and point out that Theorem B (for $E = L^1$) has been applied to the linear transport equation (see [18] or [29]).

4. Abstract Functional Differential Equations

In this section we give some applications of the previous results to a class of semigroups arising as solutions of abstract functional differential equations.

Let X be a Banach lattice and consider the space $E := C([-1, 0], X)$ of X-valued continuous functions on $[-1, 0]$. Let $(B, D(B))$ be the generator of a strongly continuous semigroup on X and L a continuous linear operator from E to X.

With the abstract delay equation

$$\dot{u}(t) = Bu(t) + Lu_t, \quad t \geqslant 0, \ u_0 = f \in E,\star \tag{4.1}$$

we can associate an operator $(A, D(A))$ defined by

$$\begin{aligned} D(A) &:= \{f \in C^1([-1,0], X) \mid f(0) \in D(B), f'(0) = Bf(0) + Lf\} \\ Af &:= f' \quad \text{for } f \in D(A). \end{aligned} \tag{4.2}$$

It is well known (see, e.g., [28]) that A, as in (4.2), is the generator of a semigroup $\mathcal{T} = (T(t))_{t \geqslant 0}$ yielding the solutions of the delay equation (4.1). Therefore, many properties of the solutions, e.g., positivity or stability, can be investigated by means of this 'solution semigroup' and its generator.

It is not too difficult to find a rather explicit representation of the resolvent $R(\lambda, A)$ of the generator A. To that purpose we take $\lambda \in \mathbb{C}$ and define the operators

$$H_\lambda \in \mathcal{L}(E) \colon H_\lambda g(x) := \int_x^0 e^{\lambda(x-y)} g(y)\, dy, \quad x \in [-1, 0], \tag{4.3}$$

$$L_\lambda \in \mathcal{L}(X) \colon L_\lambda u := L(\varepsilon_\lambda \otimes u), \quad \text{where } \varepsilon_\lambda(x) := e^{\lambda x}. \tag{4.4}$$

Here, for $f \in C[-1, 0]$ and $u \in X$, we denote by $f \otimes u$ the element of E defined by $(f \otimes u)(x) := f(x)u$, $x \in [-1, 0]$. Now the spectrum and the resolvent of A can be described by the following relations (for details see [19]):

$$\lambda \in \sigma(A) \quad \text{iff } \lambda \in \sigma(B + L_\lambda) \tag{4.5}$$

$$\begin{aligned} &\forall \lambda \in \rho(A), \quad \forall g \in E \colon \\ &\quad R(\lambda, A)g = \varepsilon_\lambda \otimes R(\lambda, B + L_\lambda)(g(0) + LH_\lambda g) + H_\lambda g. \end{aligned} \tag{4.6}$$

Next we will look for conditions that imply the positivity of the solution semigroup $\mathcal{T}$. Problems of this kind have been treated in [20] by an approach based on the notion of dispersive operators. We use the fact that $\mathcal{T}$ is positive if and only if the resolvent $R(\lambda, A)$ is positive for $\lambda > \omega(\mathcal{T})$.

PROPOSITION. *Let B be the generator of a positive semigroup on X and assume L to be a positive operator. Then the solution semigroup associated with* (4.1) *is positive.*

Proof. The operators L_λ and H_λ are positive for each $\lambda \in \mathbb{R}$ while $R(\lambda, B)$ is positive for λ sufficiently large. Hence, for large λ the operator

$$R(\lambda, B + L_\lambda) = \sum_{n=0}^{\infty} (R(\lambda, B)L_\lambda)^n R(\lambda, B)$$

is positive and so by (4.6) the resolvent $R(\lambda, A)$ is positive. □

★ Here we denote by $u_t \in E$ the segment of a function $u \colon [-1, \infty) \to X$ defined by $u_t(x) := u(t + x)$, $x \in [-1, 0)$, $t \geqslant 0$.

REMARK. If we assume, that L has *no mass in* 0 (i.e.: For every $\varepsilon > 0$ there exists $\delta > 0$ such that $\|Lf\| \leqslant \varepsilon\|f\|$ for all $f \in E$, $\operatorname{supp}(f) \subset [-\delta, 0]$) we can show that the hypotheses in the above proposition are also necessary (see [19]).

EXAMPLE. Let $X := \mathbb{R}^n$. Then L is a matrix of measures and has a decomposition in an atomic part A_0 in 0 and a matrix $\tilde{L}$ of measures, that have no mass in 0. If we denote by b_{ij} the elements of $B + A_0$, we have by the preceding: The solution semigroup $\mathscr{T}$ is positive if and only if $\tilde{L}$ is positive and $b_{ij} \geqslant 0$ for $i \neq j$ (those matrices are precisely the generators of positive semigroups on $\mathbb{R}^n$).

From the results in Section 3 we know that the spectral bound $s(A)$ of A determines, in many cases, the asymptotic behavior and, in particular, the stability of $\mathscr{T}$. In order to describe $s(A)$ we introduce the function

$$s(\lambda) := s(B + L_\lambda) \quad \text{for } \lambda \in \mathbb{R}$$

and make the following observation.

LEMMA. *Under the assumptions of the above proposition, the function* $\lambda \mapsto s(\lambda)$ *is decreasing.*

Proof. Since $L_{\lambda_1} \geqslant L_{\lambda_2}$ for $\lambda_1 \leqslant \lambda_2$ it follows by the standard perturbation formulas that the semigroups $(S_1(t))$, $(S_2(t))$ generated by $B + L_{\lambda_1}$, resp. $B + L_{\lambda_2}$ are positive and satisfy $S_1(t) \geqslant S_2(t)$, $t \geqslant 0$. Therefore, we obtain $s(B + L_{\lambda_1}) \geqslant s(B + L_{\lambda_2})$ for the corresponding spectral bounds (use the integral representation for the resolvent in the Lemma of Section 2), i.e., $s(\lambda_1) \geqslant s(\lambda_2)$. □

THEOREM. *Let B be the generator of a positive semigroup and assume L to be positive. If B has compact resolvent with $\sigma(B) \neq \emptyset$, then $\lambda \mapsto s(\lambda) := s(B + L_\lambda)$ is continuous and $s(A)$ is the unique solution of the equation $\lambda = s(\lambda)$.*

Proof. By the above lemma and the upper semicontinuity of the spectrum, $\lambda \mapsto s(\lambda)$ is continuous on the left. Suppose $s(\tau) > s_0 := \lim_{\lambda \downarrow \tau} s(\lambda)$ for some $\tau \in \mathbb{R}$. By hypothesis $B + L_\lambda$ generates a positive semigroup for every $\lambda \in \mathbb{R}$ and has compact resolvent. Therefore, $s(\tau)$ is an isolated point of $\sigma(B + L_\tau)$. Hence, there exists $s_1 \in \rho(B + L_\tau)$ with $s_0 < s_1 < s(\tau)$.

Now $R(s_1, B + L_\tau)$ is positive, since $R(s_1, B + L_\tau) = \lim_{\varepsilon \downarrow 0} R(s_1, B + L_{\tau+\varepsilon})$ and each $R(s_1, B + L_{\tau+\varepsilon})$ is positive. But then by the resolvent equation we have $R(s_1, B + L_\tau) \geqslant R(\lambda, B + L_\tau) \geqslant 0$ and, therefore, $\|R(s_1, B + L_\tau)\| \geqslant \|R(\lambda, B + L_\tau)\|$ for all $\lambda > s(\tau)$. This leads to a contradiction, since $\lim_{\varepsilon \downarrow 0} \|R(s(\tau) + \varepsilon, B + L_\tau)\| = \infty$. Consequently $\lambda \mapsto s(\lambda)$ is continuous and there exists a unique solution λ_0 of the equation $\lambda = s(\lambda)$.

By (4.5), λ_0 is the largest spectral value of A and, hence, $\lambda_0 = s(A)$. □

From the above information on the function $\lambda \mapsto s(\lambda)$, it follows that the spectral bound $s(A)$ has the same sign as the spectral bound $s(B + L_0)$ $(= s(0))$. So we can apply Theorem A of the preceding section to deduce the following.

COROLLARY. *Under the assumptions of the above theorem, the following properties are equivalent*:

(a) *The solution semigroup $\mathscr{T}$ of* (4.1) *is strongly exponentially stable.*

(*b*) *The semigroup $\mathscr{S}$ generated by $B + L_0$ in X is strongly exponentially stable.*

INTERPRETATION. Consider the equation

$$\dot{u}(t) = Bu(t) + Su(t-1), \quad S \in \mathscr{L}(X). \tag{4.7}$$

Then $L_0 = S$. The semigroup $\mathscr{S}$ generated by $B + L_0$ can be regarded as the solution of the undelayed equation

$$\dot{u}(t) = Bu(t) + Su(t). \tag{4.8}$$

So the corollary means, that for such systems strong exponential stability is independent of the delay. This is a very untypical behavior for delay equations (compare [14]).

If we take $X := C(K)$, K compact, we have coincidence of spectral bound and growth bound (see [3]), i.e., $\omega(\mathscr{T}) = s(A)$ and $\omega(\mathscr{S}) = s(B + L_0)$. Hence, we obtain an analogue to the above Corollary for uniform exponential stability. This will be used in the concluding example.

EXAMPLE. Take $X := C(0, 1]$, $D(B) := \{f \in C^2[0, 1] \mid f'(0) = f'(1) = 0\}$, $Bf = f''$ for $f \in D(B)$ and consider the equation

$$\dot{u}(t) = Bu(t) - M_d u(t) + M_b u(t-1), \tag{4.9}$$

where M_b and M_d denote the multiplication operators with the positive functions b and d respectively.

This equation can be interpreted as a linear model for the growth of a distributed population: the operator B describes migration, b and d are the distributed birth and death rates.

It is well known that B has compact resolvent, generates a positive semigroup and satisfies $s(B) = 0$.

By the above results the solution semigroup corresponding to (4.9) is positive and its asymptotic behavior can be investigated by the undelayed equation

$$\dot{u}(t) = (B + M_h)u(t), \quad \text{where } h := b - d. \tag{4.10}$$

Assume $h(x) < 0$ for all $x \in [0, 1]$, i.e., the birth rate is everywhere smaller than the death rate. Then we obtain $s(B + M_h) \leqslant \max\{h(x) : x \in [0, 1]\} < 0$. Hence, the solutions of (4.10) and consequently of (4.9) are uniformly exponentially stable.

5. Convergence to Quasi-periodic Flows

In this section we study a more complicated behavior of $T(t)$ as $t \to \infty$. In particular, we show that many positive semigroups on L^2-spaces behave asymptotically as 'rotation groups' on compact groups (see the precise statement below). Here the spaces $L^2(X, \Sigma, \mu)$ are the most appropriate, since they combine the lattice structure with that of a Hilbert space.

From now on let $\mathscr{T}$ be an irreducible strongly continuous semigroup of positive contractions on $E = L^2(X, \Sigma, \mu)$ and denote its generator by $(A, D(A))$. While in the previous section we asked whether $s(A) < 0$ implies $\lim_{t\to\infty} T(t) = 0$, we now assume that 0 is an eigenvalue of A. Then the irreducibility of $\mathscr{T}$ implies that there exists a unique normalized, strictly positive function $0 \ll u \in E$ such that $Au = 0$ and, therefore, $T(t)u = u$ for $t \geqslant 0$ (see the Theorem in Section 2). Without loss of generality, we may assume that u is the constant function $\mathbb{1}$. Since the fixed space of $T(t)$ and its (Hilbert space) adjoint $T(t)^\star$ coincide ([27], p. 185), we obtain that

$$P_1 := \mathbb{1} \otimes \mathbb{1}, \quad \text{i.e., } P_1 f = \left(\int f \, d\mu\right)\mathbb{1}$$

is the orthogonal projection onto the $\mathscr{T}$-fixed space

$$F_1 := \{f \in E : T(t)f = f \text{ for } t \geqslant 0\}.$$

Next we consider the set $\mathscr{K}$ of all purely imaginary eigenvalues of A (which by the Perron–Frobenius theorem in Section 2 is a subgroup of $i\mathbb{R}$).

To each $\lambda \in \mathscr{K}$ corresponds a normalized eigenfunction $v_\lambda \in E$ such that $|v_\lambda| = \mathbb{1}$ and $P_\lambda := v_\lambda \otimes v_\lambda$ is the orthogonal projection onto

$$F_\lambda := \{f \in E : Af = \lambda f\} = \{f \in E : T(t)f = e^{\lambda t} f, \, t \geqslant 0\}.$$

Von Neumann's mean ergodic theorem applied to the contraction semigroup $(e^{-\lambda t}\, T(t))_{t\geqslant 0}$, $\lambda \in \mathscr{K}$, yields

$$P_\lambda = \lim_{t\to\infty} \frac{1}{t} \int_0^t e^{-\lambda s}\, T(s)\, ds.$$

Therefore $P_{\lambda_1} P_{\lambda_2} = 0$ for $\lambda_1, \lambda_2 \in \mathscr{K}$, $\lambda_1 \neq \lambda_2$, and

$$Q := \sum_{\lambda \in \mathscr{K}} P_\lambda$$

exists and is the orthogonal projection onto the closed subspace generated by the eigenfunctions v_λ, $\lambda \in \mathscr{K}$.

LEMMA. *The projection Q is strictly positive and $QE \cong L^2(X, \Sigma_0, \mu)$ for some σ-algebra $\Sigma_0 \leqslant \Sigma$.*

Proof. The semigroup $\mathscr{T}$ is commutative and relatively compact with respect to the weak operator topology. Therefore the Glicksberg–de Leeuw theory (see [27], p. 214 or [17]) yields a projection $\tilde{Q}$ satisfying the following:

(i) $\tilde{Q}f$ belongs to the weak closure of $\{T(t)f : t > 0\}$ for $f \in E$.

(ii) $\tilde{Q}E$ is the closed subspace generated by all eigenfunctions v_λ, $\lambda \in \mathscr{K}$.

(iii) $\tilde{Q}f = 0$ if and only if 0 belongs to the weak closure of $\{T(t)f : t > 0\}$.

From (i) it follows that $\tilde{Q}$ is a positive projection. Moreover, $(\tilde{Q}f \mid \mathbb{1}) = (f \mid \mathbb{1}) > 0$ whenever $0 \neq f \in E_+$, i.e., $\tilde{Q}$ is strictly positive. By ([27], III. 11.5) it follows that the range of $\tilde{Q}$ is a sublattice of E and ([27], III. 11.2) shows that it is of the form $L^2(X, \Sigma_0, \mu)$.

Therefore the lemma is proved if we show $Q = \tilde{Q}$: Restrict the semigroup $\mathscr{T}$ to the invariant subspace $Q^{-1}(0)$. Since its generator $A_|$ has no point spectrum on $i\mathbb{R}$, we conclude that the Glicksberg–de Leeuw projection of $\mathscr{T}_|$ is zero, i.e., 0 belongs to the weak closure of $\{T(t)f : t > 0\}$ for every $f \in Q^{-1}(0)$. By (iii) we obtain $f \in \tilde{Q}^{-1}(0)$. Since the ranges of Q and $\tilde{Q}$ coincide (by definition and by (i)) the assertion follows. □

In the next step we analyse the semigroup $\mathscr{T}_| = (T(t)|_{QE})_{t \geqslant 0}$ of the operators $T(t)$ restricted to QE. This semigroup is still irreducible and positive, but the eigenfunctions of its generator pertaining to purely imaginary eigenvalues form a total subset (i.e., $\mathscr{T}_|$ has 'discrete spectrum', [27], p. 208). Therefore, $\mathscr{T}_|$ is relatively compact in the strong operator topology and the Halmos–von Neumann theorem ([27], III. 10.5) asserts that it is isomorphic to a rotation group. More precisely, consider the dual group $\mathscr{G}$ of the discrete group $\mathscr{K}$. Then $\mathscr{G}$ is compact and solenoidal, i.e., contains a dense homomorphic image $\{g_t : t \in \mathbb{R}\}$ of $\mathbb{R}$ (see [16]). The induced operators

$$\tilde{R}(t)f(h) := f(g_t h)$$

for $f \in L^2(\mathscr{G}, m)$, m Haar measure, $h \in \mathscr{G}$, $t \geqslant 0$, define a strongly continuous group on $L^2(\mathscr{G}, m)$ called *rotation group* or *quasi-periodic flow*. We refer to [16], Section 25, for a classification of quasi-periodic flows and to [9] for concrete examples and collect the previous information in the following lemma.

LEMMA. *There exists a positive unitary operator*

$$V: L^2(X, \Sigma_0, \mu) \longrightarrow L^2(\mathscr{G}, m)$$

such that the diagram

$$\begin{array}{ccc} L^2(X, \Sigma_0, \mu) & \xrightarrow{\;T(t)\;} & L^2(X, \Sigma_0, \mu) \\ V \big\downarrow & & \big\uparrow V^{\star} \\ L^2(\mathscr{G}, m) & \xrightarrow[\tilde{R}(t)]{} & L^2(\mathscr{G}, m) \end{array}$$

commutes for every $t \geqslant 0$.

Thus we have singled out a subspace of E on which the behavior of $\mathscr{T}$ can be described quite explicitly and, in fact, is determined by the eigenvalues in $\mathscr{K} = P\sigma(A) \cap i\mathbb{R}$. For greater convenience, we introduce the following terminology.

DEFINITION. A family of positive operators $(R(t))_{t \geqslant 0}$ on $E = L^2(X, \Sigma, \mu)$ is called a *partially quasi-periodic flow* if $Q = R(0)$ is a positive projection onto a sublattice of E and $(R(t)|_{QE})_{t \geqslant 0}$ is (isomorphic to) a quasi-periodic flow.

We now arrive naturally at the following question: Under which conditions converges a positive semigroup $\mathscr{T}$ to a quasi-periodic flow as $t \to \infty$, i.e., does there exist a partially quasi-periodic flow $(R(t))_{t \geqslant 0}$ such that

$$\lim_{t \to \infty} (T(t) - R(t)) = 0?$$

It is clear that the partially quasi-periodic flow in question is determined by the set of eigenvalues $\mathscr{K}$ of A and must be obtained as $R(t) := T(t) \circ Q$, where Q is the projection onto $L^2(X, \Sigma_0, \mu)$ (use the lemmas above). On the other hand it follows from the Glicksberg–de Leeuw theory of weakly compact operator semigroups that we always have some kind of *weak* convergence (see [17] or [22]).

Here we concentrate on convergence with respect to the operator norm. Typically for semigroups of positive operators it turns out that some information on the spectrum of A near 0 will allow the desired conclusion.

In order to formulate the result we introduce one more notation: For $\varepsilon \geqslant 0$ we consider the eigenvalues $\lambda \in \mathscr{K}$ such that $|\lambda| \leqslant \varepsilon$ and define the associated projection

$$P_\varepsilon := \sum_{|\lambda| \leqslant \varepsilon} P_\lambda .$$

THEOREM. *Let $\mathscr{T} = T(t))_{t \geqslant 0}$ be an irreducible strongly continuous semigroup of positive contractions on $E = L^2(X, \Sigma, \mu)$ and assume that there exists at least one nonzero, purely imaginary eigenvalue of the generator A, i.e., $|\mathscr{K}| > 1$. If there exists $\varepsilon > 0$ such that*

$$\lim_{\alpha \downarrow 0} \| \alpha R(\alpha, A)(\mathrm{Id} - P_\varepsilon) \| = 0,$$

then $\mathscr{T}$ converges uniformly to a partially quasi-periodic flow $(R(t))_{t \geqslant 0}$, i.e.,

$$\lim_{t \to \infty} \| T(t) - R(t) \| = 0.$$

REMARK. The formally weaker hypothesis $\lim_{\alpha \downarrow 0} \| \alpha R(\alpha, A)(\mathrm{Id} - Q) \| = 0$ immediately implies the convergence of $\alpha R(\alpha, A)$ on the larger space $P_\varepsilon^{-1}(0) = Q^{-1}(0) \oplus \overline{\mathrm{lin}}\{v_\lambda : \lambda \in \mathscr{K}, |\lambda| > \varepsilon\}$.

Proof. From the assumptions follows that $\mathscr{K} = P\sigma(A) \cap i \cdot \mathbb{R}$ is an infinite subgroup of $i\mathbb{R}$. We distinguish two cases:

1° *$\mathscr{K} = i\alpha_0\mathbb{Z}$ for some $\alpha_0 > 0$*: Then P_ε is a finite sum of one-dimensional projections and, therefore, the hypothesis on the resolvent $R(\alpha, A)$ implies

$$\lim_{\alpha \downarrow 0} \| \alpha R(\alpha, A) - P_0 \| = 0.$$

This means that $\mathscr{T}$ is uniformly ergodic and 0 is a pole of first order of $R(., A)$. Now the assertion follows from Greiner's Niiro–Sawashima theorem ([10], 2.5) and arguments as in the case below (see also [23]).

2° *$\mathscr{K}$ is dense in $i\mathbb{R}$*: We restrict $\mathscr{T}$ to the invariant subspace

$$H_0 := Q^{-1}(0) = \{f \in E : \textstyle\int f \bar{v}_\lambda \, \mathrm{d}\mu = 0 \text{ for all eigenfunctions } v_\lambda, \lambda \in \mathscr{K}\}.$$

The hypothesis implies that the restricted semigroup is uniformly ergodic with projection 0, hence there exists a neighborhood of 0 on which the norm of $R(., A|_{H_0})$ is bounded by some constant c.

Next, it follows from the second Lemma above that the product of two eigenfuntions $v_\varkappa$, v_λ for $\varkappa, \lambda \in \mathscr{K}$ is again an eigenfunction (pertaining to the eigenvalue $\varkappa + \lambda$). Therefore, H_0 is invariant under the multiplication operators $M_{v_\lambda} f := v_\lambda \cdot f$, $\lambda \in \mathscr{K}$. But for every $\lambda \in \mathscr{K}$, the operators A and $A + \lambda$, resp. $R(\alpha, A)$ and $R(\alpha - \lambda, A)$ are conjugate via these multiplication operators (see the Remark following the Theorem in Section 2).

Therefore the estimate of the norm of $R(.,A|_{H_0})$ in the above neighborhood of 0 translates to an estimate near each $\lambda \in \mathcal{K}$. Consequently we have

$$s(A|_{H_0}) < 0 \quad \text{and} \quad \| R(\alpha + i\beta, A|_{H_0}) \| \leqslant c$$

for small α and all $\beta \in \mathbb{R}$. Since $\mathcal{T}|_{H_0}$ is a strongly continuous semigroup on the Hilbert space H_0, these two conditions allow us to conclude

$$\lim_{t \to \infty} \| T(t)|_{H_0} \| = 0$$

(see [15, 26]).

Since $E = QE \oplus Q^{-1}(0)$ and $T(t) = T(t)|_{QE} \oplus T(t)|_{H_0}$ the assertion now follows from the preceding lemmas. □

COROLLARY 1 [23]. *If under the assumptions of the above theorem, 0 is a pole of the resolvent $R(.,A)$ then $\mathcal{T}$ converges uniformly to a partially quasi-periodic flow defined by the rotation on the unit circle.*

Proof. In fact this case has been discussed separately in part 1° of the proof above. □

COROLLARY 2. *Let $\mathcal{T} = (T(t))_{t \geqslant 0}$ be an irreducible strongly continuous semigroup of positive contractions on $E = L^2(X, \Sigma, \mu)$ and assume $\mathcal{K} = \{0\}$. If 0 is a pole of $R(.,A)$ and $\{R(\alpha + i\beta, A) : \beta \in \mathbb{R}\}$ is uniformly bounded for each α near 0, then $\mathcal{T}$ converges uniformly to the projection P_0 onto the fixed space F_1, i.e., $\lim_{t\to\infty} \| T(t) - P_0 \| = 0$.*

Proof. Greiner's theorem ([10], 2.5) implies $\sigma(A) = \{0\} \cup \sigma_2$ where $\sup\{\text{Re}\,\lambda : \lambda \in \sigma_2\} < 0$. Therefore, the spectral bound of the semigroup $\mathcal{T}$ restricted to $P_0^{-1}(0)$ is smaller than zero. The boundedness condition for the resolvent in the Hilbert space $P_0^{-1}(0)$ implies $\omega(\mathcal{T}_|) < 0$ by [15] or [26]. Hence we proved the assertion. □

We conclude with a simple example satisfying the assumptions of Corollary 1: Take $E = L^2([0,1]^2, m)$,

$$Af := f_x(x,y) + f_{yy}(x,y)$$

for

$$f \in D(A) := \{ f \in E : f(.,y) \in AC[0,1],\ f(0,y) = f(1,y) \text{ for all } y \in [0,1] \text{ and } f_x \in E;$$

$$f(x,.) \in C^1[0,1],\ f_y(x,.) \in AC[0,1], \quad f_y(x,0) = f_y(x,1) = 0 \text{ for all } x \in [0,1] \text{ and } f_{yy} \in E \}.$$

Here f_x, f_y and f_{yy} denote the partial derivatives of f.

References

1. Arendt, W.: 'The Generator of Positive Semigroups', preprint, 1983.
2. Batty, C. J. K. and Davies, E. B.: 'Positive Semigroups and Resolvents', *J. Operator Theory* **10** (1982), 357 – 363.
3. Batty, C. J. K. and Robinson, D. W.: 'Positive One-Parameter Semigroups on Ordered Spaces', *Acta Appl. Math.* **2** (1984), 221–296 (this issue).
4. Berg, C. and Forst, G.: *Potential Theory on Locally Abelian Groups*, Springer-Verlag, Berlin, Heidelberg, New York, 1975.
5. Braun, M.: *Differential Equations and their Applications*, Springer-Verlag, Berlin, Heidelberg, New York, 1975.
6. Davies, E. B.: *One-parameter Semigroups*, Academic Press, London, 1980.
7. Derndinger, R.: 'Über das Spektrum positiver Generatoren', *Math. Z.* **172** (1980), 281–293.
8. Fattorini, H. O.: *The Cauchy Problem*, Addison-Wesley, London, 1983.
9. Galavotti, G.: *The Elements of Mechanics*, Springer-Verlag, Berlin, Heidelberg, New York, 1983.
10. Greiner, G.: 'Zur Perron-Frobenius Theorie stark stetiger Halbgruppen', *Math. Z.* **177** (1981), 401–423.
11. Greiner, G.: 'Spektrum und Asymptotik stark stetiger Halbgruppen positiver Operatoren', *Sitzungsber, Heidelberger Akad. Wiss. Math. Nat. Kl.* **3**, Abh. 1982.
12. Greiner, G. and Nagel, R.: 'On the Stability of Strongly Continuous Semigroups of Positive Operators on $L^2(\mu)$, *Ann. Scuola Normale Sup. Pisa* **10** (1983), 257–262.
13. Greiner, G., Voigt, J. and Wolff, M.: 'On the Spectral Bound of the Generator of Semigroups of Positive Operators', *J. Operator Theory* **5** (1981), 245–256.
14. Hale, J.: *Theory of Functional Differential Equations*, Springer-Verlag, Berlin, Heidelberg, New York, 1977.
15. Herbst, I.: 'The Spectrum of Hilbert Space Semigroups', *J. Operator Theory* **10** (1983), 87–94.
16. Hewitt, E. and Ross, K. A.: *Abstract Harmonic Analysis*, Vol. I, Springer-Verlag, Berlin, Heidelberg, New York, 1979.
17. Hiai, F.: 'Weakly Mixing Properties of Semigroups of Linear Operators', *Kodai Math J.* **1** (1978), 376–393.
18. Kaper, H. G. and Hejtmanek, J.: 'Recent Progress in the Reactor Problem of Linear Transport Theory', preprint.
19. Kerscher, W.: 'Positive Lösungen retardierter Cauchyprobleme', *Semesterbericht Funktionalanalysis Tübingen* (1983), 139–163.
20. Kunisch, K. and Schappacher, W.: 'Order Preserving Evolution Operators, *Boll. U.M.I.* **16B** (1979).
21. Nagel, R.: 'Zur Charakterisierung stabiler Operatorhalbgruppen', *Semesterber. Funktionalanalysis Tübingen* 1982, 99–119.
22. Nagel, R.: 'Ergodic and Mixing Properties of Linear Operators', *Proc. Royal Irish Acad. Sect. A* **74** (1974), 245–261.
23. Nagel, R.: 'What Can Positivity Do for Stability?', in *Functional Analysis: Surveys and Recent Results III*, K. D. Bierstedt and B. Fuchssteiner (eds.), North-Holland, 1984, pp. 145–154.
24. Neubrander, F.: 'Laplace Transform and Asymptotic Behavior of Strongly Continuous Semigroups', *Semesterber. Funktionalanalysis, Tübingen* 1983, 139–161.
25. Pazy, A.: *Semigroups of Linear Operators and Applications to Partial Functional Differential Equations*, Springer-Verlag, Berlin, Heidelberg, New York, 1983.
26. Prüss, J.: 'On the Spectrum of C_0-semigroups', preprint, 1983.
27. Schaefer, H. H.: *Banach Lattices and Positive Operators*, Grundl. Math. Wiss. 215, Springer-Verlag, Berlin, Heidelberg, New York, 1974.
28. Travis, C. C. and Webb, G. F.: 'Existence and Stability for Partial Functional Differential Equations', *Trans. Amer. Math. Soc.* **200** (1974).
29. Voigt, J.: 'Positivity in Time-dependent Linear Transport Theory', *Acta Appl. Math.* **2** (1984), 311–331 (this issue).

Acta Applicandae Mathematicae **2**, 311–331. 0167–8019/84/0024–0311\$03.15

Positivity in Time Dependent Linear Transport Theory

JÜRGEN VOIGT
Fach Mathematik, Universität Trier, F.R. Germany★

(Received: 22 December 1983)

Abstract. We present methods using positive semigroups and perturbation theory in the application to the linear Boltzmann equation. Besides being a review, this paper also presents generalizations of known results and develops known methods in a more abstract setting.

In Section 1 we present spectral properties of the semigroup operators $W_a(t)$ of the absorption semigroup and its generator T_a. In Section 2 we treat the full semigroup $(W(t); t \geqslant 0)$ as a perturbation of the absorption semigroup. We discuss part of the problems (perturbation arguments and existence of eigenvalues) which have to be solved in order to obtain statements about the large time behaviour of $W(\cdot)$. In Section 3 we discuss irreducibility of $W(\cdot)$.

In four appendices we present abstract methods used in Sections 1, 2 and 3.

AMS (MOS) subject classifications (1980). 47D05, 82A70, 47B55, 47G05, 35K22, 35F15, 35C15.

Key words. linear transport theory, neutron transport equation, Boltzmann equation, semigroups, perturbations, irreducibility, collision, absorption, scattering, spectrum, Dyson–Phillips expansion, flows.

0. Introduction

In this paper we discuss properties of the C_0-semigroup associated with the linear Boltzmann equation (neutron transport equation), abbreviated LBE,

$$\frac{\partial f}{\partial t}(t, x, \xi) = -\xi \cdot \operatorname{grad}_x f(t, x, \xi) - h(x, \xi) f(t, x, \xi) + \\ + \int k(x, \xi, \xi') f(t, x, \xi')\, \mathrm{d}\rho(\xi').$$

Here $x \in D$, and $D = \mathring{D} \subset \mathbb{R}^n$ is the set where neutron transport takes place in space, and $\xi \in V$, where $V \subset \mathbb{R}^n$ is the set of velocities neutrons may assume, and ρ is a measure on V. For suitable assumptions on V, ρ, h, k we refer to Sections 1 and 2.

The LBE arises in the following way. The volume D is filled with material, and neutrons migrate in this volume, are scattered and absorbed by the material. If $f(t, ., .)\colon\ D \times V \to [0, \infty)$ describes the neutron density (in configuration-velocity

★ Permanent address: Mathematisches Institut der Universität München, Theresienstraße 39, D-8000 München 2, Federal Republic of Germany.

space!) at time $t \geqslant 0$, then certain assumptions (among which the assumption that neutrons do not interact with each other) imply that f should (formally) satisfy the LBE, which thus is an equation expressing balance of mass. The first term on the right-hand side corresponds to the motion of neutrons between collisions with the background material. The second term corresponds to collisions (including absorption). The third term corresponds to scattering of neutrons: particles at the point x with velocity ξ' 'generate' particles at x with new velocity ξ, and the transition is governed by $k(x, \xi, \xi')$. For more information, motivation, and for the derivation of the LBE we refer to [6], sect. 1.3, [42, 19].

The appropriate functional analytical tool for the treatment of the initial value problem for the LBE is the theory of C_0-semigroups. Since $f(t, ., .)$ has the meaning of a density it seems most appropriate to consider the LBE as an evolution equation in the Banach space $L_1(D \times V)$. We consider more generally the Banach spaces $L_p(D \times V)$, where $1 \leqslant p < \infty$. Historically, the case $p = 2$ has been treated first (cf. [25, 17]), mainly because the authors preferred to stay in the safe realm of Hilbert space. Let $(W(t); t \geqslant 0)$ denote the C_0-semigroup on $L_p(D \times V)$ associated with the LBE. The problem of main interest is the question of the behaviour of $W(t)$ for large t. This behaviour is reflected by spectral properties of $W(\cdot)$, and (partly) of the generater T of $W(\cdot)$. For more information and the precise formulation of the problem we refer to Section 2.

The aim of this paper is to review some methods used in time-dependent linear transport theory, in particular methods which rely on the positivity of $W(t)$ $(t \geqslant 0)$. At the same time, we want to put some of the methods into a general context. We feel that this will make it clearer what are the essential ingredients needed for the proofs of different facts. Owing to this aim we have organized the paper as follows. The main body of the paper, Sections 1, 2, 3, deals strictly with the LBE. The more abstract tools have been collected in Appendices A, B, C, D:

1. The free semigroup and the absorption semigroup
2. The full semigroup, spectral properties
3. Irreducibility of the full semigroup

Appendices:

A. Partial flows and generated semigroups
B. Approximate point spectrum by invariant subspaces
C. Asymptotic behaviour of (positive, irreducible) C_0-semigroups
D. Irreducibility of positively perturbed positive C_0-semigroups.

In 'Notes and Remarks' at the end of Sections 1, 2, and 3, we add additional references and general pertinent remarks.

In this last part of the introduction we recall some notations and facts.

Let X be a complex Banach space. If $A \in L(X)$ (= bounded linear operators $X \to X$), then $r(A)$ denotes the spectral radius of A, and $r_e(A)$ denotes the *essential spectral radius*,

$$r_e(A) := \sup\{|\lambda|;\ \lambda \in \sigma(A),\ \lambda \text{ not an eigenvalue of finite algebraic multiplicity}\}.$$

Let $(U(t);\, t \geqslant 0) = U(\cdot)$ be a C_0-semigroup on X. Then the type

$$\omega_0 := \inf_{t>0} t^{-1} \log \| U(t) \| = \lim_{t\to\infty} t^{-1} \log \| U(t) \| \in [-\infty, \infty)$$

of $U(\cdot)$ satisfies

$$r(U(t)) = e^{\omega_0 t} \quad (t > 0)$$

(cf. [7]). Further, there exists $\omega_e \in [-\infty, \omega_0]$, the *essential type* of $U(\cdot)$, satisfying

$$r_e(U(t)) = e^{\omega_e t} \quad (t > 0)$$

(cf. [38, Lemma 2.1]). The *generator* T of $U(\cdot)$ is defined by

$$D(T) := \{x \in X;\ Tx := \lim_{t\to 0} t^{-1}(U(t)x - x) \text{ exists}\}.$$

We denote by

$$s(T) := \sup\{\operatorname{Re} \lambda;\ \lambda \in \sigma(T)\}$$

the *spectral bound* of T. If $A \in L(X)$, then $T + A$ is the generator of a C_0-semigroup $(V(t);\, t \geqslant 0)$ which can be obtained by the *Dyson–Phillips expansion* (*of* $V(\cdot)$ *with respect to* $U(\cdot)$)

$$V(t) = \sum_{j=0}^{\infty} U_j(t)$$

(converging in operator norm, uniformly on bounded intervals), where

$$U_0(t) := U(t),\ U_j(t) = \int_0^t U(t-s) A U_{j-1}(s)\, ds \quad (t \geqslant 0, j \in \mathbb{N})$$

(cf. [7], Eqn. (3.1), p. 68, [20], IX, §1.7). We also note the *Duhamel formula*

$$V(t) = U(t) + \int_0^t U(t-s) A V(s)\, ds$$

(cf. [20], *loc. cit.*).

For an operator S in X, the *approximate point spectrum* is defined by

$$\sigma_{ap}(S) := \{\lambda \in \mathbb{C};\ \text{there exists } (x_n) \subset D(S),$$
$$\| x_n \| = 1 (n \in \mathbb{N}), \text{ such that } (\lambda - S) x_n \to 0\}.$$

1. The Free Semigroup and the Absorption Semigroup

Let $D = \mathring{D} \subset \mathbb{R}^n$, and let $V \subset \mathbb{R}^n$, V locally compact under the topology induced by $\mathbb{R}^n$. (We note that V is countable at infinity, in particular a Borel subset of $\mathbb{R}^n$.) On D we consider the restriction of the Lebesgue–Borel measure λ^n. Let ρ be a locally integrable Borel measure on V. We denote by $\mu = \lambda^n \times \rho$ the product measure on $D \times V$.

We define the functions $t_\pm : D \times V \to (0, \infty]$,

$$t_\pm(x, \xi) := \inf\{s > 0;\ x \pm s\xi \notin D\} \tag{1.1}$$

($= \infty$ if $x \pm s\xi \in D$ for all $s > 0$). The number $t_\pm(x, \xi)$ indicates the time a particle starting at x with velocity $\pm\xi$ needs until it reaches the boundary of D. We define the partial flow

$$u: \{(t, (x, \xi)) \in \mathbb{R} \times D \times V;\ -t_-(x, \xi) < t < t_+(x, \xi)\} \to D \times V,$$

$$u_t(x, \xi) = u(t, (x, \xi)) := (x + t\xi, \xi).$$

If $(x, \xi) \in D \times V$, $t \in \mathbb{R}$ *are such that* $-t_-(x, \xi) < t < t_+(x, \xi)$, then $x + t\xi \in D$, and

$$t_\pm(u_t(x, \xi)) = t_\pm(x, \xi) \mp t.$$

For $t \geqslant 0$, the sets

$$\Omega_{\pm t} := \{(x, \xi) \in D \times V;\ t_\pm(x, \xi) > t\}$$

are open, i.e., the functions $t_\pm$ are lower semicontinuous, in particular they are measurable (cf. [39], Lemma 1.5). It is not difficult to verify that $(u_t)_{t \leqslant 0}$ is a partial backward flow in the sense of Appendix A.

For the remainder of this section we fix $p \in [1, \infty)$. We define the *free semigroup* (or *semigroup of free streaming*) $(W_0(t); t \geqslant 0)$ on $L_p(D \times V)$ $(= L_p(D \times V, \mu))$ as $(U_0(t); t \geqslant 0)$ in Appendix A,

$$W_0(t)f(x, \xi) = \begin{cases} f(x - t\xi, \xi) & \text{if } t_-(x, \xi) > t, \\ 0 & \text{if } t_-(x, \xi) \leqslant t \end{cases}$$

$$= \chi_t(x, \xi) f(x - t\xi, \xi), \tag{1.2}$$

where

$$\chi_t = \chi_{\Omega_t}. \tag{1.3}$$

(Here and in the sequel we use the convention that all functions defined on $D \times V$ are extended by zero to $\mathbb{R}^n \times \mathbb{R}^n$.)

It is not difficult to show that $(W_0(t); t \geqslant 0)$ is a C_0-semigroup (cf. [39]). The generator T_0 of $W_0(\cdot)$ is a realization of the differential expression $-\xi \cdot \mathrm{grad}_x$ $(= -\mathrm{div}_x(\xi \cdot))$ on a suitable domain. In this paper we are not interested in the description of the domain of T_0. We refer to [41], sect. 1, [19], Ch. 12.1, for a core for T_0 (in special cases), and to [39], Thm 1.11, p. 36, for a precise description.

Let $h: D \times V \to \mathbb{C}$ be measurable. We want to associate a C_0-semigroup with the differential expression $-\xi \cdot \mathrm{grad}_x - h$, where the last term should be multiplication by h. Let first $h \in L_\infty(D \times V)$, and define $A \in L(L_p(D \times V))$ by $Af := -hf$. Then $T_a := T_0 + A$ is the generator of a C_0-semigroup which can be described explicitly by

$$W_a(t)f(x, \xi) = \begin{cases} \exp\left(-\int_{-t}^{0} h(x + s\xi, \xi)\, ds\right) f(x - t\xi, \xi) & \text{if } t_-(x, \xi) > t, \\ 0 & \text{if } t_-(x, \xi) \geqslant t \end{cases}$$

$$= m_t(x, \xi) f(x - t\xi, \xi), \tag{1.4}$$

where

$$m_t(x, \xi) = \chi_t(x, \xi) \exp\left(- \int_{-t}^{0} h(x + s\xi, \xi)\, ds \right) \tag{1.5}$$

(cf. [41, 19]).

In order to define $W_a(\cdot)$ by (1.4) for more general h it is sufficient to find conditions on h such that $(m_t)_{t \geqslant 0}$ defined by (1.5) satisfied (A4), (A5), (A6) of Appendix A. A set of such conditions is given by

$$\int_{-t}^{0} |h(x + s\xi, \xi)|\, ds < \infty \text{ a.e. on } \Omega_{-t}, \quad \text{for all } t \geqslant 0, \tag{1.6}$$

$$\inf \left\{ \operatorname*{ess\,inf}_{(x, \xi) \in \Omega_{-t}} \operatorname{Re} \int_{-t}^{0} h(x + s\xi, \xi)\, ds;\ 0 < t \leqslant 1 \right\} > -\infty. \tag{1.7}$$

In the case of neutron transport, the function h is assumed to be nonnegative. In this case (A4), (A5), (A6) are satisfied if

$$\int_{-t}^{0} h(x + s\xi, \xi)\, ds \to 0 \quad (t \to 0) \text{ a.e. on } D \times V \tag{1.8}$$

(where $\int_{-t}^{0} h(x + s\xi, \xi)\, ds$ is defined for $0 < t < t_-(x, \xi)$, with values in $[0, \infty]$). The function h is the *collision frequency*. We call the C_0-semigroup $(W_a(t), t \geqslant 0)$, defined by (1.4), the *absorption semigroup*. (We use this terminology although it may be misleading, since h also includes the losses resulting from scattering collisions.) By T_a we denote the generator of $W_a(\cdot)$.

1.1. THEOREM★. *Assume that* (1.6), (1.7) *are satisfied, or assume* $h \geqslant 0$ *and* (1.8).

(a) *We define*

$$\lambda^* := \sup_{t > 0} \left\{ \operatorname*{ess\,inf}_{(x, \xi) \in \Omega_{-t}} \operatorname{Re} \int_{-t}^{0} h(x + s\xi, \xi)\, ds \right\}$$

$$\left(= \lim_{t \to \infty} \{\ldots\}\right).$$

Then $r(W_a(t)) = e^{-\lambda^* t}$ $(t > 0)$, i.e., $\omega_0(W_a(\cdot)) = -\lambda^*$.

(b) *Assume additionally* $\rho(V \cap \{0\}) = 0$. *Then* $\sigma(W_a(t)) = \sigma(W_a(t)) \cdot \Gamma$ $(t > 0)$, *where* $\Gamma = \{z \in \mathbb{C}, |z| = 1\}$, $\sigma(T_a) = \sigma(T_a) + i\mathbb{R}$.

(c) *Assume additionally* $\rho(V \cap \{0\}) = 0$, *that h is real, and* $t_+(x, \xi) < \infty$ *a.e. on* $D \times V$. *Then*

$$\sigma(W_a(t)) = \sigma_{ap}(W_a(t)) = \{\mu \in \mathbb{C};\ |\mu| \leqslant e^{-\lambda^* t}\} \qquad (t > 0),$$

$$\sigma(T_a) = \sigma_{ap}(T_a) = \{\lambda \in \mathbb{C};\ \operatorname{Re} \lambda \leqslant s(T_a)\}.$$

Further, if $p = 1$ *or* $p = 2$, *then* $s(T_a) = -\lambda^*$.

★ cf. 'Added in proof'.

Proof. (a) is the same as for [41], Lemma 1.1. (b) is a consequence of Proposition A1, since the function $\alpha: D \times V \to \mathbb{R}$,

$$\alpha(x, \xi) := \begin{cases} \dfrac{x \cdot \xi}{|\xi|^2} & \text{if } \xi \neq 0, \\ 0 & \text{if } \xi = 0, \end{cases}$$

satisfies (A.7).

(c) follows from Proposition A4, since $t_+(x, \xi) < \infty$ a.e. is equivalent to $\mu(\bigcap_{t>0} \Omega_t) = 0$.

For the final statement, cf. Remark A5. □

NOTES AND REMARKS. First, we comment shortly on our assumptions on D, V, and ρ. In many references, D is assumed to be either bounded and convex ('reactor problem'; cf. [19], Ch. 12, [18, 13], or $D = \mathbb{R}^n$ ('multiple scattering problem'; cf. [19], Ch. 13). Further, V is assumed to be an annular domain $\{\xi \in \mathbb{R}^n;\ v_{\min} \leqslant |\xi| < v_{\max}\}$, where $0 \leqslant v_{\min} \leqslant v_{\max} \leqslant \infty$, and ρ is Lebesgue measure or weighted Lebesgue measure; in the latter case, mostly $V = \mathbb{R}^n$, and the weight is some Maxwellian (cf. [33]). Or else, in the 'multigroup case', V is a finite union of spheres centered at the origin, and ρ is the surface measure (cf. [22]). (Our references are not intended to be complete in any sense, at this point.) The assumptions we start with cover the cases mentioned above.

The inclusion of unbounded collision frequencies is motivated by the collision frequency obtained for a free gas moderator (cf. [27, 1, 33]).

Theorem 1.1(a) is a generalization of [41], Lemma 1.1. Theorem 1.1(b) is new. Theorem 1.1(c) generalizes [12], Beispiel 2.7, p. 15. The result of Theorem 1.1(c) concerning $\sigma(T_a)$ is known under special assumptions on h. We refer to [19], sect. 12.2, as well as to [41], Remarks 1.2.

2. The Full Semigroup, Spectral Properties

We use the definitions and notations introduced in Section 1. We consider $p \in [1, \infty)$ fixed throughout this section. We assume that h is nonnegative and satisfies (1.8).

Further we assume that

$$k: D \times V \times V \to [0, \infty)$$

is measurable. If $p > 1((1/p') + (1/p) = 1)$, then we assume

$$\begin{aligned} c_p(k) := &\left(\operatorname*{ess\,sup}_{(x, \xi) \in D \times V} \int_V k(x, \xi, \xi')\, \mathrm{d}\rho(\xi') \right)^{1/p'} \cdot \\ &\left(\operatorname*{ess\,sup}_{(x, \xi') \in D \times V} \int_V k(x, \xi, \xi')\, \mathrm{d}\rho(\xi) \right)^{1/p} < \infty. \end{aligned} \tag{2.1}$$

If $p = 1$, then we assume

$$c_1(k) := \operatorname*{ess\,sup}_{(x, \xi') \in D \times V} \int_V k(x, \xi, \xi')\, \mathrm{d}\rho(\xi) < \infty. \tag{2.2}$$

This implies that the operator K defined by

$$Kf(x, \xi) = \int_V k(x, \xi, \xi')f(x, \xi')\,d\rho(\xi')$$

($(x, \xi) \in D \times V$, $f \in L_p(D \times V)$) is a bounded operator in $L_p(D \times V)$, $\| K \| \leqslant c_p(k)$. Therefore $T := T_a + K$ is the generator of a C_0-semigroup $(W(t); t \geqslant 0)$, the *full semigroup*. Since obviously $K \geqslant 0$ holds, the Dyson–Phillips expansion implies

$$0 \leqslant W_a(t) \leqslant W(t) \quad (t \geqslant 0). \tag{2.3}$$

Our interest concentrates on spectral properties of $W(\cdot)$, more precisely on such properties which concern the large time behaviour of $W(\cdot)$. In order to make the results of Appendix C applicable to $W(\cdot)$, the following three problems are of interest.

2.1. PROBLEM. Find conditions implying $\omega_e \leqslant \lambda^*$ (where ω_e denotes the essential type of $W(\cdot)$, and λ^* is as in Theorem 1.1).

Loosely speaking, Problem 2.1 asks to find conditions implying that the addition of K to T_a can augment the spectrum of $W_a(t)$ only by eigenvalues of finite algebraic multiplicity.

The next problem concerns the existence of eigenvalues for the case that $\omega_e \leqslant -\lambda^*$ is known.

2.2. PROBLEM. Find conditions implying $\omega_0 > -\lambda^*$ (ω_0 = type of $W(\cdot)$).

2.3. PROBLEM. Find conditions implying that $(W(t); t \geqslant 0)$ is irreducible.

We shall discuss Problems 2.1 and 2.2 in this section, whereas Problem 2.3 will be treated in Section 3.

The only way we are aware of to attack Problem 2.1 is to look at the Dyson–Phillips expansion of $W(\cdot)$ with respect to $W_a(\cdot)$ and to show that some remainder has a certain compactness property. We refer to [38], Thm 2.2, for a precise statement.

The first result concerning Problem 2.1 was obtained by Jörgens [17]. We note that this is the only result for a case other than $V = \mathring{V}$ and $\rho = \lambda^n$.

2.4. EXAMPLE (cf. [17]). Assume $n = 3$, D bounded, $V \subset \{\xi \in \mathbb{R}^3; v_{\min} \leqslant |\xi| \leqslant v_{\max}\}$, with $0 < v_{\min} < v_{\max} < \infty$. Assume either (i) $V = \mathring{V}$ and $\rho = \lambda^n$ or (ii) V the finite union of spheres centered at the origin and ρ the surface measure (multigroup case). Assume h bounded and k bounded, and $p = 2$.

Then $\omega_e \leqslant -\lambda^*$, which in this case means $\omega_e = -\lambda^* = -\infty$.

For $p \neq 2$ it was only recently ([13], $p = 1$) that a fairly general result concerning Problem 2.1 was proved. This result is covered by the following result.

2.5. EXAMPLE (cf. [41], Thms 2.1 and 3.1). Assume $V = \mathring{V}$, $\rho = \lambda^n$, h bounded (the latter can be dropped). Assume that there exists $\kappa \in L_p(V)$ such that

$$k(x, \xi, \xi') \leqslant \kappa(\xi) \quad \text{for all } (x, \xi, \xi') \in D \times V \times V.$$

Further a second condition is assumed which we do not want to state here. This second condition follows from the remaining conditions for $p = 1$ if D is bounded and for $1 < p < \infty$ if D and V are bounded. Then $\omega_e \leqslant -\lambda^*$.

2.6. REMARKS. (a) In the proof of the result mentioned in Example 2.5 the compactness property of the second-order remainder is proved. It should be mentioned that the method of proof used in [41] may also be applicable under different conditions.

(b) Vidav [36] indicates sufficient conditions in an abstract context for a remainder to be compact. He derives a result for $p = 2$ and homogeneous medium, i.e., h and k do not depend on x. We are not aware that Vidav's abstract conditions have been verified for any class of h and k, in the literature, for $p \neq 2$ (cf. [19], Sect. 12.3, in particular lower part of p. 286, [41], Remark 2.2 (d)).

(c) Problem 2.1 is also treated in [2].

Finally we are going to make several remarks concerning Problem 2.2.

There are qualitative approaches to this problem, based on comparison arguments. For this we refer to [17], Thm 6.1, Thm 6.2, [1], Thm 2, p. 32, [10] (cf. also [18], sect. 4).

There is also a recent result of Schaefer [32], Thm B, p. 23, which is applicable in a rather general context, as is illustrated in the following example.

2.7. EXAMPLE. Let $p = 1$, D bounded, $V = \mathring{V} \subset \{\xi \in \mathbb{R}^n;\ |\xi| > v_{\min}\}$ for some $v_{\min} > 0$, $\rho = \lambda^n$. Assume that $W(\cdot)$ is irreducible, and that the second-order remainder $R_2(t)$ in the Dyson–Phillips expansion of $W(t)$ with respect to $W_a(\cdot)$ is weakly compact for $t \geqslant 0$. (The latter is the case, for instance, if the condition stated in Example 2.5 is satisfied.)

Then $\omega_0 > \omega_e\ (= -\infty)$.

Proof. The assumptions imply $W_a(t_0) = 0$ for suitable $t_0 > 0$. This implies $\int_0^t W_a(t-s)KW_a(s)\,ds = 0$ for $t \geqslant 2t_0$, and therefore $W(t) = R_2(t)$ for $t \geqslant 2t_0$. Now [32], *loc. cit.*, implies the desired statement. □

Further, for many specific geometries and (or) functions h and k, calculations have been made. We refer to [25, 29, 1, 34, 35, 28, 26].

We finally want to point out a problem (stated already in [24, 17] for the cases considered there) concerning the eigenfunction expansion of $W(t)$.

2.8. PROBLEM. Assume $\omega_e \leqslant -\lambda^*$. Then $\{\lambda \in \sigma(T);\ \mathrm{Re}\,\lambda > -\lambda^*\}$ is a countable (possible finite) set $\{\lambda_k;\ k \in N\}$, where $N \subset \mathbb{N}_0$, $\lambda_k \neq \lambda_j\ (k \neq j)$. Let P_k be the spectral projection corresponding to λ_k and T.

Find conditions implying that $\Sigma_{k \in N} W(t)P_k$ converges (in a suitable sense) for all $t \geqslant 0$, or at least for t large enough. If this is solved, does it then follow that the remainder

$$Z(t) = W(t) - \sum_{k \in N} W(t)P_k$$

obeys $\|Z(t)\| = O(e^{(-\lambda^* + \varepsilon)t})$, for all $\varepsilon > 0$?

2.9. REMARKS. (a) We note that the answer to Problem 2.8 is affirmative if N is finite. This follows from formula (C1). In fact, there are examples where it is shown that N is finite; cf. [25, 1], [30], sect. 4.

(b) The calculations made in [28] aim at a series expansion for a different quantity. It seems rather probable that for the specific situation considered there, Problem 2.8 can be answered affirmatively.

(c) It seems too early to judge the significance of [9] for Problem 2.8.

NOTES AND REMARKS. Since one is interested in the (time dependent) behaviour of solutions of the LBE, it is of primary importance to study the semigroup $(W(t); t \geqslant 0)$ directly. As long as there is no additional information enabling to translate spectral properties of the generator T into spectral properties of $W(\cdot)$ (which in general is not possible), the study of the spectrum of T can only be considered as a preliminary step. We therefore refer to [23] concerning this aspect, where an extensive list of references can be found; for more recent references we refer to [19]. We note, however, that, if Problem 2.1 is answered affirmatively for a system, then Problem 2.2 can be attacked by studying the generator T.

3. Irreducibility of the Full Semigroup

We use the definitions and notations introduced in Sections 1 and 2. We consider $p \in [1, \infty)$ fixed throughout this section.

3.1. REMARK. Intuitively, irreducibility of $W(\cdot)$ means that, given any (x', ξ'), $(x, \xi) \in D \times V$, a particle starting at (x', ξ') can be transported to (x, ξ) by alternatingly travelling on classical paths and being scattered at a fixed point in space to a new velocity. Unfortunately, because of the sets of measure zero involved, it is not easy to translate this concept into mathematics.

The following result reflects the idea sketched in Remark 3.1. In its proof, however, this idea is realized only indirectly.

Before stating the result we introduce the notion of *path diameter* p-diam(D), for an open connected set $D \subset \mathbb{R}^n$. For $x, x' \in D$ we define the *path distance*

$$p\text{-dist}(x, x') := \inf\{\text{length of } C;\ C \text{ polygonal path in } D \text{ connecting } x \text{ and } x'\},$$

$$p\text{-diam}(D) := \sup\{p\text{-dist}(x, x');\ x, x' \in D\}.$$

3.2. THEOREM. *Let D be connected, $V = \mathring{V} \subset \mathbb{R}^n$, $\rho = \lambda^n$. Assume that h satisfies* (1.6). *Assume that there exist $0 \leqslant v_1 < v_2 \leqslant \infty$ such that*

$$V_0 := \{\xi \in \mathbb{R}^n;\ v_1 < |\xi| < v_2\} \subset V,$$

and $k(x, \xi, \xi') > 0$ a.e. on $(D \times V_0 \times V) \cup (D \times V \times V_0)$.

Then $(W(t); t \geqslant 0)$ is irreducible. Moreover, if $v_2 = \infty$, then $W(t)$ is positivity improving for all $t > 0$. If $v_2 < \infty$ and p-diam$(D) < \infty$, then $W(t)$ is positivity improving for all

$t > t' := v_2^{-1}$ (p-diam(D)); if additionally D is convex, then $W(t')$ is also positivity improving.

The result stated in Theorem 3.2 is certainly not surprising if one adopts the idea expressed in Remark 3.1. It is therefore rather distressing to see its tedious proof. In [2], [19], Chap. 12, Sect. 3, p. 288, [18], the special case 'D bounded (and convex), $V = V_0$, h bounded' of Theorem 3.2 is stated without proof or reference. It is claimed there that the result is obvious from the Dyson–Phillips expansion.

The essential piece of analysis for the proof of Theorem 3.2 is contained in the following lemma.

3.3. LEMMA. *Let the assumptions be as in Theorem* 3.2, *with $v_2 < \infty$. Let $x_0 \in D$, and let $r > 0$ be such that*

$$\hat{B} := B(x_0, 3r)\ (:= \{x \in \mathbb{R}^n;\ |x - x_0| < 3r\})$$

is contained in D. Let $B := B(x_0, r)$.

Then for each $f \in L_p(D \times V)_+$ with $f | B \times V \neq 0$ we have $W(t)f(x, \xi) > 0$ a.e. on $B \times V$, for all $t \geqslant t_0 := 2r/v_2$.

Proof. (i) We note first that it is sufficient to prove the statement for $t = t_0$.

Assume that this is proved, and let $t > t_0$. Choose $m \in \mathbb{N}$ such that $(t - t_0)/m \leqslant t_0$, and define $r' := v_2(t - t_0)/2m$ ($\leqslant r$). Then the result already proved can be applied m times to each ball B' of radius r' contained in B, and we obtain

$$W(t)f(x, \xi) = W((t - t_0)/m)^m W(t_0)f(x, \xi) > 0$$

a.e. on $B' \times V$ for all $f \in L_p(D \times V)_+$ such that $f | B \times V \neq 0$. This implies $W(t)f(x, \xi) > 0$ a.e. on $B \times V$.

(ii) We now show the statement for $t = t_0$. This will be done by showing that $W(t_0)$ dominates an integral operator whose kernel is positive a.e. on $(B \times V) \times (B \times V)$. This integral operator is the second-order term $U_2(t_0)$ in the Dyson–Phillips expansion of $W(t_0)$ with respect to $W_a(\cdot)$. (This term, in turn, corresponds to the second generation of successors of particles present initially.)

Denote $S := \{(s', s) \in (0, \infty)^2;\ s' + s < t_0\}$. A straightforward calculation yields that the second-order term is given by

$$U_2(t_0)f(x, \xi) = \int_S W_a(s)KW_a(t_0 - s' - s)KW_a(s')f \, \mathrm{d}s \, \mathrm{d}s'(x, \xi)$$

$$= \int_{D \times V} k(x, \xi, x', \xi')f(x', \xi') \, \mathrm{d}x' \, \mathrm{d}\xi',$$

with

$$k(x, \xi, x', \xi')$$

$$= \int_S m_s(x, \xi) m_{t_0 - s' - s}\left(x - s\xi, \frac{x - s\xi - x' - s'\xi'}{t_0 - s' - s}\right) m_{s'}(x' + s'\xi', \xi') \times$$

$$\times (t_0 - s' - s)^{-n} k\left(x - s\xi, \xi, \frac{x - s\xi - x' - s'\xi'}{t_0 - s' - s}\right) \times$$

$$\times k\left(x' + s'\xi', \frac{x - s\xi - x' - s'\xi'}{t_0 - s' - s}, \xi'\right) ds'\, ds$$

$$= \int_S m_s(x, \xi) m_{t_0 - s' - s}(\ldots) m_{s'}(\ldots)(t_0 - s' - s)^{-n} \times$$

$$\times \tilde{k}\left(x - s\xi, x' + s'\xi', \xi, \frac{x - x\xi - x' - s'\xi'}{t_0 - s' - s}, \xi'\right) ds'\, ds,$$

where $\tilde{k}: D \times D \times V \times V \times V \to [0, \infty)$ is defined by

$$\tilde{k}(x_1, x_2, \xi_1, \xi_2, \xi_3) := k(x_1, \xi_1, \xi_2) k(x_2, \xi_2, \xi_3).$$

From the assumption we know $\tilde{k}(x_1, x_2, \xi_1, \xi_2, \xi_3) > 0$ a.e. on $\hat{B} \times \hat{B} \times V \times V_0 \times V$. Without restriction we may drop 'a.e.'. We define $\alpha: D \times V \times D \times V \times S \to \mathbb{R}^{5n}$,

$$\alpha(x, \xi, x', \xi', s', s) := \left(x - s\xi, x' + s'\xi', \xi, \frac{x - s\xi - x' - s'\xi'}{t_0 - s' - s}, \xi'\right).$$

It is now sufficient to show that for a.e. $(x, \xi, x', \xi') \in B \times V \times B \times V$ the (open!) set

$$\{(s', s) \in S;\ \alpha(x, \xi, x', \xi', s', s) \in \hat{B} \times \hat{B} \times V \times V_0 \times V\}$$

is nonempty. (For these (s', s) it is easy to see that the first three factors under the integral for $\hat{k}$ are positive as a consequence of (1.6), and therefore $\hat{k}(x, \xi, x', \xi') > 0$.)

Let $(x, \xi, x', \xi') \in B \times V \times B \times V$, $\xi' \neq \xi$ or $x - x' - t_0\xi \neq 0$. (The set of these points obviously has full measure in $B \times V \times B \times V$.) Then $|(x - x')/t_0| < 2r/t_0 = v_2$; therefore there exists $s' \in (0, t_0)$ such that

$$\left|\frac{x - x' - s'\xi'}{t_0 - s'}\right| < v_2, \quad x' + s'\xi' \in B,\ x - x' - t_0\xi + s'(\xi - \xi') \neq 0.$$

Then either $x - s\xi \in \hat{B}$ for all $s \in (0, t_0 - s')$, or else there exists $s_0 \in (0, t_0 - s')$ such that $x - s\xi \in \hat{B}$ for all $s \in (0, s_0)$, but $x - s_0\xi \notin \hat{B}$. In the first case we have

$$\left|\frac{x - s\xi - x' - s'\xi'}{t_0 - s' - s}\right| \to \infty \quad (s \to t_0 - s');$$

in the second we obtain

$$\left|\frac{x - s_0\xi - (x' + s'\xi')}{t_0 - s' - s_0}\right| > \frac{2r}{t_0 - s' - s_0} > \frac{2r}{t_0} = v_2\,.$$

In both cases there exists $s \in (0, t_0 - s')$ such that

$$x - s\xi \in \hat{B}, \quad v_1 < \left|\frac{x - s\xi - x' - s'\xi'}{t_0 - s' - s}\right| < v_2. \qquad \square$$

3.4. REMARK. It seems worth while to note that, in the proof of Lemma 3.3 we have shown in step (ii) that $U_2(t_0)$ acts 'positivity improving on the set $B \times V$'. For $t > t_0$, this is not true in general. In fact, if D is bounded and V is bounded away from zero, then $U_2(t) = 0$ for t sufficiently large. Therefore, it is a necessity to use $W(\cdot)$ itself and not $U_2(\cdot)$ in step (i) of the proof.

Proof of Theorem 3.2. We only comment shortly on the case $v_2 < \infty$, p-diam$(D) < \infty$; the proof is easily adopted to the other case.

Let $f \in L_p(D \times V)_+$, $f \neq 0$. Then there exists $x_0 \in D$ such that for all $\varepsilon > 0$ with $B(x_0, \varepsilon) \subset D$ we have $f | B(x_0, \varepsilon) \times V \neq 0$.

Let $t > 0$, $x' \in D$ such that p-dist$(x_0, x') < v_2 t$. Then there exists a polygonal path C connecting x_0 with x', of length $< v_2 t$. There exists a covering of C by balls $\{B(x_j, r_j); j = 0, \ldots, m\}$ such that $x_m = x'$, $B(x_j, r_j) \cap B(x_{j-1}, r_{j-1}) \neq \emptyset$ $(j = 1, \ldots, m)$, $B(x_j, 3r_j) \subset D (j = 0, \ldots, m)$, $2\Sigma_{j=0}^m r_j \leqslant v_2 t$. Repeated application of Lemma 3.3 yields $W(t)f(x, \xi) > 0$ a.e. on $B(x', r_m) \times V$.

This implies the desired result. □

It seems to the author that it should be possible to develop a geometrically more intuitive method of proof for irreducibility, in the spirit of Remark 3.1. Since probably the results of Appendix D will be useful for this approach we shall illustrate some of the concepts introduced in Appendix D.

Due to Proposition D3, it is of interest to study the closed solid subcones of $L_p(D \times V)_+$ invariant under $W_a(\cdot)$ and K separately. We note that the closed solid subcones of $L_p(D \times V)_+$ are of the form $L_p(M)_+$ (canonically injected), for measurable $M \subset D \times V$.

The proof of the following proposition is obvious. For '$\prec$' and '$\sim$' we refer to Appendix D.)

3.5. PROPOSITION. (a) *$W_a(t) \prec W_0(t)$ for all $t \geqslant 0$.* (b) *If additionally* (1.6) *holds, then $W_a(t) \sim W_0(t)$ for all $t \geqslant 0$.*

We show next that the invariant cones for K are completely determined by

$$[k > 0] := \{(x, \xi, \xi') \in D \times V \times V;\ k(x, \xi, \xi') > 0\}.$$

Let $k' : D \times V \times V \to [0, \infty)$ be measurable and satisfy $c_p(k') < \infty$ (cf. (2.1), (2.2)), and define K' correspondingly. Let $(W'(t); t \geqslant 0)$ be the C_0-semigroup generated by $T_a + K'$.

3.6. PROPOSITION. *Assume*

$$(\lambda^n \times \rho \times \rho)([k > 0] \backslash [k' > 0]) = 0.$$

Then $K \prec K'$, $\{W(t); t \geqslant 0\} \prec \{W'(t); t \geqslant 0\}$.

Proof. For $j \in \mathbb{N}$ we define $k_j := k \wedge (jk')$, and K_j correspondingly. Then $K_j \leqslant jK'$ $(j \in \mathbb{N})$, $K_j \to K$ in the strong operator topology, and therefore $K \prec K'$ follows from Remarks D1 (b, e, f).

The second statement follows then from Proposition D3 and Remark D1 (c). □

3.7. REMARKS. (a) In [21], 4.2.3, Satz, the following criterion for irreducibility is stated. Let $n = 3$, D convex, $0 \leqslant v_1 < v_2 < \infty$, $V = \{\xi \in \mathbb{R}^3; v_1 \leqslant |\xi| \leqslant v_2\}$, $\rho = \lambda^3$, $p = 1$, h bounded. Assume that

$$M_0 := \{x \in D;\ \exists S_x \subset V \text{ measurable, } 0 < \rho(S_x) < \rho(V), \text{ such that } k(x, \xi, \xi') = 0 \text{ for a.e. } (\xi, \xi') \in S_x \times (V \backslash S_x)\}$$

is a null set. Then $W(\cdot)$ is irreducible.

It is first shown in [21], 4.2.2, that each K-invariant closed solid subcone C of $L_1(D \times V)_+$ is of the form $L_1(D_C \times V)_+$, with measurable $D_C \subset D$. In view of Proposition D3, Remark D1 (g), and Proposition 3.5 it is now sufficient to show that $L_1(D_C \times V)_+$ can be $W_0(\cdot)$-invariant only in the trivial cases $\lambda^3(D_C) = 0$, $\lambda^3(D \backslash D_C) = 0$. This statement, which is definitely plausible, should nevertheless be provided with a proof. (In [21] it is claimed that this follows by simple geometrical considerations.)

(b) From [13], Prop. 2.5, we take the following criterion for irreducibility. Let $n = 3$, D bounded and convex, $v_2 < \infty$, $V = \{\xi \in \mathbb{R}^3; |\xi| \leqslant v_2\}$, $\rho = \lambda^3$, h bounded, k bounded, and assume that there exists $\varepsilon > 0$ such that $k(x, \xi, \xi') > 0$ for all $x \in D_\varepsilon := \{x \in D; \operatorname{dist}(x, \mathbb{R}^3 \backslash D) < \varepsilon\}$, $\xi, \xi' \in V$. Then $W(\cdot)$ is irreducible.

We note, however, that in the proof in [13] a simplifying assumption is made which would need justification.

(c) The criteria presented in Theorem 3.2 and in (a), (b) above are, on the one hand, intuitively convincing but, on the other hand, not easy to prove. We think that an analysis of the $W_0(\cdot)$-invariant closed solid subcones of $L_p(D \times V)_+$, combined with the results of Appendix D, should make the proofs easier and more natural.

NOTES AND REMARKS. Theorem 3.2 and Lemma 3.3 are essentially contained in [37]. The idea to discuss the irreducibility of $W(\cdot)$ by discussing $W_a(\cdot)$ and K separately goes back to [21]. Proposition 3.6 is contained in [13], Prop. 2.4.

Appendix A: Partial Flows and Generated Semigroups

Let $(\Omega, \mathscr{A}, \mu)$ be a measure space. Assume that a family $(\Omega_t)_{t \in \mathbb{R}}$ of measurable subsets of Ω is given, satisfying

$$\Omega = \Omega_0 \supset \Omega_{\pm t} \supset \Omega_{\pm s} \quad (0 \leqslant t \leqslant s).$$

A partial backward flow is a family $(u_t)_{t \leqslant 0}$ of bijective and measure preserving (for μ) mappings

$$u_t := \Omega_t \to \Omega_{-t} \quad (t \leqslant 0),$$

satisfying

$$u_0 = \mathrm{id}_\Omega,\ u_{t+s} = u_t \circ u_s \quad (t, s \leqslant 0), \tag{A1}$$

in the sense that

$$\Omega_{t+s} = \{w \in \Omega_s;\ u_s(w) \in \Omega_t\},$$
$$u_{t+s}(w) = u_t(u_s(w)) \quad (w \in \Omega_{t+s}). \tag{A2}$$

Let $p \in [1, \infty)$. With the partial backward flow we associate for each $t \geqslant 0$ an operator $U_0(t) \in L(L_p(\mu))$,

$$U_0(t)f := \begin{cases} f \circ u_{-t} & \text{on } \Omega_{-t}, \\ 0 & \text{on } \Omega \backslash \Omega_{-t}, \end{cases}$$

for $f \in L_p(\mu)$. It is easy to show $\| U_0(t) \| \leqslant 1$ $(t \geqslant 0)$, $U_0(0) = I$, $U_0(t+s) = U_0(t)U_0(s)$ $(t, s \geqslant 0)$. We now assume that

$$U_0(t)f \to f \ (t \to 0), \quad \text{for all } f \in L_p(\mu); \tag{A3}$$

thus $U_0(\cdot)$ is a C_0-semigroup. This condition requires an additional property of $(u_t)_{t \leqslant 0}$ which we do not want to state in more explicit form.

We want to derive spectral properties of a more general semigroup. We assume that a family $(m_t)_{t \geqslant 0}$ of $L_\infty(\mu)$-functions is given, satisfying

$$m_0 = 1,\ m_{t+s} = \left.\begin{cases} m_t(m_s \circ u_{-t}) & \text{on } \Omega_{-t} \\ 0 & \text{on } \Omega \backslash \Omega_{-t} \end{cases}\right\} (t, s \geqslant 0), \tag{A4}$$

$$\sup\{\| m_t \|_\infty;\ 0 \leqslant t \leqslant 1\} < \infty, \tag{A5}$$

$$m_{t_j} \to 1 \text{ a.e., for each null sequence } (t_j) \subset (0, \infty). \tag{A6}$$

This implies that a C_0-semigroup $(U(t); t \geqslant 0)$ on $L_p(\mu)$ is given by $U(t)f := m_t U_0(t) f (f \in L_p(\mu), t \geqslant 0)$. Let T denote its generator.

For all $t \geqslant 0$, the operator $U(t)$ is a Lamperti operator (cf. [4]). If additionally $m_t \geqslant 0$ $(t \geqslant 0)$ then $U(t)$ is a lattice homomorphism for all $t \geqslant 0$. It might be of interest to apply known decomposition theorems (cf. [3]) in order to translate spectral properties of the semigroup or its generator into properties of the partial backward flow. Our emphasis here goes in the converse direction.

A1. PROPOSITION. *Additionally to the assumptions made so far we assume that there exists a measurable function* $\alpha: \Omega \to \mathbb{R}$ *satisfying*

$$\alpha(u_t(\cdot)) = \alpha(\cdot) + t \tag{A7}$$

for all $t \leqslant 0$, *a.e. on* Ω_t.

Then for $\eta \in \mathbb{R}$ *the mapping* $M_\eta \in L(L_p(\mu))$,

$$M_\eta f := \mathrm{e}^{-i\eta\alpha(\cdot)} f,$$

is an isometric isomorphism such that

$$M_\eta^{-1} U(t) M_\eta = \mathrm{e}^{i\eta t} U(t) \quad (t \geqslant 0),$$
$$M_\eta^{-1} T M_\eta = T + i\eta I. \tag{A8}$$

As a consequence,

$$\sigma(U(t)) = \sigma(U(t)) \cdot \Gamma \quad (t > 0),$$

where $\Gamma := \{z \in \mathbb{C}; |z| = 1\}$, $\sigma(T) = \sigma(T) + i\mathbb{R}$. *The latter equalities hold also for all other kinds of spectra, e.g.*

$$\sigma_{\text{ap}}(U(t)) = \sigma_{\text{ap}}(U(t)) \cdot \Gamma \quad (t > 0),$$

$$\sigma_{\text{ap}}(T) = \sigma_{\text{ap}}(T) + i\mathbb{R}.$$

Proof. Obviously it is sufficient to prove (A8), and this property follows by a straightforward calculation. □

A2. REMARK. We note that the existence of the function in Proposition A1 excludes the existence of (too many) periodic orbits.

The following lemma is a preparation for the subsequent proposition. We omit its straightforward proof.

A3. LEMMA. *In addition to the assumptions made initially we assume*

$$\mu\left(\bigcap_{t \geq 0} \Omega_t\right) = 0. \tag{A9}$$

Then the function $\alpha\colon \Omega \to (-\infty, 0]$,

$$\alpha(w) := \begin{cases} -\sup\{t \geq 0;\ w \in \Omega_t\} & \text{if } w \in \Omega \setminus \bigcap_{t \geq 0} \Omega_t, \\ 0 & \text{if } w \in \bigcap_{t \geq 0} \Omega_t \end{cases}$$

is measurable and satisfies (A7).

A4. PROPOSITION. *In addition to the assumptions made initially we assume (A9) and*

$$m_t \geq 0 \ (t \geq 0). \tag{A10}$$

Then

$$\sigma(U(t)) = \sigma_{\text{ap}}(U(t)) = \{\mu \in \mathbb{C};\ |\mu| \leq r(U(t))\} \quad (t > 0),$$

$$\sigma(T) = \sigma_{\text{ap}}(T) = \{\lambda \in \mathbb{C};\ \operatorname{Re} \lambda \leq s(T)\}.$$

Proof. In view of Lemma A3 and Proposition A1 it is sufficient to show

$$[0, r(U(t))] \subset \sigma_{\text{ap}}(U(t)) \quad (t > 0), \ (-\infty, s(T)] \subset \sigma_{\text{ap}}(T).$$

For the proof of these inclusions we want to apply Proposition B1 and Proposition B2.

Because of (A10) the semigroup $U(\cdot)$ is positive. For $n \in \mathbb{N}$ we define

$$X_n := \{f \in L_p(\mu);\ f | \Omega_n = 0\}.$$

Then X_n is a closed ideal of $L_p(\mu)$, and (A9) implies that $\bigcup_{n \in \mathbb{N}} (X_n)_+$ is dense in $L_p(\mu)_+$.

Next we show that X_n is invariant under $U(t)$ $(t \geqslant 0)$. For $t \geqslant 0$, (A2) can be rewritten as $\Omega_n \cap \Omega_{-t} = u_{-n}(\Omega_{-n-t})$, which implies $u_{-t}(\Omega_n \cap \Omega_{-t}) = u_{-n-t}(\Omega_{-n-t}) = \Omega_{n+t} \subset \Omega_n$. Thus for $f \in X_n$ we have $f \circ u_{-t} = 0$ on $\Omega_n \cap \Omega_{-t}$, and therefore $U(t)f = U_0(t)f = 0$ on Ω_n, i.e., $U(t)f \in X_n$.

In order to show that Propositions B1 and B2 yield the desired result, it is now sufficient to show that the semigroups $(U(t)|X_n; t \geqslant 0)$ on X_n are nilpotent $(n \in \mathbb{N})$. The latter, however, is true since obviously $U(n)f = 0$ holds for all $f \in X_n$ $(n \in \mathbb{N})$. □

A5. REMARK. If assumption (A10) is made (and thus $U(t) \geqslant 0$ for all $t \geqslant 0$), then for $p = 1$ and $p = 2$ we know

$$\omega_0(U(\cdot)) = s(T) \tag{A11}$$

(cf. [8] for $p = 1$, [14] for $p = 2$). It is an open question whether (A11) holds for positive C_0-semigroups on L_p for p other than 1 or 2. We were not able to decide this even for the special case considered here, or for the even more special case considered in Section 1★.

Appendix B: Approximate Point Spectrum by Invariant Subspaces

Throughout this section let X be a complex ordered Banach space, which means that X is the complexification of a (real) ordered Banach space, and assume that the positive cone X_+ is normal and generating (cf. [5]).

For $n \in \mathbb{N}$ let X_n be a closed subspace of X, with $X_n \subset X_{n+1}$ $(n \in \mathbb{N})$, and such that

$$X_+ = \overline{\bigcup_{n \in \mathbb{N}} (X_n \cap X_+)}. \tag{B1}$$

B1. PROPOSITION. *Let $A \in L(X)$ be a positive operator, and assume $A(X_n) \subset X_n$ $(n \in \mathbb{N})$, $A_n := A|X_n$. Let $0 \leqslant r < r(A)$ be such that $r(A_n) \leqslant r$ for all $n \in \mathbb{N}$.*

Then $[r, r(A)] \subset \sigma_{ap}(A)$. For all $\lambda \in [r, r(A)]$, a sequence $(x_n) \subset X$ can be found such that $x_n \in X_n$, $\|x_n\| = 1$ $(n \in \mathbb{N})$, $\|(\lambda - A_n)x_n\| \to 0$ $(n \to \infty)$.

We omit the proof which is rather similar to the proof of the following proposition.

B2. PROPOSITION. *Let $(U(t); t \geqslant 0)$ be a positive C_0-semigroup on X, with generator T. For all $n \in \mathbb{N}$ let X_n be invariant under $U(\cdot)$, and denote by $U_n(\cdot)$ the restriction of $U(\cdot)$ to X_n, and by T_n the generator of $U_n(\cdot)$. (Note $T_n \subset T_{n+1} \subset T$ $(n \in \mathbb{N})$.) Assume that there exists $s < s(T)$ such that*

$$\int_0^\infty e^{-\lambda t} U_n(t)\, dt = \lim_{a \to \infty} \int_0^a e^{-\lambda t} U_n(t)\, dt \tag{B2}$$

exists for all $\lambda > s$ and all $n \in \mathbb{N}$.

Then $[s, s(T)] \subset \sigma_{ap}(T)$. For $\lambda \in [s, s(T)]$, a sequence $(x_n) \subset D(T)$ can be found such that $x_n \in D(T_n)$, $\|x_n\| = 1 (n \in \mathbb{N})$,

$$\|(\lambda - T_n)x_n\| \to 0 \quad (n \to \infty).$$

★ cf. 'Added in proof'.

Proof. We define the operator S by $S := T|\bigcup_{n\in\mathbb{N}} D(T_n)$ and note that T is the closure of S. This follows since, by (B1), $\bigcup_{n\in\mathbb{N}} X_n$ is dense in X, and since, for $\mu > \omega_0(U(\cdot))$, we have $(\mu - T)^{-1} = \int_0^\infty e^{-\mu t} U(t)\, dt$.

The second statement of the proposition implies the first, and it is sufficient to prove the second statement for $\lambda \in (s, s(T)]$.

Let $\lambda \in (s, s(T)]$. From (B2) we have $\lambda \in \rho(T_n)$,

$$(\lambda - T_n)^{-1} = \int_0^\infty e^{-\lambda t} U_n(t)\, dt \tag{B3}$$

($n \in \mathbb{N}$). (Cf. [40], Thm 1.1; the existence of the limit in (B2) implies that the abscissa of convergence for the Laplace transform of $U_n(\cdot)$ is $\leqslant s$; for large λ, (B3) holds.) It obviously suffices to show

$$\sup_{n\in\mathbb{N}} \|(\lambda - T_n)^{-1}\| = \infty.$$

Assume $c := \sup_{n\in\mathbb{N}} \|(\lambda - T_n)^{-1}\| < \infty$. We define $B\colon \bigcup_{n\in\mathbb{N}} X_n \to X$ by $B|X_n := (\lambda - T_n)^{-1}$. Then $B = (\lambda - S)^{-1}$ is bounded and densely defined. This implies $\lambda \in \rho(T)$, $(\lambda - T)^{-1} = \bar{B}$. Also by (B1) and (B3), $(\lambda - T)^{-1} = \bar{B}$ is positive. For $\mu > s(T)$ we have $\mu \in \rho(T)$, $(\mu - T)^{-1} \geqslant 0$ (cf. [15]), and the resolvent equation implies

$$(\mu - T)^{-1} = (\lambda - \mu)(\mu - T)^{-1}(\lambda - T)^{-1} + (\lambda - T)^{-1} \leqslant (\lambda - T)^{-1}. \tag{B4}$$

Since $L(X)_+$ is normal in $L(X)$ (cf. [5], Thm 1.7.3, (B4) implies that the set $\{(\mu - T)^{-1}; \mu > s(T)\}$ is bounded in $L(X)$. This is a contradiction to $s(T) \in \sigma(T)$ (cf. [15], Thm 3.3a). □

B3. REMARK. The condition that the limit in (B2) exists implies $s(T_n) \leqslant s (n \in \mathbb{N})$ (cf. the above proof). The latter condition is equivalent to the former in the following case: For each $n \in \mathbb{N}$, X_n is a complex ordered Banach space with generating positive cone $X_n \cap X_+$ (cf. [15], Thm 3.3b).

Appendix C: Asymptotic Behaviour of (Positive, Irreducible) C_0-semigroups

Let X be a complex Banach space. Let $(U(t); t \geqslant 0)$ be a C_0-semigroup on X, with generator T, type ω_0, and essential type ω_e. Then for each $\omega > \omega_e$ there exists a finite rank projection P_ω in X, commuting with $U(t)$ $(t \geqslant 0)$, and a constant $C_\omega \geqslant 0$ such that

$$\|U(t)(I - P_\omega)\| \leqslant C_\omega e^{t\omega} \quad (t > 0). \tag{C1}$$

In fact, P_ω can be taken as the spectral projection corresponding to $\{\lambda \in \sigma(T); \operatorname{Re}\lambda \geqslant \omega\}$ and T (which coincides with the spectral projection corresponding to $\{\mu \in \sigma(U(t)); |\mu| \geqslant e^{\omega t}\}$ and $U(t)$, for all $t > 0$).

If $\omega_e < \omega < \omega_0$, then $P_\omega \neq 0$, and the asymptotic behaviour of $U(\cdot)$ for $t \to \infty$ is determined by the part of the semigroup $U(\cdot)$ in the finite dimensional space $R(P_\omega)$ (cf. [36], Sect. III, p. 272, [38], proof of Lemma 2.1, p. 157). (As a consequence, we note that $\omega_e < \omega_0$ implies $s(T) = \omega_0$.)

C1. PROPOSITION. *In addition to the previous assumptions we assume that X is a Banach lattice and that $U(t) \geqslant 0$ $(t \geqslant 0)$ holds. Further we assume $\omega_e < \omega_0$.*

Then $\sigma_{\text{per}}(U(t)) = \{r(U(t))\}$ $(= \{e^{\omega_0 t}\})$ holds for all $t > 0$ (where $\sigma_{\text{per}}(B) := \{\mu \in \sigma(B); |\mu| = r(B)\}$ – the peripheral spectrum – for $B \in L(X)$). Also, ω_0 is an eigenvalue of T, and there exists $\varepsilon > 0$ such that

$$\{\lambda \in \sigma(T);\ \text{Re}\,\lambda \geqslant \omega_0 - \varepsilon\} = \{\omega_0\},$$

i.e., ω_0 is a strictly dominant eigenvalue of T.

Proof. (In [13], proof of Cor. 2.2, this result is derived (for a special context) from [11], Thm (2.5) (a). We are going to reproduce a simple proof from [21], 3.1.5, avoiding the sophisticated result of [11].) It is easy to see that it is sufficient to show

$$\{\lambda \in \sigma(T);\ \text{Re}\,\lambda = \omega_0\} \subset \{\omega_0\}.$$

Let $b \in \mathbb{R}$ be such that $\omega_0 + ib \in \sigma(T)$. This implies $e^{t(\omega_0 + ib)} \in \sigma_{\text{per}}(U(t))$ for all $t \geqslant 0$. By [31], Ch. V, Cor. of Thm 4.9, $\sigma_{\text{per}}(U(t))$ is cyclic for $t > 0$. Therefore $t(b/2\pi)$ is rational for all $t > 0$, which implies $b = 0$. □

C.2. PROPOSITION. *Let the assumptions be as in Proposition* 1, *and let P_0 be the spectral projection corresponding to ω_0 and T. Then the following statements are equivalent:*

(i) *$(U(t); t \geqslant 0)$ is irreducible,*

(ii) *$U(t)$ is irreducible for some (all) $t > 0$,*

(iii) dim $R(P_0) = 1$, *and P_0 is strictly positive (= positivity improving).*

Proof. (i)⇒(iii). By [15], Prop. 3.9 and its proof, we have dim $R(P_0) = 1$ and $P_0 \geqslant 0$ irreducible. Clearly, for a positive one-dimensional projection irreducibility and strict positivity are equivalent.

(iii)⇒(ii) (with 'all'). Let $x \in X_+$, $0 < x' \in X'_+$. Then $\langle P_0 x, x' \rangle > 0$ and therefore

$$\begin{aligned} &e^{-\omega_0 t} \langle U(t)x, x' \rangle \\ &= \langle P_0 x, x' \rangle + \langle e^{-\omega_0 t} U(t)(I - P_0)x, x' \rangle \to \\ &\to \langle P_0 x, x' \rangle \ (t \to \infty), \end{aligned}$$

because of (C1). Therefore $\langle U(t)x, x' \rangle > 0$ for all large enough t. This implies that $U(t)$ is irreducible for all $t > 0$.

(ii) (with 'some')⇒(i) is trivial. □

Appendix D: Irreducibility of (Positively Perturbed) Positive Semigroups

Throughout this section let X be a real or complex ordered Banach space, and assume that the positive cone X_+ is normal and generating.

Let $\mathcal{S}, \mathcal{S}_1 \subset L(X)_+$. We write $\mathcal{S} \prec \mathcal{S}_1$ if each closed solid (= hereditary) $\mathcal{S}_1$-invariant subcone of X_+ is also $\mathcal{S}$-invariant. We write $\mathcal{S} \sim \mathcal{S}_1$ if $\mathcal{S} \prec \mathcal{S}_1$ and $\mathcal{S}_1 \prec \mathcal{S}$. We collect several evident properties concerning this relation.

D1. REMARKS. Let $\mathscr{S}, \mathscr{S}_1, \mathscr{S}_2 \subset L(X)_+$.

(a) $\mathscr{S} \prec \mathscr{S}_1$, $\mathscr{S}_1 \prec \mathscr{S}_2 \Rightarrow \mathscr{S} \prec \mathscr{S}_2$. This implies: '$\sim$' is an equivalence relation on Pot $(L(X)_+)$, whose equivalence classes are ordered by '$\prec$'.

(b) $\mathscr{S} \subset \mathscr{S}_1 \Rightarrow \mathscr{S} \prec \mathscr{S}_1$.

(c) $\mathscr{S} \prec \mathscr{S}_2$, $\mathscr{S}_1 \prec \mathscr{S}_2 \Rightarrow (\mathscr{S} \cup \mathscr{S}_1) \prec \mathscr{S}_2$. In particular: $\mathscr{S} \sim \mathscr{S}_1 \Rightarrow (\mathscr{S} \cup \mathscr{S}_2) \sim (\mathscr{S}_1 \cup \mathscr{S}_2)$.

(d) $\mathrm{co}\,(\bigcup_{\lambda \geq 0} \lambda \mathscr{S}) \sim \mathscr{S}$.

(e) $\{A \in L(X)_+;\ \exists A_1 \in \mathscr{S}\colon A \leqslant A_1\} \sim \mathscr{S}$.

(f) $\mathscr{S}^{\mathfrak{T}_s} \sim \mathscr{S}$, where $\mathfrak{T}_s$ is the strong operator topology on $L(X)$.

(g) $\mathscr{S} \prec \mathscr{S}_1$, $\mathscr{S}$ irreducible $\Rightarrow \mathscr{S}_1$ irreducible.

In particular: $\mathscr{S} \sim \mathscr{S}_1 \Leftrightarrow (\mathscr{S}$ irreducible $\Leftrightarrow \mathscr{S}_1$ irreducible).

The proof of the following proposition is contained in [15], proof of Thm 3.8.

D2. PROPOSITION. *Let $(U(t);\ t \geqslant 0)$ be a positive C_0-semigroup on X, with generator T. Then $\{U(t);\ t \geqslant 0\} \sim (\lambda - T)^{-1}$, for all $\lambda > s(T)$.*

The following proposition was inspired by [21], 4.2.1.

D3. PROPOSITION. *Let $(U(t);\ t \geqslant 0)$ be a positive C_0-semigroup on X, with generator T. Let $A \in L(X)_+$, and denote by $(V(t);\ t \geqslant 0)$ the C_0-semigroup generated by $T + A$. Then $\{U(t);\ t \geqslant 0\} \cup \{A\} \sim \{V(t);\ t \geqslant 0\}$.*

Proof. '$\prec$'. $\{U(t);\ t \geqslant 0\} \prec \{V(t);\ t \geqslant 0\}$ follows from $U(t) \leqslant V(t)\ (t \geqslant 0)$ and Remark D1 (b, e). In view of Remark D1 (c) it remains to show $\{A\} \prec \{V(t);\ t \geqslant 0\}$.

Let C be a norm closed solid subcone of X_+, invariant under $\{V(t);\ t \geqslant 0\}$, and let $x \in C$. The Duhamel formula implies

$$0 \leqslant \int_0^t U(t-s)AV(s)x\,\mathrm{d}s \leqslant V(t)x \in C, \quad \int_0^t U(t-s)AV(s)x\,\mathrm{d}s \in C,$$

and therefore, for $t \to 0$,

$$t^{-1} \int_0^t U(t-s)AV(x)x\,\mathrm{d}s \to Ax \in C.$$

'$\succ$'. Let C be a norm closed solid subcone of X_+, invariant under $U(t)\ (t \geqslant 0)$ and A. Then the Dyson–Phillips expansion implies that C is also invariant under $V(t)\ (t \geqslant 0)$. □

The following corollary is an immediate consequence of Remark D1 (c), Propositions D2, and D3.

D4. COROLLARY. *Let all the notations be as in Proposition D3. Then $\{(\lambda - T)^{-1}, A\} \sim \{U(t);\ t \geqslant 0\} \cup \{A\} \sim \{V(t);\ t \geqslant 0\} \sim (\lambda' - T - A)^{-1}$, for all $\lambda > s(T)$, $\lambda' > s(T + A)$.*

Added in Proof

It follows from [J. Voigt: 'Interpolation for (Positive) C_0-semigroups on L_p-Spaces', preprint, Cor. 2 (b)] that in Theorem 1.1, in the case that h is real, one has $s(T_a) = -\lambda^*$ for all $p \in [1, \infty)$.

Also, it follows that in the situation described in Appendix A, one has $\omega_0(U(\cdot)) = s(T)$ for all $p \in [1, \infty)$ if (A10) is assumed; cf. Remark A5. It is sufficient to show this for $1 < p < \infty$. We note first $\| U(t) \| = \| m_t \|_\infty$ $(t \geqslant 0)$, and therefore $\omega_0(U(\cdot))$ is independent of p. The continuity of $U(\cdot)$ in the strong operator topology for one $p \in [1, \infty)$ implies that $U(\cdot)$ is continuous in the weak operator topology for all $p \in (1, \infty)$, and therefore a C_0-semigroup on $L_p(\mu)$ for all $p \in (1, \infty)$; cf. [K. Yosida: *Functional Analysis*. 2nd edn. Springer-Verlag, New York, 1968, Ch. IX, Sect. 1, Thm, p. 233]. Now [Voigt, *loc. cit.*] implies the desired assertion.

References

1. Albertoni, S. and Montagnini, B.: 'On the Spectrum of Neutron Transport Equation in Finite Bodies', *J. Math. Anal. Appl.* **13** (1966), 19–48.
2. Angelescu, N. and Protopopescu, V.: 'On a Problem in Linar Transport Theory', *Rev. Roum. Phys.* **22** (1977), 1055–1061.
3. Arendt, W.: 'Kato's Equality and Spectral Decomposition for Positive C_0-groups', *Manuscripta Math.* **40** (1982), 277–298.
4. Arendt, W.: 'Spectral Properties of Lamperti Operators', *Indiana Univ. Math. J.* **32** (1983), 199–215.
5. Batty, C. J. K. and Robinson, D.W.: 'Positive One-parameter Semigroups on Ordered Banach Spaces', *Acta Appl. Math.* **2** (1984), 221–296 (this issue).
6. Case, K. M. and Zweifel, P. F.: *Linear Transport Theory*. Addison-Wesley, Reading, Mass. 1967.
7. Davies, E. B.: *One-parameter Semigroups*, Academic Press, London, 1980.
8. Derndinger, R.: Über das Spektrum positiver Generatoren', *Math. Z.* **172** (1980), 281–293.
9. Emamirad, H.: 'Generalized Eigenfunction Expansions in Transport Theory', Preprint, 1983.
10. Fuß, J.: 'Über die Spektralschranke des linaren Transportoperators', Dissertation, Wien, 1982.
11. Greiner, G.: 'Zur Perron-Frobenius-Theorie stark stetiger Halbgruppen', *Math. Z.* **177** (1981), 401–423.
12. Greiner, G.: *Spektrum und Asymptotik start stetiger Halbgruppen positiver Operatoren*, Sitzungsber. Heidelb. Akad. Wiss., Math.-naturwiss. Kl., Springer Verlag, Berlin, 1982.
13. Greiner, G.: 'Asymptotics in Linear Transport Theory'. Semesterbericht Funktionalanalysis, Tübingen, Sommersemester 1982.
14. Greiner, G. and Nagel, R.: 'On the Stability of Strongly Continuous Semigroups of Positive Operators on $L^2(\mu)$', *Annali Scuola Normale Sup.* **10** (1983), 257–262.
15. Greiner, G., Voigt, J. and Wolff, M.: 'On the Spectral Bound of the Generator of Semigroups of Positive Operators', *J. Operator Theory* **5** (1981), 245–256.
16. Huber, A.: 'Spectral Properties of the Linear Multiple Scattering Operator in L^1-Banach Lattices', *Integral Equations and Operator Theory* **6** (1983), 357–371.
17. Jörgens, K.: 'An Asymptotic Expansion in the Theory of Neutron Transport', *Commun. Pure Appl. Math.* **11** (1958), 219–242.
18. Kaper, H. G. and Hejtmanek, J.: 'Recent Progress on the Reactor Problem of Linear Transport Theory, Preprint, 1983.
19. Kaper, H. G., Lekkerkerker, C. G. and Hejtmanek, J.: *Spectral Methods in Linear Transport Theory*, Birkhäuser-Verlag, Basel, 1982.
20. Kato, T.: *Perturbation Theory for Linear Operators*. Springer-Verlag, Berlin, 1966.
21. Koschat, M.: 'Die lineare Boltzmanngleichung im Banachverband $L^1(D \times W)$', Dissertation, Wien, 1979.
22. Larsen, E. W.: 'The Spectrum of the Multigroup Neutron Transport Operator for Bounded Spatial Domains', *J. Math. Phys.* **20** (1979), 1776–1782.

23. Larsen, E. W. and Zweifel, P. F.: 'On the Spectrum of the Linear Transport Operator', *J. Math. Phys.* **15** (1974), 1987–1997.
24. Lehner, J.: 'An Unsymmetric Operator Arising in the Theory of Neutron Diffusion', *Commun. Pure Appl. Math.* **9** (1956), 487–497.
25. Lehner, J. and Wing, G. M.: 'On the Spectrum of an Unsymmetric Operator Arising in the Transport Theory of Neutrons', *Commun. Pure Appl. Math.* **8** (1955), 217–234.
26. Montagnini, B.: 'Existence of Complex Eigenvalues for the Mono-energetic Neutron Transport Equation', *Transport Theory Stat. Phys.* **5** (1976), 127–167.
27. Montagnini, B. and Demuru, M. L.: 'Complete Continuity of the Free Gas Scattering Operator in Neutron Thermalization Theory', *J. Math. Anal. Appl.* **12** (1965), 49–57.
28. Montagnini, B. and Pierpaoli, V.: 'The Time-dependent Rectilinear Transport Equation', *Transport Theory Stat. Phys.* **1** (1971), 59–75.
29. van Norton, R.: 'On the Real Spectrum of a Mono-energetic Neutron Transport Operator', *Commun. Pure Appl. Math.* **15** (1962), 149–158.
30. Ribarič, M. and Vidav, I.: 'Analytic Properties of the Inverse $A(z)^{-1}$ of an Analytic Linear Operator Valued Function $A(z)$', *Arch. Rational Mech. Anal.* **32** (1969), 298–310.
31. Schaefer, H. H.: *Banach Lattices and Positive Operators*, Springer-Verlag, Berlin, 1974.
32. Schaefer, H. H.: 'On the Spectral Bound of Irreducible Semi-groups'. Semesterber. Funktionalanalysis, Tübingen, Sommersemester, 1983, 21–28.
33. Suhadolc, A.: 'Linearized Boltzmann Equation in L^1 Space', *J. Math. Anal. Appl.* **35** (1971), 1–13.
34. Ukai, S.: 'Real Eigenvalues of the Monoenergetic Transport Operator for a Homogeneous Medium', *J. Nuclear Sci. Technol.* **3** (1966), 263–266.
35. Ukai, S.: 'Eigenvalues of the Neutron Transport Operator for a Homogeneous Finite Moderator', *J. Math. Anal. Appl.* **18** (1967), 297–314.
36. Vidav, I.: 'Spectra of Perturbed Semigroups with Applications to Transport Theory', *J. Math. Anal. Appl.* **30** (1970), 264–279.
37. Voigt, J.: 'Spectral properties of the linear transport operator', Talk given at a workshop on 'Transport Theoretical Methods in Physics', Graz, 1977, unpublished.
38. Voigt, J.: 'A Perturbation Theorem for the Essential Spectral Radius of Strongly Continuous Semi-groups', *Mh. Math.* **90** (1980), 153–161.
39. Voigt, J.: 'Functional Analytic Treatment of the Initial Boundary Value Problem for Collisionless Gases', Habilitationsschrift, Universität München, 1981.
40. Voigt, J.: 'On the Abscissa of Convergence for the Laplace Transform of Vector Valued Measures', *Arch. Math.* **39** (1982), 455–462.
41. Voigt, J.: 'Spectral Properties of the Neutron Transport Equation', *J. Math. Anal. Appl.*, to appear.
42. Williams, M. M. R.: *Mathematical Methods in Particle Transport Theory*, Butterworths, London, 1971.

Acta Applicandae Mathematicae **2**, 333–352. 0167–8019/84/0024–0333$03.00.

Quantum Dynamical Semigroups, Symmetry Groups, and Locality

DAVID E. EVANS
Mathematics Institute, University of Warwick, Coventry CV4 7AL, England

(Received: 20 August 1983; revised: 24 October 1983)

Abstract. A survey of the recent work on the infinitesimal generators of one-parameter semigroups of positivity preserving maps on operator algebras, in the presence of compact symmetry groups or flows.

AMS (MOS) Subject classifications (1980). 46L10, 46L55, 46L60, 43A35, 47C10, 47D05, 82A15.

Key words. Infinitesimal generator, derivation, dissipation positive semigroups, compact groups, locality.

1. Introduction

Here we survey recent work on the infinitesimal generators of quantum dynamical semigroups on operator algebras which commute with certain compact groups of symmetries or flows. A quantum dynamical semigroup is a strongly continuous semigroup of completely positive maps on a C^*-algebra; this terminology arising from their use to describe irreversible Markovian dynamics in open quantum systems. For the physical background, I refer the reader to [36, 39, 45] which also survey the theory of dilations, spatial structure of generators, the quasi-free theory and certain ergodic aspects. We will not enter into a discussion of these topics here.

Let us adopt the view [19, 20] that reversible dynamics in a quantum system is described by a flow in the Heisenberg picture by a strongly continuous group $\{\tau_t : t \in \mathbb{R}\}$ of $*$-automorphisms of a C^*-algebra $\mathscr{A}$. Then the infinitesimal generator

$$\delta = \frac{\mathrm{d}}{\mathrm{d}t}\bigg|_{t=0} \tau_t$$

is a derivation, i.e.,

$$\text{the domain } D \text{ of } \delta \text{ is a } *\text{-subalgebra of } \mathscr{A} \tag{1.1}$$

$$\delta(fg) = \delta(f)g + f\delta(g), \qquad \delta(f^*) = \delta(f)^*, \quad f, g \in D. \tag{1.2}$$

One can then ask which derivations give rise to a one-parameter group of automorphisms. Moreover, if one also has a compact group acting on $\mathscr{A}$, e.g., the d-dimensional torus $\mathbb{T}^d$ with derivations $\{\partial/\partial t_i : i = 1, \ldots, d\}$ which are the generators arising from the canonical one-parameter subgroups of $\mathbb{T}^d$, one can ask whether any derivation δ, defined say on the C^∞ vectors for $\mathbb{T}^d$, can be expressed as

$$\delta = \Sigma\lambda_i \frac{\partial}{\partial t_i} + \text{perturbation} \tag{1.3}$$

where λ_i are suitable operator coefficients and the perturbation small in some sense, e.g., approximately inner. Such questions are familiar for derivations and have been studied extensively ([10, 12, 14–18, 30, 46, 47, 51, 52, 56, 58, 66–70]). For example, if $\mathbb{T}^d$ acts ergodically on a simple C^*-algebra, then Bratteli, Elliott and Jørgensen [12] showed that any derivation δ on the C^∞ vectors has such a decomposition, with $\{\lambda_i\}_{i=1}^d$ scalars, and $\Sigma\lambda_i(\partial/\partial t_i)$ is the part of δ which commutes with the $\mathbb{T}^d$-action and generates a one-parameter subgroup of that action. Here we go further and look at some problems for irreversible dynamics described by a strongly continuous semigroup $\{\alpha_t : t \geqslant 0\}$ of completely positive maps on a C^*-algebra $\mathscr{A}$. One first needs a workable algebraic alternative to (1.1) and (1.2). This turns out to be restrictive because the domain of the infinitesimal generator of a dynamical semigroup need not, in general, even have a core which is a subalgebra (see the example in [21]).

In Section 2 we give a brief account of characterisations of norm continuous positivity preserving maps on C^*-algebras. Then we use this as a guide for the minimal algebraic type conditions which a candidate for a dynamical semigroup would need. Then in Sections 3 and 4 we give some precise sufficient conditions for obtaining generators in the presence of a compact group or flow. One could also hope to obtain an analogue of (1.3) for semigroups. It was shown by Bratteli and Evans [13] that if a generator L of a dynamical semigroup commutes with a continuous action of $\mathbb{T}^d$, with simple fixed point algebra and L is trivial on the fixed point algebra, then it is a combination of three terms:

(1.4) a linear combination of the canonical derivations $\dfrac{\partial}{\partial t_i}$

(1.5) an elliptic operator in $\dfrac{\partial}{\partial t_i}$

(1.6) a superposition of $\{\tau_t - 1 : t \in \mathbb{T}^d\}$.

For a C^*-algebra $\mathscr{A}$, which possesses a flow described by a strongly continuous group $\{\tau_t : t \in \mathbb{R}\}$ of $*$-automorphisms, a concept of locality was introduced by Bratelli, Elliott and Evans in [11]. This gave a characterisation of certain second-order differential operators of the form

$$M\delta + N\delta^2 \tag{1.7}$$

where δ is the generator of τ, and M, N certain unbounded multipliers. They then gave sufficient conditions under which these operators generated dynamical semigroups. Thus for a periodic action τ, the locality condition eliminated the term (1.6), and so the locality is much stronger than $L|_{\mathscr{A}^\tau} = 0$. (For non-periodic flows τ, the fixed point algebra $\mathscr{A}^\tau$ usually does not contain enough information.) A related locality property had previously been employed by Bratteli, Digernes and Robinson [10], (following Batty [5]). In particular they used locality to characterise those derivations ∂ which could be expressed as $\partial = M\delta$, where M is again an unbounded multiplier. This locality condition is much closer related to $\partial|_{\mathscr{A}^\tau} = 0$. In Section 4 we review these locality results in some detail.

2. Bounded Generators

If $\mathscr{H}$ is a Hilbert space, $\mathscr{B}(\mathscr{H})$ will denote the algebra of all bounded linear operators on $\mathscr{H}$. Recall that a C^*-algebra $\mathscr{A}$ is a norm closed $*$-subalgebra of $\mathscr{B}(\mathscr{H})$ or, equivalently, it can be characterised abstractly as a Banach $*$-algebra $\mathscr{A}$, where the norm is related to the involution by $\| x^*x \| = \| x \|^2$, for all x in $\mathscr{A}$. The positive cone $\mathscr{A}_+$ of $\mathscr{A}$ is defined by

$$\begin{aligned}\mathscr{A}_+ &= \{a \in \mathscr{A} : \langle a\xi, \xi \rangle \geqslant 0, \quad \forall \xi \in \mathscr{H}\}\\ &= \{b^*b : b \in \mathscr{A}\}\end{aligned}$$

If $S \subset A$, we let $S^* = \{s^* : s \in S\}$, and $S_+ = S \cap \mathscr{A}_+$. We say that $S \subset \mathscr{A}$ is symmetric if $S = S^*$. A linear map T defined on a domain $D = D(T) \subset \mathscr{B}$, between C^*-algebras $\mathscr{B}$ and $\mathscr{A}$ is said to be symmetric if D is symmetric and $T(x^*) = T(x)^*$, for all $x \in D$, and positive if $T(D_+) \subset \mathscr{A}_+$. The positive linear functionals of $\mathscr{A}$ are denoted by $\mathscr{A}^*_+$. For non-commutative C^*-algebras, the concept of complete positivity has been found to be more useful than positivity [22] although for commutative C^*-algebras the concepts coincide [65, 74, 75]. If $\mathscr{A}$ is a C^*-algebra, then $M_n(\mathscr{A})$, the $*$-algebra of $n \times n$ matrices over $\mathscr{A}$ is a C^*-algebra, since it can be represented on $\mathscr{H} \oplus \cdots \oplus \mathscr{H}$ (n copies), if $\mathscr{A}$ is represented on $\mathscr{H}$, by

$$[a_{ij}]^n_{i,j=1} \cdot [\xi_j]^n_{j=1} = \left[\sum_{j=1}^{n} a_{ij}\xi_j \right]_{i=1}^{n}, \qquad [a_{ij}] \in M_n(A), [\xi_j] \in \mathscr{H} \oplus \cdots \oplus \mathscr{H}.$$

We also identify $M_n(\mathscr{A})$ with $\mathscr{A} \otimes M_n$ if $M_n = M_n(\mathbb{C})$. An element $[b_{ij}]$ of $M_n(\mathscr{A})$ is positive if and only if it is a finite sum of matrices $[a_i^* a_j]$ or, equivalently, if and only if $\Sigma b_i^* b_{ij} b_j \geqslant 0$ for all sequences $b_i, \ldots b_n$ in $\mathscr{A}$ [45], Lemma 4.1. Then a linear map T between C^*-algebras $\mathscr{B}$ and $\mathscr{A}$ is said to be n-positive if $T_n \equiv T \otimes 1_n : \mathscr{B} \otimes M_n \to \mathscr{A} \otimes M_n$ is positive, where 1_n denotes the identity operator on M_n, i.e., $[T(b_i^* b_j)] \in M_n(\mathscr{A})_+$ for all sequences $b_1, \ldots, b_n$ in $\mathscr{B}$. Note that if T is n-positive then

$$\| T \| = \| T_n \|. \tag{2.1}$$

[Observe that T is n-positive if and only if T^{**} is n-positive on the bidual, and $\| T \| = \| T^{**}(1) \|$ if T is positive. Thus $\| T_n \| = \| T_n^{**}(1) \| = \| T^{**}(1) \| = \| T \|$.] We record, for comparison with the infinitesimal versions later that T is n-positive if and only if one of the following equivalent conditions hold:

$$(2.2)\quad \sum_{i,j=1}^{n} \langle T(b_i^* b_i)\xi_j, \xi_i \rangle \geqslant 0, \quad \forall b_1, \ldots, b_n \in \mathscr{B}, \xi_1, \ldots, \xi_n \in \mathscr{H}.$$

$$(2.3)\quad \sum_{i,j=1}^{n} a_j^* T(b_i^* b_j) a_j \geqslant 0, \quad \forall b_1, \ldots, b_n \in \mathscr{B}, a_1, \ldots, a_n \in \mathscr{B}.$$

We say that T is completely positive if it is n-positive for all n.

It was shown by Evans [33] that if a linear map T between C^* algebras $\mathscr{B}$ and $\mathscr{A} \subseteq \mathscr{B}(\mathscr{H})$ is $(n+1)$-positive then the following holds:

$$(2.4)\quad [T(b_i^* b_j)]_{i,j=1}^{n} \geqslant [T(b_i^* y) T(y^* y)^{-1} T(y^* b_j)]$$

in $M_n(\mathscr{B})$, for all sequences $b_1, \ldots, b_n \in \mathscr{B}$, with a suitable interpretation of the right-hand side (see [25] and [53] for the case $n = 1$, and also [59] and [45, Theorem 1.4]). In particular:

$$(2.5)\quad \|T\| [T(b_i^* b_j)] \geqslant [T(b_i^*) T(b_j)] \quad \text{for all } b_1, \ldots, b_n \in \mathscr{B},$$

$$2.6)\quad \|T\| \sum_{i,j=1}^{n} \langle T(b_i^* b_j)\xi_i, \xi_j \rangle \geqslant \left\| \sum_{j=1}^{n} T(b_j)\xi_j \right\|^2$$

for all $b_1, \ldots, b_n \in \mathscr{B}, \xi_1, \ldots, \xi_n \in \mathscr{H}$,

$$(2.7)\quad \|T\| \sum_{i,j=1}^{n} a_i^* T(b_i^* b_j) a_j \geqslant \left(\sum_i T(b_i) a_i \right)^* \left(\sum_j T(a_j) b_j \right)$$

for all $b_1, \ldots, b_n \in \mathscr{B}, a_1, \ldots, a_n \in \mathscr{A}$.

A bounded linear map T between C^*-algebras $\mathscr{B}$ and $\mathscr{A}$ is said to be $(n+\frac{1}{2})$-positive if one of the equivalent conditions (2.5)–(2.7) hold. In particular we say that T is sesqui-positive if it is $1\frac{1}{2}$ positive. Note that it follows from (2.1) that T is $(n+\frac{1}{2})$ positive if and only if $T_n = T \otimes 1_n$ is sesqui-positive. Thus [33] tells us that

$$(n+1)\text{-positive} \Rightarrow (n+\tfrac{1}{2}) \text{ positive} \Rightarrow n\text{-positive}. \tag{2.8}$$

This trail of implications in (2.8) is the subject of a careful analysis by Takasaki and Tomiyama [78, 79]. We are particularly interested here in the infinitesimal generators of strongly continuous semigroups on C^*-algebras, with one of these positivity properties, and with particular emphasis on the generators of dynamical semigroups. First however, let us recall the situation for norm continuous semigroups, beginning with the following lemma.

If X is a set, and $\mathscr{A}$ a C^*-algebra, a map $K: X \times X \to \mathscr{A}$ is called a kernel, and a kernel is said to be positive definite if

$$[K(x_i, x_j)]_{i,j=1}^{n} \in M_n(A)_+, \quad \text{for all } x_1 \ldots x_n \in X, [45].$$

LEMMA 2.1. *Let $\mathscr{B}$ be a C^*-subalgebra of a C^*-algebra $\mathscr{A} \subseteq \mathscr{B}(\mathscr{H})$, and $L: \mathscr{B} \to \mathscr{A}$ a symmetric bounded linear map.*

Then the following conditions are equivalent:

(2.9) *For all b in* $\mathscr{B}$*, the kernels*

$$(s,t) \to L(s^*b^*bt) + s^*L(b^*b)t - L(s^*b^*b)t - s^*L(b^*bt)$$

are positive definite on $\mathscr{B} \times \mathscr{B}$.

(2.10) *The following holds for all n in* $\mathbb{N}$:

$$\sum \langle L(b_i^*b_j)\xi_j,\ \xi_i\rangle \geq 0$$

for all $b_1, \ldots, b_n \in \mathscr{B}$, $\xi_1, \ldots, \xi_n \in \mathscr{H}$ *which satisfy* $\Sigma_{i=1}^n b_i\xi_i = 0$.

(2.11) *The following holds for all n in* $\mathbb{N}$

$$\sum a_i^* L(b_i^*b_j)a_j \geq 0$$

for all $b_1, \ldots, b_n \in \mathscr{B}$, $a_1, \ldots, a_n \in \mathscr{A}$ *which satisfy* $\Sigma_{i=1}^n b_i a_i = 0$.

(2.12) *The following holds for all n in* $\mathbb{N}$

$$f[L(b_i^*b_j)] \geq 0$$

for all $f \in M_n(\mathscr{A})_+^*$, $b_1, \ldots, b_n \in \mathscr{B}$ *which satisfy* $f([b_i^*b_j]) = 0$.

If L satisfies one of the equivalent conditions of Lemma 2.1, we say it is conditionally completely positive [35]. The equivalence between (2.9) and (2.11) is in [45], Lemma 14.5. That (2.12) implies (2.10) is trivial. Suppose (2.10) holds, and $b_1, \ldots, b_n \in \mathscr{B}$, $a_1, \ldots, a_n \in \mathscr{A}$ satisfy $\Sigma_{i=1}^n b_i a_i = 0$. Then $\Sigma_{i=1}^n b_i(a_i\eta) = 0$ for all $\eta \in \mathscr{H}$. Hence, by (2.10)

$$\langle \sum a_i^* L(b_i^*b_j)a_j\eta, \eta\rangle = \sum_{i,j} \langle L(b_i^*b_j)(a_j\eta), (a_i\eta)\rangle \geq 0,$$

and so (2.11) holds. Suppose (2.9) holds, and $b_1, \ldots, b_n \in \mathscr{B}$, $f \in M_n(\mathscr{A})_+^*$ with $f[(b_i^*b_j)] = 0$. Let

$$b_0 = \begin{pmatrix} b_1 & \ldots & b_n \\ & 0 & \end{pmatrix} \in M_n$$

so that $f(b_0^*b_0) = 0$ and so $f(xb_0) = 0 = f(b_0^*x)$ for all $x \in M_n(\mathscr{A})$ by the Schwarz inequality. In particular

$$f([x_ib_j]) = 0 = f([b_i^*x_j]) \quad \text{for all } x_1, \ldots, x_n \in \mathscr{A}.$$

But by (2.9), for any $b \in \mathscr{B}$:

$$[L(b_i^*b^*bb_j) + b_i^*L(b^*b)b_j] \geq [L(b_i^*b^*b)b_j] + [b_i^*L(b^*bb_j)].$$

Then applying the state $f([L(b_i^*b^*bb_j)]) \geq 0$ and, hence, $f([L(b_i^*b_j)]) \geq 0$ by taking an approximate identity for $\mathscr{B}$, so that (2.12) holds.

Conditions (2.9), (2.10) and (2.11) can be regarded as the infinitesimal analogues of (2.4), (2.2) and (2.3) respectively.

LEMMA 2.2. *Let* $\mathscr{B}$ *be a* C^**-subalgebra of a* C^**-algebra* $\mathscr{A}$ *and* $L = \mathscr{B} \to \mathscr{A}$ *a symmetric bounded linear map. Then the following conditions are equivalent*

(2.13) *L is conditionally completely positive and* $1L^{**}(1)1 \leqslant 0$, *where* 1 *denotes the identity of* $\mathscr{B}^{**}$.

(2.14) *The kernel*

$$(s, t) \to L(s^*t) - L(s)^*t - s^*L(t)$$

is positive definite on $\mathscr{B} \times \mathscr{B}$.

Proof. Suppose L is conditionally completely positive. Then taking b in (2.9) to be an approximate identity for $\mathscr{B}$, converging to 1 in $\mathscr{B}^{**}$ we see that (2.14) holds. Conversely suppose (2.14) holds. Then clearly (see [45], p. 71) L is conditionally completely positive. Moreover for all self adjoint b in $\mathscr{B}$, $L(b^2) \geqslant L(b)b + bL(b)$, and so

$$L^{**}(1) \geqslant L^{**}(1)1 + 1L^{**}(1) \text{ and finally } 0 \geqslant 1L^{**}(1)1.$$

THEOREM 2.3. *Let L be a bounded symmetric linear map on a* C^**-algebra* $\mathscr{A} \subseteq \mathscr{B}(\mathscr{H})$. *Then the following conditions are equivalent*:

(2.15) e^{tL} *is positive for all positive t.*

(2.16) $(\lambda - L)^{-1}$ *is positive for all positive* λ.

(2.17) *For any commutative* C^**-subalgebra* $\mathscr{B}$ *of* $\mathscr{A}$, *the map* $L|_{\mathscr{B}}: \mathscr{B} \to \mathscr{A}$ *is conditionally completely positive.*

(2.18) $L(x^2y^2) + xL(y^2)x \geqslant L(xy^2)x + xL(xy^2)$ *for all commuting pairs of self adjoint elements x, y in* $\mathscr{A}$.

(2.19) $f(L(b^*b)) \geqslant 0$ *for all* $f \in \mathscr{A}^*_+$, $b \in \mathscr{A}$ *which satisfy* $f(b^*b) = 0$.

(2.20) $\langle L(b^*b)\xi, \xi\rangle \geqslant 0$ *for all* $b \in \mathscr{A}, \xi \in H$ *which satisfy* $b\xi = 0$.

(2.21) $a^*L(b^*b)a \geqslant 0$ *for all* $a, b \in \mathscr{A}$ *which satisfy* $ba = 0$.

If the algebra is unital, then these conditions are also equivalent to

(2.22) $L(x^2) + xL(1)x \geqslant L(x)x + xL(x)$ *for all self-adjoint x in* $\mathscr{A}$.

(2.23) $L(1) + u^*L(1)u \geqslant L(u^*)u + u^*L(u)$ *for all unitaries u in* $\mathscr{A}$.

The crunch here comes in showing that (2.21) implies (2.16) and is due to Evans and Hanche Olsen [41]. That (2.17)⇒(2.18)⇒(2.19)⇒(2.20)⇒(2.21) is contained in Lemma 2.1 and its proof (or see [45]). Suppose (2.15) holds and $\mathscr{B}$ is a commutative subalgebra, then $e^{tL}|_{\mathscr{B}}$ is completely positive for all positive t. Hence, if $b_1, \ldots, b_n \in \mathscr{B}$, $a_1, \ldots, a_n \in \mathscr{A}$ with $\Sigma b_i a_i = 0$ we have

$$\sum_{i,j} a_i^* L(b_i^* b_j) a_j = \lim_{t \downarrow 0} \sum a_i^* [e^{tL}(b_i^* b_j) - b_i^* b_j] a_j / t$$
$$= \lim_{t \downarrow 0} \sum a_i^* e^{tL}(b_i^* b_j) a_j / t \geqslant 0.$$

Thus (2.17) holds. In the unital case, that (2.15) implies (2.23) is due to Tsui [81] (see also [41]), who also observed that (2.23) implies (2.15) is implicit in the work of Lindblad [60].

COROLLARY 2.4. *Let L be a bounded symmetric linear map on a C*-algebra $\mathscr{A}$. Then the following conditions are equivalent*

(2.24) e^{tL} *is a positive contraction for all positive t.*

(2.25) $\lambda(\lambda - L)^{-1}$ *is a positive contraction for all large positive* λ.

(2.26) $L^{**}(1_{A^{**}}) \leqslant 0$, *and for any commutative C*-subalgebra* $\mathscr{B}$ *of* $\mathscr{A}$, $L|_{\mathscr{B}}$ *is conditionally completely positive.*

(2.27) *For any commutative C*-subalgebra* $\mathscr{B}$ *of* $\mathscr{A}$, *the kernel* $(s, t) \to L(s^*t) - L(s)^*t - s^*L(t)$ *is positive definite on* $\mathscr{B} \times \mathscr{B}$.

This is essentially contained in Evans [35] and Evans and Hanche Olsen [41]. To show that (2.24) and (2.26) are equivalent, it is enough by Theorem 2.3 to show that if e^{tL} is a positive semigroup, then e^{tL} is contraction semigroup if and only if $L^{**}(1) \leqslant 0$. If e^{tL} are contractions, then $e^{tL^{**}}(1) \leqslant \|e^{tL^{**}}\| \leqslant 1$, and so $L^{**}(1) \leqslant 0$ by differentiating. Conversely if $L^{**}(1) \leqslant 0$, then $(\lambda - L^{**}) \geqslant \lambda 1$, and so $\lambda(\lambda - L^{**})^{-1}(1) \leqslant 1$ and, hence, $\lambda(\lambda - L)^{-1}$ is a contraction for $\lambda \geqslant 0$. If (2.24) holds then (2.27) follows by differentiating the associated Schwarz inequality (c.f. (2.5)). If (2.27) holds, then one can easily see that $L|_{\mathscr{B}}$ is conditionally completely positive for any commutative C*-subalgebra $\mathscr{B}$ of $\mathscr{A}$. Thus, by Theorem 2.3, e^{tL} is a positive semigroup. Adjoin an identity 1 to $\mathscr{A}$, letting $\tilde{\mathscr{A}} = \mathscr{A} + \mathbb{C}1$, and define $\tilde{L}(x + \lambda 1) = L(x)$, $x \in \mathscr{A}$. Then it is easy to see that condition (2.22) of Theorem 2.3 holds for $\tilde{L}$ on $\tilde{\mathscr{A}}$. Hence, e^{tL} is a positive semigroup. Since $\tilde{L}(1) = 0$, $e^{t\tilde{L}}$ is norm one, and so $e^{tL} = e^{t\tilde{L}}|_{\mathscr{A}}$ is a positive contraction for t positive.

From this one can proceed and obtain the following result of Evans and Hanche Olsen [41], which was first proved by Lindblad [60] for identity preserving semigroups on unital algebras by a different method (see also [34]).

THEOREM 2.5 [41]. *Let L be a symmetric bounded linear map on a C*-algebra* $\mathscr{A}$. *Then the following conditions are equivalent:*

(2.28) $e^{tL}(x^*x) \geqslant e^{tL}(x)^* e^{tL}(x)$ *for all* $x \in \mathscr{A}$, $t \geqslant 0$.

(2.29) $L(x^*x) \geqslant L(x)^*x + x^*L(x)$ *for all x in* $\mathscr{A}$.

The following characterisation of the generators of dynamical semigroups is due to Evans and Lewis [45] based on Evans [35].

THEOREM 2.6 *Let L be a bounded symmetric linear map on a C*-algebra* $\mathscr{A}$. *Then the following conditions are equivalent.*

(2.30) e^{tL} *is completely positive for all positive t.*

(2.31) *L is conditionally completely positive on* $\mathscr{A}$.

In which case e^{tL} *is a semigroup of contractions if and only if* $L^{**}(1_{\mathscr{A}^{**}}) \leqslant 0$.

Furthermore, we have the following spatial description of the generators of norm continuous dynamical semigroups due to Christenssen and Evans [26]. It was first shown for finite-dimensional matrix algebras by Gorini, Kossakowski and Sudarshan [50] and for normal norm continuous dynamical semigroups on a hyperfinite Von Neumann algebra by Lindblad [60]. It should be compared with the situation with reversible processes, where every bounded derivation is weakly inner [54, 72].

THEOREM 2.7 [26]. *Let L be a bounded symmetric map on a* C^**-algebra* $\mathscr{A}$, *faithfully represented on a Hilbert space H. Then the following conditions are equivalent:*

(2.32) $\{e^{tL} : t \geqslant 0\}$ *is a dynamical semigroup.*

(2.33) *There exists a norm continuous semigroup* $\{G_t : t \geqslant 0\}$ *in the weak closure* $\mathscr{A}''$ *such that* $x \to e^{tL}(x) - G_t x G_t^*$ *is completely positive from* $\mathscr{A}$ *into* $\mathscr{A}''$, *for all* $t \geqslant 0$.

(2.34) *There exists a K in* $\mathscr{A}''$ *such that* $\psi \equiv L - K(\cdot) - (\cdot)K^*$ *is completely positive from* $\mathscr{A}$ *into* $\mathscr{A}''$.

If $\{G_t : t \geqslant 0\}$ is as in (2.33), then the K in (2.34) can be taken to be the infinitesimal generator of $\{G_t : t \geqslant 0\}$, simply by differentiation of $t \to e^{tL} - e^{Kt}(\cdot)\, e^{K^*t}$. The formulation of this spatial description as in (2.33) is particularly useful when L is unbounded so as to avoid domain problems. We now wish to consider unbounded operators which satisfy some of the infinitesimal conditions of the previous results, in the hope that they and some additional verifiable criteria are enough to generate dynamical semigroups. Such questions have been looked at in detail for certain algebras, e.g., the case of the algebra $\mathscr{B}(\mathscr{H})$ by Davies [28, 29] and for the C^*-algebra of the canonical anti-commutation relations by Evans [36–38] and are reviewed in [39]. But here we want to take advantage of a symmetry group or flow, to give algebraic or order theoretic conditions on an unbounded L which ensure that it, or its closure generates a dynamical semigroup, and not have to check directly any dissipativity condition or norm estimate on the operators $\{(\lambda - L) : \lambda \geqslant 0\}$. However, domain problems arise in using (2.9) or (2.14), because the generator of a dynamical semigroup can fail to have a core which is a subalgebra, even for commutative C^*-algebras [21]. These conditions are, however, at least formally satisfied by the following non-commutative elliptic type operator:

$$L = \delta + \sum_{\alpha} \delta_{\alpha}^2 \tag{2.35}$$

where δ, δ_α are symmetric derivations. For then

$$L(s^*a^*at) + s^*L(a^*a)t - L(s^*a^*a)t - s^*L(a^*at)$$
$$= \Sigma\, 2\, \delta_\alpha(s)^* a^* a \delta_\alpha(t)$$

which is positive definite.

That squares of derivations should give rise to generators of dynamical semigroups can also be seen as follows. Let δ be the infinitesimal generator of a strongly continuous one-parameter group of *-automorphisms of a C^*-algebra $\mathscr{A}$. Then

$$e^{t\delta^2} = \int_{-\infty}^{\infty} e^{-s^2/4t}(4\pi t)^{-1/2}\, e^{s\delta}\, ds \tag{2.36}$$

and, hence, is completely positive, being an average of the completely positive automorphisms $\{e^{s\delta} : s \in \mathbb{R}\}$. Alternatively one sees from

$$(\lambda - \delta)^{-1} = \int_{0}^{\infty} e^{t\delta}\, e^{-\lambda t}\, dt \tag{2.37}$$

that $(\lambda \pm \delta)^{-1}$ is completely positive for all real $\lambda \geqslant 0$. Hence, $(\lambda - \delta^2)^{-1} = (\sqrt{\lambda} - \delta)^{-1}(\sqrt{\lambda} + \delta)^{-1}$ is also completely positive for $\lambda \geqslant 0$. Hence, again

$$e^{t\delta^2} = \lim_{n \to \infty} (1 - t\delta^2/n)^{-n}$$

is completely positive for $t \geqslant 0$.

For norm-continuous semigroups, the generators of the type (2.35) do not exhaust all possibilities. Before proceeding to study unbounded generators, let us pause and examine the connection between the spatial decomposition of Theorem 2.7 for a Von Neumann algebra and the elliptic operators of the form (2.35). Suppose $\mathscr{M}$ is a Von Neumann algebra on a Hilbert space $\mathscr{H}$ and L a generator of a norm-continuous normal dynamical semigroup of unital maps on $\mathscr{M}$. Then by Theorem 2.7

$$L = \psi + K(\cdot) + (\cdot)K^* \tag{2.38}$$

where $K \in \mathscr{M}$, and $\psi : \mathscr{M} \to \mathscr{M}$ is completely positive and normal. By the Stinespring decomposition [74, 45], $\psi = V^*\pi(\cdot)V$, where π is a normal representation of $\mathscr{M}$ on a Hilbert space $\mathscr{K}$ and $V : \mathscr{H} \to \mathscr{K}$ is a bounded operator. But then $\pi = W^*\theta(\cdot)W$ where θ is a normal representation of $\mathscr{B}(\mathscr{H})$ on a Hilbert space $\mathscr{L}$ and $W : \mathscr{R} \to \mathscr{L}$ is an isometry. Since every normal representation of $\mathscr{B}(\mathscr{H})$ is a multiple of the identity representation, it follows that $\psi = V^*W^*\theta(\cdot)WV$ can be decomposed as

$$\psi = \sum_{\alpha} V_\alpha^*(\cdot)V_\alpha \tag{2.39}$$

where $V_\alpha \in \mathscr{B}(\mathscr{H})$ (cf [57] and [45], Theorem 4.6). Hence

$$L(x) = \sum V_\alpha^* x V_\alpha + Kx + xK^*, \quad x \in \mathscr{M} \tag{2.40}$$

where $K \in \mathscr{M}$, and $\Sigma V_\alpha^* V_\alpha$ converges weakly. Since the semigroup is unital, or equivalently $L(1) = 0$, we see that

$$K = ih - \sum_{\alpha} V_\alpha^* V_\alpha / 2 \tag{2.41}$$

where $h \in \mathscr{M}$ is self adjoint.

Then taking (2.35) and formally writing

$$\delta = i[h, \cdot], \qquad \delta_\alpha = i[h_\alpha, \cdot] \tag{2.42}$$

where h, h_α are self-adjoint, one has

$$L(x) = \sum 2h_\alpha x h_\alpha - [\sum h_\alpha^2, x]_+ + i[h, x]_- \tag{2.43}$$

where we write $[a, b]_\pm = ab \pm ba$. Thus to write (2.40) as (2.43) it would be sufficient to have the V_α's self-adjoint. More generally, it would be enough to have

$$\sum V_\alpha^* x V_\alpha = \sum V_\alpha x V_\alpha^* \quad \text{for all } x \in \mathcal{M} \tag{2.44}$$

for then

$$\begin{aligned}\psi(x) &= \sum (V_\alpha^* x V_\alpha + V_\alpha x V_\alpha^*)/2 \\ &= \sum [(V_\alpha^* + V_\alpha)x(V_\alpha^* + V_\alpha) + (iV_\alpha^* - iV_\alpha)x(iV_\alpha^* - iV_\alpha)]/4 \\ &= \sum d_\gamma x d_\gamma\end{aligned}$$

where the d_γ are self-adjoint.

Condition (2.44) is a detailed balance condition. It says, formally at least, that

$$\operatorname{tr} \psi(x)y = \operatorname{tr} x\psi(y) \tag{2.45}$$

for a trace. Note that an inner derivation δ is skew-symmetric with respect to a trace $\operatorname{tr} \delta(x)y = -\operatorname{tr} x\,\delta(y)$ so that its square is symmetric $\operatorname{tr} \delta^2(x)y = +\operatorname{tr} x\,\delta^2(y)$. Thus (2.43) amounts to a decomposition into symmetric and skew-symmetric parts. For more precise information on the concept of detailed balance, we refer the reader to [1, 77].

3. Dynamical Semigroups and Gauge Actions

As promised, we finally begin to consider the structure of generators of dynamical semigroups when there is a symmetry group G acting on the C^*-algebra $\mathcal{A}$ in question. Let us first describe the set-up. If G is a compact abelian group, an action τ of G on a C^*-algebra $\mathcal{A}$ will be a homomorphism τ from G into $\operatorname{Aut}(\mathcal{A})$, the group of all $*$-automorphisms of $\mathcal{A}$, which is strongly continuous in the sense that $g \to \tau(g)(a)$ is norm continuous for all a in $\mathcal{A}$. If $\gamma \in \Gamma = \hat{G}$, the dual group of G, P_γ denotes the norm-one projection $\int_G dg\,\overline{\langle \gamma, g \rangle}\tau(g)$ on $\mathcal{A}$ so that $P_\gamma P_{\gamma'} = 0$ if $\gamma \neq \gamma'$. The ranges are denoted by $\mathcal{A}^\tau(\gamma) = P_\gamma \mathcal{A}$, the spectral subspace corresponding to γ, and we often write P for P_0 and $\mathcal{A}^\tau = \mathcal{A}^\tau(0)$, the fixed point algebra. Then

$$\mathcal{A}^\tau(\gamma) = \{a \in \mathcal{A} : \tau(g)(a) = \langle \gamma, g \rangle a, g \in G\}$$

and $\mathcal{A}^\tau(\gamma_1)\mathcal{A}^\tau(\gamma_2) \subseteq \mathcal{A}^\tau(\gamma_1 + \gamma_2)$, $\mathcal{A}^\tau(\gamma)^* = \mathcal{A}^\tau(-\gamma)$. An element $x \in \mathcal{A}$ lies in the linear span $\{\mathcal{A}^\tau(\gamma) : \gamma \in \Gamma\}$ if and only if the linear span $\{\tau(g)(x) : g \in G\}$ is finite-dimensional. Such elements are called G-finite or trigonometric polynomials, and the G-finite elements $\mathcal{A}_F^\tau = \mathcal{A}_F$ form a dense symmetric subalgebra of $\mathcal{A}$.

We now look in detail at some motivating examples:

3.1. SEMIGROUPS ON M_2.

We analysis ergodic actions of $G = \mathbb{Z}_2 \times \mathbb{Z}_2$ on a simple C^*-algebra $\mathscr{A}$. Up to isomorphism, there is only one such action: $\mathscr{A} = M_2$, and if

$$\sigma_0 = \begin{pmatrix} 1 & 0 \\ 0 & 1 \end{pmatrix}, \quad \sigma_1 = \begin{pmatrix} 1 & 0 \\ 0 & -1 \end{pmatrix}, \quad \sigma_2 = \begin{pmatrix} 0 & 1 \\ 1 & 0 \end{pmatrix}, \quad \sigma_3 = \begin{pmatrix} 0 & i \\ -i & 0 \end{pmatrix}$$

are the Pauli matrices, and g_1, g_2 are the canonical generators of G, define

$$\tau(g_i) = \mathrm{Ad}\,\sigma_i, \quad i = 1, 2.$$

Then $\hat{G} = G$ and generated by the two characters γ_1, γ_2 given by $\langle \gamma_i, g_j \rangle = (-1)^{i-j}$. Put $\gamma_0 = 0$, $\gamma_3 = \gamma_1 + \gamma_2$. The spectral subspaces $\mathscr{A}^\tau(\gamma_i)$ are one-dimensional and spanned by the unitaries σ_i for each $i = 0, 1, 2, 3$. If $\{e^{tL} : t \geqslant 0\}$ is a semigroup commuting with τ, then the generator L must map the spectral subspaces into themselves, i.e. there exist scalars λ_i such that

$$L\sigma_i = \lambda_i \sigma_i, \quad i = 0, 1, 2, 3. \tag{3.1}$$

Without loss of generality we can assume $L(1) = 0$, i.e. $\lambda_0 = 0$. Then:

(3.2) $\{e^{tL} : t \geqslant 0\}$ is a positive semigroup if and only if $\lambda_i \geqslant 0$, $i = 1, 2, 3$. [13],

(3.3) $\{e^{tL} : t \geqslant 0\}$ is a semigroup of sesqui-positive maps if and only if $\lambda_i \geqslant 0$ and $4\lambda_i\lambda_j - t_k^2 \geqslant 0$ for $i \neq j \neq k \neq i$,

where $t_i = -\lambda_i + \Sigma_{j \neq} \lambda_j$, $i = 1, 2, 3$ [18]).

(3.4) $\{e^{tL} : t \geqslant 0\}$ is a semigroup of 2-positive maps if and only if $t_i \geqslant 0$, $i = 1, 2, 3$

where t_i are as above ([13], see also [50]).

We see from the above, that even for semigroups commuting with ergodic actions, the concepts of positivity, sesqui-positivity and 2-positivity are distinct. Note that on M_2, the concept of 2-positivity and complete positivity are equivalent [22].

3.2. DERIVATIONS AND SYMMETRY GROUPS

Let $(\mathscr{A}, G, \tau)$ be a C^*-dynamical system where G is compact abelian. We review some of the work on the structure of derivations with domain the G-finite or C^∞ vectors and generation problems for derivations that commute with the symmetry group.

Suppose G is a compact abelian group acting ergodically on a C^*-algebra $\mathscr{A}$. Then Bratteli, Elliott and Jørgensen [12] have shown that if δ is a derivation defined on the G-finite elements, then δ has an unique decomposition

$$\delta = \delta_0 + \tilde{\delta} \tag{3.5}$$

where δ_0 generates a one-parameter subgroup of G, and $\tilde{\delta}$ is approximately inner. Note that it is the assumed that δ commutes with the action of G. In fact $\delta_0 = \int \tau(g)\delta\tau(-g)\mathrm{d}g$, the G-invariant part of δ, and $\tilde{\delta}$ consists of the remaining Fourier components of δ. Thus if $G = \mathbb{T}^d$ $(d \leqslant 2)$, the d-dimensional torus

$$\delta_0 = \sum_{i=1}^{d} \lambda_i \frac{\partial}{\partial t_i} \tag{3.6}$$

where $\lambda_i \in \mathbb{R}$, and $\{\partial/\partial t_i\}_{i=1}^{d}$ are the generators of the canonical one-parameter subgroups of $\mathbb{T}^d$.

Suppose δ is a closed densely defined derivation on $\mathscr{A}$ commuting with the group G. If $\delta|_{\mathscr{A}^\tau} = 0$, it was shown by Bratteli and Jørgensen [16] that δ is then automatically the generator of a strongly continuous group of *-automorphisms. One cannot, in general, weaken the hypothesis by merely assuming that $\delta|_{\mathscr{A}^\tau}$ is a generator on $\mathscr{A}^\tau$ and expect to get the same result that δ is a generator. See [16] for a counter-example with commutative C^*-algebras. However, if one has lots of projections in $\mathscr{A}^\tau$ (to be precise, if the ideals $(\mathscr{A}^\tau(\gamma)\mathscr{A}^\tau(\gamma)^*)^-$ in $\mathscr{A}^\tau$ have approximate identities consisting entirely of projections, which would be the case were $\mathscr{A}^\tau$ an AF algebra) then one can get away with assuming that $\delta|_{\mathscr{A}^\tau}$ is a generator. This was shown by Bratteli and Jørgensen [17], following on earlier work by Kishimoto and Robinson [56] in this direction.

Before looking at the analogous questions for semigroups, let us finally note that Batty *et al.* [6] have looked at the problem of deciding when a given derivation on the fixed point algebra actually extends to a derivation on the given algebra which commutes with the group action, see also [69].

THEOREM 3.1 [6]. *Let τ be a continuous action of a compact abelian group on a unital C^*-algebra $\mathscr{A}$. Let $\mathscr{D}$ be a symmetric subalgebra of $\mathscr{A}$ which contains a unitary $u(\gamma)$ in each spectral subspace $\mathscr{A}^\tau(\gamma)$, $\gamma \in \Gamma$, and such that δ_0 is a densely defined derivation on $\mathscr{D} \cap \mathscr{A}^\tau$. Suppose that there exists a family of traces on $\mathscr{A}$ which separate its center. Then δ_0 extends to a derivation on $\mathscr{D}$ if and only if both the following conditions hold:*

(3.7) *$u(\gamma)\delta_0(u(\gamma)^*(\cdot)u(\gamma))u(\gamma)^* - \delta_0(\cdot)$ is a bounded inner derivation on $\mathscr{A}$ for all γ in Γ.*

(3.8) *$\phi[\delta_0(u(\gamma_1)^*u(\gamma_2)^*u(\gamma_1)u(\gamma_2))u(\gamma_2)^*u(\gamma_1)^*u(\gamma_2)u(\gamma_1)] = 0$ for any trace ϕ on $\mathscr{A}$, $\gamma_1, \gamma_2 \in \Gamma$.*

In [13] Bratteli and Evans began a programme for analysing conditionally completely positive maps defined on the G-finite vectors by first looking at those which commute with the group action. To describe their results, we need some more notation. In particular we need to make sense of

$$\int d\mu_t(g)\tau_t(g)(x), \quad x \in \mathscr{A} \tag{3.9}$$

for a probability measure μ_t taking values in a C^*-subalgebra of the centre of the multiplier algebra of $\mathscr{A}$. (The multiplier algebra $M(A) = \{T \in \mathscr{A}'' : T\mathscr{A} \subset \mathscr{A}, \mathscr{A}T \subset \mathscr{A}\}$ is a certain unital C^*-algebra containing $\mathscr{A}$, and corresponds to taking the Stone–Cech compactification for commutative algebras.) Thus, let *$\mathscr{C}$ be a C^*-subalgebra* of the center of $M(\mathscr{A})$ containing the unit of $M(\mathscr{A})$. Then a $\mathscr{C}$-valued probability measure μ on G is a positive unital linear map from $C(G)$ into $\mathscr{C}$. By [40], Prop. 4.7, and nuclearity of $\mathscr{C}$, there exists an unique *-homomorphism ε of $\mathscr{C}\otimes\mathscr{A}$ onto $\mathscr{A}$ such that $\varepsilon(c\otimes a)=ca$, $c\in\mathscr{C}$, $a\in\mathscr{A}$. Then define $\hat{\mu}:c(G)\otimes\mathscr{A}\to\mathscr{A}$ by $\hat{\mu}=\varepsilon(\mu\otimes 1)$, and write

$$\int d\mu(g)X(g)=\hat{\mu}(X) \quad \text{for } X\in C(G,\mathscr{A}). \tag{3.10}$$

A version of Bochner's theorem [13], Lemma 3.1, says that a map Z from Γ into $\mathscr{C}$ is positive definite if and only if there exists an unique $\mathscr{C}$-valued probability measure μ on G such that

$$Z(\gamma)=\mu(\langle\gamma,\cdot\rangle)=\int d\mu(g)\langle\gamma,g\rangle. \tag{3.11}$$

We refer to [13] for details. In particular if μ_1,μ_2 are $\mathscr{C}$-valued probability measures on G, we can define their convolution $\mu_1*\mu_2$ to be unique positive linear functional such that

$$(\mu_1*\mu_2)(\langle\gamma,\cdot\rangle)=\mu_1(\langle\gamma,\cdot\rangle)\mu_2(\langle\gamma,\cdot\rangle),\quad \gamma\in L^1.$$

We can now state the main result of [13]:

THEOREM 3.2 [13]. *Let τ be an action of a compact abelian group G on a C^*-algebra A such that either*

(a) *$Z(M(\mathscr{A}^\tau))\subset Z(M)(\mathscr{A}))$ and each ideal $A^\tau(\gamma)\mathscr{A}^\tau(\gamma)^*$, $\gamma\in\Gamma$, is either zero or dense in $\mathscr{A}^\tau$: or*

(b) *there exists a faithful representation π_0 of $\mathscr{A}^\tau$ such that $\pi(\mathscr{A}^\tau)''$ is a factor, if π is the Stinespring representation of $\pi_0\circ P$;*

holds. Suppose $L:\mathscr{A}_F\to\mathscr{A}$ is a symmetric linear map such that $L|_{\mathscr{A}^\tau}=0$, $L\tau(g)=\tau(g)L$, $g\in G$ and $(s,t)\to L(s^*t)-L(s^*)t-s^*L(t)$ is positive definite on $\mathscr{A}_F\times\mathscr{A}_F$.

Then L is closable and its closure $\bar{L}$ generates a strongly continuous dynamical semigroup. Moreover, there exists a convolution semigroup μ of probability measures, which are $Z(M(\mathscr{A}^\tau))$-valued in case (a), *and scalar-valued in case* (b) *such that*

$$e^{t\bar{L}}(a)=\int d\mu_t(g)\tau(g)(a),\quad a\in\mathscr{A},\ t\geqslant 0. \tag{3.12}$$

The crunch here is that L is closable and its closure is a generator. Note that hypotheses (a) holds if $\mathscr{A}^\tau$ is simple. In this case if δ is a derivation on $\mathscr{A}_F$ commuting with G, then (3.12) says that $\{e^{t\delta}:t\in\mathbb{R}\}$ is a one parameter subgroup of G as one would expect from looking at δ_0 in (3.5). The result expressed by (3.12) is closely related to Pontryagin duality. In fact by Araki *et al.* [3], Appendix C (although they actually work with a W^*-dynamical system such that the fixed-point algebra is a factor) if τ is an action of a compact abelian group G on a C^*-algebra $\mathscr{A}$ such that the fixed-point algebra is simple, one has the following result which is usually called Roberts' formulation of Pontryagin duality: If α is an automorphism of $\mathscr{A}$ satisfying

$$\alpha\tau(g) = \tau(g)\alpha \tag{3.13}$$

$$\alpha(x) = x, \quad x \in \mathscr{A}^\tau \tag{3.14}$$

then there exists a $g \in G$ such that $\alpha = \tau(g)$. In [13] it was shown that if S is a completely positive map on $\mathscr{A}$ satisfying (3.13) and (3.14) then

$$S = \int \mathrm{d}\mu(g)\tau(g) \tag{3.15}$$

where μ is a probability measure on G, showing that the extremal completely positive maps satisfying (3.13) and (3.14) are the automorphism $\tau(G)$.

By using the Levy Khinchine formula in (3.12) one can get further information on the operator L. For example, suppose $G = \mathbb{T}^d$, the d-dimensional torus, and τ an action of $\mathbb{T}^d$ on a C^*-algebra $\mathscr{A}$ such that either $\mathscr{A}^\tau$ is simple or (b) holds. Let $\{\partial/\partial t_i\}_{i=1}^d$ be the generators of the actions of the canonical one-parameter subgroups of $\mathbb{T}^d$ on $\mathscr{A}$. Then if L is as in Theorem 3.2, there is [13] a triple (b, a, μ) where $b = (b_1, \ldots, b_d)$ is a d-tuple of real numbers, $a = [a_{ij}]$ is a real positive $d \times d$ matrix and μ is a non-negative bounded measure on

$$\mathbb{T}^d \backslash \{0\} = \{x = (x_i) : -\pi < x_i \leqslant \pi, x \neq 0\}$$

such that

$$L(X) = \left\{ \sum_{k=1}^{d} b_k \frac{\partial}{\partial t_k} + \sum_{ij} a_{ij} \frac{\partial^2}{\partial t_i \partial t_j} - \right.$$

$$\left. - \int_{T^d\backslash\{0\}} \frac{\mathrm{d}\mu(x)}{\|x\|^2} \left[1 + \sum_k x_k \frac{\partial}{\partial t_k} - \exp\left(\sum_k x_k \frac{\partial}{\partial t_k} \right) \right] \right\} X \tag{3.16}$$

where $\|x\|^2 = \Sigma_{i=1}^d x_1^2$.
The triple (b, a, μ) is arbitrary within the constraints given there and is uniquely determined by L.

Bratteli *et al.* [18] have weakened some of the hypotheses of Theorem 3.2 yet still show that one gets a generator:

THEOREM 3.3 [18]. *Let τ be an action of a compact abelian group G on a C^*-algebra $\mathscr{A}$, and D a τ-invariant dense *-subalgebra of $\mathscr{A}$ such that $\mathscr{A}^\tau \subset \mathscr{D}$, and D is the linear span of $D \cap \mathscr{A}^\tau(\gamma)$, $\gamma \in \Gamma$. Suppose $L: D \to \mathscr{A}$ is a symmetric linear map such that $L|_{\mathscr{A}^\tau} = 0$, $L\tau(g) = \tau(g)L$, $g \in G$ and $(s, t) \to L(s^*t) - L(s^*)t - s^*L(t) - s$ is positive definite on $D \times D$.*

Then L is closable and its closure $\bar{L}$ generates a strongly continuous dynamical semigroup.

A classification is also given in [18] of all strongly continuous dynamical semigroups which commute with G and restrict to the identity on $\mathscr{A}^\tau$, in terms of maps from Γ to unbounded operators affiliated with the centres of the multiplier algebras of the ideals $\mathscr{A}^\tau(\gamma)\mathscr{A}^\tau(\gamma)^*$ in $\mathscr{A}^\tau, \gamma\in\Gamma$. To describe this, suppose for brevity that $\mathscr{A}^\tau(\gamma)\mathscr{A}^\tau(\gamma)^*$ is always dense in $\mathscr{A}^\tau$. Then there exists an action α of Γ on $Z = Z(M(\mathscr{A}^\tau))$ such that $\alpha_\gamma(a)x = xa$ for all $x\in\mathscr{A}^\tau(\gamma)$, $a\in Z$. Then given a dynamical semigroup $\{T_t : t\geqslant 0\}$ as above there exists a map $\gamma\in\Gamma\to e^{-t\mathscr{L}(\gamma)}$ such that

(3.17) $e^{-t\mathscr{L}(\gamma)}\in Z$.

(3.18) $t\to e^{-t\mathscr{L}(\gamma)}$ is a contraction semigroup continuous in the strict topology on $M(\mathscr{A}^\tau)$.

(3.19) $(\gamma_1,\gamma_2)\to \alpha_{\gamma_1}(\exp - t\mathscr{L}(\gamma_2-\gamma_1)) - \exp(-t\mathscr{L}(\gamma_1))^*\exp(-t\mathscr{L}(\gamma_2))$ is positive definite on $\Gamma\times\Gamma$.

(3.20) $e^{-t\mathscr{L}(0)} = 1$.

In which case $T_t(x) = e^{-t\mathscr{L}(\gamma)}x$, $x\in\mathscr{A}^\tau(\gamma)$, $\gamma\in\Gamma$, $t\geqslant 0$. Conversely a map $\mathscr{L}$ on Γ satisfying (3.17)–(3.20) gives rise to a dynamical semigroup with the required properties. If α is trivial, or equivalently, if $Z(M(\mathscr{A}^\tau))\subset Z(M(\mathscr{A}))$ then (3.19) reduces to $(\gamma_1,\gamma_2)\to e^{-t\mathscr{L}(\gamma_2-\gamma_1)}$ being positive definite. Thus Bochner's theorem as in [13] gives (3.12) for a Z-valued convolution semigroup μ_t.

4. Dynamical Semigroups, Flows and Locality

Suppose $\{\tau_t : t\in\mathbb{R}\}$ is a strongly continuous group of *-automorphisms on a C^*-algebra $\mathscr{A}$ with infinitesimal generator δ. Using a concept of locality, Bratteli, Elliott and Evans [11] have characterised those linear operators on a suitable domain, which can be expressed

$$L = M\delta + N\delta^2 \tag{4.1}$$

with certain unbounded operator coefficients. This work was motivated by a related concept of relative locality for derivations, introduced by Batty [5] in the commutative case, and developed by Bratteli, Digernes and Robinson [10], which we first describe.

A map ∂ is said to be strongly δ-local if

$$\omega(\delta(a)^*\,\delta(a)) = 0$$

implies

$$\omega(\partial(a)^*\,\partial(a)) = 0$$

for each a in $D(\delta)\cap D(\delta)$, and each pure state ω on $\mathscr{A}$. Moreover, ∂ is said to be n-strongly δ-local if $\partial\otimes 1_n$ is strongly $\delta\otimes 1_n$ local on $\mathscr{A}\otimes M_n$, and completely strongly δ-local if it is n-strongly δ-local for all n. For abelian $\mathscr{A}$, the concepts of strong locality and complete strong locality coincide [10].

To describe the main result of [10], we first develop some notation which will be used throughout this section. Let $\mathscr{B}$ denote the closed ideal $(\mathscr{A}\,\delta(D(\delta))\mathscr{A})^-$ in $\mathscr{A}$ and Prim($\mathscr{B}$), the primitive spectrum of $\mathscr{B}$ viewed as an open subset of Prim($\mathscr{A}$), the primitive spectrum of $\mathscr{A}$. If M is a multiplier of the minimal dense ideal Ped($\mathscr{B}$), and Δ is a map on $\mathscr{B}$ with domain $D(\Delta)$, we let

$$D(M\Delta) = \{b\in\mathscr{B} : M(\Delta(b))\in\mathscr{B}\}$$

and

$$M\Delta(b) = M(\Delta(b)), \quad b \in D(M\Delta).$$

THEOREM 4.1 [10]. *Let $\mathscr{A}$ be a C^*-algebra and τ a strongly continuous one-parameter group of *-automorphisms of $\mathscr{A}$ with infinitesimal generator δ, and ∂ a closed *-derivation of $\mathscr{A}$. Suppose that ∂ comutes with τ, is completely strongly δ-local and*

$$(\text{center } (M(\mathscr{B})^{\tau})(D(\partial) \cap \mathscr{B}) \subset D(\partial).$$

*Then ∂ generates a strongly continuous one-parameter group of *-automorphisms of $\mathscr{A}$, and there is a continuous real valued function μ on* Prim$(\mathscr{B})$ *such that*

$$(4.2) \quad \pi e^{t\partial} = \begin{cases} \pi\tau_{\mu(\pi)t}, & \pi \in \mathrm{Prim}(\mathscr{B}) \\ \pi, & \pi \in \mathrm{Prim}(\mathscr{A}) \backslash \mathrm{Prim}(\mathscr{B}). \end{cases}$$

Moreover there exists a τ-invariant self-adjoint central multiplier M of Ped$(\mathscr{B})$ such *that*

$$(4.3) \quad \pi(M) = \mu(\pi), \quad \pi \in \mathrm{Prim}(\mathscr{B}).$$

(4.4) *$M\delta$ is closable and densely defined.*

(4.5) *$D(M\delta) = D(\partial) \cap D(\delta)$ is a core for ∂ and $\partial(a) = M\delta(a)$, $a \in D(\partial) \cap D(\delta)$.*

$$(4.6) \quad D(\overline{M\delta}) = D(\partial), \qquad \partial(a) = \overline{M\delta}(a), \qquad a \in D(\partial).$$

The proof of this relies heavily on ∂ being a derivation.

Following this, Bratteli, Elliott and Evans [11] characterised some second-order differential operators on a C^*-algebra using another concept of locality.

A linear map L with domain $D(L) \subseteq \mathscr{A}$ is local with respect to (δ, δ^2) if

$$\omega(\delta(a)^*b^*b\,\delta(a)) = 0$$

and

$$\omega(\delta^2(a)^*b^*b\,\delta^2(a)) = 0$$

imply

$$\omega(L(a)^*b^*bL(a)) = 0$$

for each $a \in D(L) \cap D(\delta^2)$, $b \in \mathscr{A}$ and pure state ω on $\mathscr{A}$. Note that this is somewhat related to the concept of 2-strong δ-locality. Suppose that ∂ is a derivation which is 2-strongly δ-local, and that $\omega(\delta(a)^*b^*b\,\delta(a)) = 0$ for some $a, b \in \mathscr{D}(\delta) \cap \mathscr{D}(\delta)$, and a (pure) state on $\mathscr{A}$. Then if π is a representation of $\mathscr{A}$ on a Hilbert space $\mathscr{H}$ with cyclic vector Ω such that $\omega = \langle \pi(\cdot)\Omega, \Omega \rangle$, we have

$$\pi(b\,\delta(a))\Omega = 0$$

i.e.

$$\pi(\delta(ba))\Omega - \pi(\delta(b))\pi(a)\Omega = 0.$$

Hence, by 2-strong locality,

$$\pi(\partial(ba))\Omega - \pi(\partial(b))\pi(a)\Omega = 0$$

i.e.

$$\pi(b\,\partial(a))\Omega = 0 \quad \text{or} \quad w(\partial(a)^* b^* b\,\partial(a)) = 0.$$

So that 2-strong locality in the sense of [10] almost implies locality in the sense of [11].

Then the main result of [11] is

THEOREM 4.2 [11]. *Let τ be a strongly continuous one-parameter group of *-automorphism of a C^*-algebra $\mathscr{A}$. Let L be a closed, densely defined *-linear map on $\mathscr{A}$ which commutes with τ and satisfies*

(4.7) *L is local with respect to (δ, δ^2)*

(4.8) centre $(M(\mathscr{B})^\tau)(D(L) \cap \mathscr{B}) \subset D(L)$.

(4.9) *if $a = a^* \in D(L)$, $a^2 \in D(L)$, then $L(a^2) \geqslant L(a)a + aL(a)$*

then L generates a strongly continuous dynamical semigroup of contractions on $\mathscr{A}$. There exist continuous real-valued functions μ, ν on Prim$(\mathscr{B})$ *such that $\nu \geqslant 0$ and*

$$\pi_0(e^{Lt}(a)) = \int_{-\infty}^{\infty} ds (4\pi\nu(\pi_0)t)^{-1/2}\, e^{-s^2/4\nu(\pi_0)t}\pi_0(\tau_{s+\mu(\pi_0)t}(a))$$

for $a \in \mathscr{A}$, $\pi_0 \in$ Prim$(\mathscr{B})$,

$$\pi_0\, e^{Lt} = \pi_0,\ \pi_0 \in \text{Prim}(\mathscr{A})\backslash\text{Prim}(\mathscr{B}).$$

Moreover there exist τ-invariant self-adjoint central multipliers M, N of Ped$(\mathscr{B})$ such that $N \geqslant 0$, *and*:

(4.10) $\pi(M) = \mu(\pi), \qquad \pi(N) = \nu(\pi), \quad \pi \in \text{Prim}(B)$

(4.11) *$M\delta$, $N\delta^2$ are closable and densely defined.*

(4.12) $D(L) \cap D(\delta^2) = D(M\delta) \cap D(N\delta^2)$ is a core for L and
$L(a) = M\delta(a) + N\delta^2(a), \quad a \in D(L) \cap D(\delta^2)$.

(4.13) $D(L) = D(\overline{M\delta}) \cap D(\overline{N\delta^2})$ *is a *-algebra* and
$L(a) = \overline{M\delta}(a) + \overline{N\delta^2}(a), \quad a \in D(L)$.

The programme of the proof is as follows. First it is shown that the decomposition in (4.12) exists in an irreducible representation π of $\mathscr{A}$:

$$\pi L(a) = \mu\pi(a) + \nu\pi^2(a), \quad a \in D(L) \cap D(\delta^2)$$

for some scalars (μ, ν). For this one does not need that L commutes with τ, or that the domain of L is saturated as in (4.8) or even the infinitesimal Schwarz inequality (4.9). (Note however, in the proof of (4.5) in [10] it was crucial that ∂ was a derivation.) Then the (μ, ν) are pieced together by the Dauns Hofman theorem to obtain the unbounded multipliers (M, N), so that

$$L(a) = M\delta(a) + N\delta^2(a), \qquad a \in D(L) \cap D(\delta^2) \tag{4.14}$$

in the sense that $L(a)$ is the sum of $M\delta(a)$ and $N\delta^2(a)$ as multipliers of $\mathrm{Ped}(\mathscr{B})$. Next using the fact that L commutes with τ, it was shown that the decomposition actually holds in a much stronger form, namely that

$$D(L) \cap D(\delta^2) \subset D(M\delta) \cap D(N\delta^2)$$

and (4.14) holds as operators in $\mathscr{B}$. In particular, $M\delta$ and $N\delta^2$ are densely defined, and are, in fact, closable. Next if $D(L)$ is saturated in the sense of (4.8) then one shows that (4.13) holds. The infinitesimal Kadison Schwarz inequality in (4.9) is only used to show that $N \geqslant 0$ and, hence, that $\overline{N\delta^2}$ generates a dynamical semigroup.

Acknowledgements

This paper was written when the author was visiting the University of Ottawa, and supported by the Natural Sciences and Engineering Research Council of Canada, through grants to George A. Elliott and David E. Handelman. I would like to express my gratitude to G. A. Elliott for his kind invitation to Ottawa.

References

1. Alicki, R.: *Rep. Math. Phys.* **10** (1976), 249.
2. Araki, H.: *Publ. RIMS Kyoto Univ.* **8** (1972–73), 439.
3. Araki, H., Haag, R., Kastler, D., and Takesaki, M.: *Commun. Math. Phys.* **53** (1977), 97.
4. Arendt, W., Chernoff, P. R., and Kato, T. *J. Operator Theory* **8** (1982), 167.
5. Batty, C. J. K.: *Proc. London Math. Soc.* (3) **42** (1981), 299.
6. Batty, C. J. K., Carey, A. L., Evans, D. E., and Robinson, D. W.: Publ. *RIMS Kyoto Univ.* (to appear).
7. Berg, C. and Forst, G.: *Potential Theory on Locally Compact Abelian Groups*, Springer-Verlag, Berlin, Heidelberg, New York, 1975.
8. Bratteli, O.: *On Dynamical Semigroups and Compact Group Actions*. Springer Lecture Notes in Mathematics. (ed. L. Accardi), to appear.
9. Bratteli, O., Digernes, T., and Robinson, D. W.: Positive semigroups on ordered Banach spaces'.
10. Bratteli, O., Digernes, T., Robinson, D. W.: 'Relative locality of derivations'.
11. Bratteli, O., Elliott, G. A., and Evans, D. E.: 'Locality and differential operators on C^*-algebras'.
12. Bratteli, O., Elliott, G. A., and Jørgensen, P. E. T.,: *J. Reine Angew. Math.* **346** (1984), 166–193.
13. Bratteli, O. and Evans, D. E.: *Ergodic Theory and Dynamical Systems* **3** (1983), 187–217.
14. Bratteli, O., Goodman, F., and Jørgensen, P. E. T.: 'Unbounded derivations tangential to compact groups of automorphisms III'. Preprint.
15. Bratteli, O. and Jørgensen, P. E. T.: *Proc. Symp. in Pure Math. Part 2*, AMS, Providence RI, 1980, pp. 353–365.
16. Bratteli, O. and Jørgensen, P. E. T.: *J. Funct. Anal.* **48** (1982), 107.
17. Bratteli, O. and Jørgensen, P. E. T.: *Commun. Math. Phys.* **87** (1982), 353.

18. Bratteli, O., Jørgensen, P. E. T., Kishimoto, A. and Robinson, D. W.: 'A C^*-algebraic Schoenberg Theorem'. Preprint 1983. To appear in *Ann. Inst. Fourier (Grenoble)*.
19. Bratteli, O. and Robinson, D. W.: *Operator Algebras and Quantum Statistical Mechanics. I*, Springer-Verlag, New York, 1979.
20. Bratteli, O. and Robinson, D. W.: *Operator Algebras and Quantum Statistical Mechanics. II*. Springer-Verlag, New York, 1981.
21. Bratteli, O. and Robinson, D. W.: *Math. Scand.* **49** (1981), 259.
22. Choi, M. D.: *Can. J. Math.* **24** (1972), 520.
23. Choi, M. D.: *Illinois J. Math.* **18** (1974), 18.
24. Choi, M. D.: *Linear Algebra and Appl.* **10** (1974), 285.
25. Choi, M. D.: *J. Operator Theory* **4** (1980), 271.
26. Christensen, E. and Evans, D. E.: *J. London Math. Soc.* (2) **20**, (1978), 358.
27. Davies, E. B.: *Quantum Theory of Open Systems*, Academic Press, London, 1976.
28. Davies, E. B.: *Rep. Math. Phys.* **11** (1977), 169.
29. Davies, E. B.: *J. Funct. Anal.* **34** (1979), 421.
30. Davies, E. B.: 'A generation theorem for operators commuting with group actions'.
31. Demoen, B., Vanheuverzwijn, P., and Verbeure, A.: *Lett. Math. Phys.* **2** (1977), 161
32. Demoen, B., Vanheuverzwijn, P., and Verbeure, A.: *Rep. Math. Phys.*
33. Evans, D. E.: *Commun. Math. Phys.* **48** (1976), 15.
34. Evans, D. E.: *Commun. Math. Phys.* **54** (1977), 293.
35. Evans, D. E.: *Quart. J. Math. Oxford* (2), **28** (1977), 369.
36. Evans, D. E.: 'Mathematical problems in the quantum theory of irreversible processes', (eds. L. Accardi, V. Gorini, G. Parravicini), in *Proceedings of Arco Felice Conference*, 1978, pp. 136–162.
37. Evans, D. E.: *Commun. Math. Phys.* **70** (1979), 53.
38. Evans, D. E.: *J. Funct. Anal.* **37** (1980), 318.
39. Evans, D. E.: *Lecture Notes in Physics* **116** (1980).
40. Evans, D. E.: 'Operator algebras and applications', in R. V. Kadison (ed.) *Proc. Symp. Pure Math*, Vol. 38, Amer. Math. Soc., Providence, RI, 1982, part 2, pp. 377–379.
41. Evans, D. E. Hanche-Olsen, H.: *J. Funct. Anal.* **32** (1979), 207.
42. Evans, D. E. and Hoegh-Krohn, R.: *J. London Math. Soc.* **17** (1978), 345.
43. Evans, D. E. and Lewis, J. T.: *Commun. Math. Phys.* **50** (1976), 219.
44. Evans, D. E. and Lewis, J. T.: *J. Funct. Anal.* **26** (1977), 369.
45. Evans, D. E. and Lewis, J. T.: 'Dilations of irreversible evolutions in algebraic quantum theory', *Comm. Dubl. Inst. Adv. Studies, Ser A*, **24**, 1977.
46. Goodman, F. and Jørgensen, P. E. T.: *Commun. Math. Phys.* **82** (1981), 399.
47. Goodman, F. and Wasserman, A. J.: 'Unbounded derivations commuting with compact group actions II', *J. Funct. Anal.* **55** (1984), 389–397.
48. Gorini, V., Frigerio, A., Verri, M., Kossakowski, A., and Sudarshan, E. C. G., *Rep. Math. Phys.* **12** (1977), 359.
49. Gorini, V., Frigerio, A., Verri, M., and Kossakowski, A.: *Commun. Math. Phys.* **57** (1977), 97.
50. Gorini, V., Kossakowski, A., and Sudarshan, E. C. G.: *J. Math. Phys.* **17** (1976), 821.
51. Ikunishi, A.: 'Derivations in C^*-algebras commuting with compact actions', *Publ. RIMS* **19** (1983), 99–106.
52. Jørgensen, P. E. T.: *Z. Wahrscheinlickstheorie Verw Gebiete.* **63** (1983), 17.
53. Kadison, R. V.: *Ann. Math.* (2) **56** (1952), 494.
54. Kadison, R. V.: *Ann. Math.* **83** (1966), 280.
55. Kishimoto, A.: *Commun. Math. Phys.* **47** (1976), 25.
56. Kishimoto, A. and Robinson, D. W.: *Publ. RIMS Kyoto Univ.*
57. Kraus, K.: *Ann. Phys.* **64** (1971), 311.
58. Kumjian, A.: *Semestericht Funktionalysis*, Tübingen Wintersemester 1982/83. pp. 179–191.
59. Lieb, E. H. and Ruskai, M. B.: *Adv. Math.* **12** (1974), 269.
60. Lindblad, G.: *Commun. Math. Phys.* **48** (1976), 119.
61. Lindblad, G.: *Lett. Math. Phys.* **1**. (1976), 219.
62. Longo, R. and Peligrad, C.: 'Non-commutative topological dynamics and compact actions on C^*-algebras' Preprint, 1983.
63. Nagel, R. and Uhlig, H.: *J. Operator Theory*. **6** (1981), 113.

64. Naimark, M. A.: *C. R. (Doklady) Acad. Sci. URSS (N. S.)* **41** (1943) 359.
65. Naimark, M. A.: *Bull. Acad. Sci. URSS. Ser. Math.* **1** (1943), 237.
66. Peligrad, C.: *Topics in Modern Operator Theory*, OT Series Vol. 2, Birkhauser-Verlag, Basle, 1981, pp. 259–268.
67. Peligrad, C.: OT Series Vol. 6, Birkhauser-Verlag Basle, 1982, pp. 181–194.
68. Powers, R. T. and Price, G. L.: *Commun. Math. Phys.* **84** (1982) 439.
69. Price, G. L.: *Publ. RIMS Kyoto Univ.* **19** (1983), 345.
70. Price, G. L.: 'On derivations annihilating a maximal abelian subalgebra'.
71. Robinson, D. W.: *Commun. Math. Phys.* **85** (1982), 129.
72. Sakai, S.: *Ann. Math.* **83** (1966), 273.
73. Simon, B.: *Indiana Univ. Math. J.* **26** (1977), 1067.
74. Stinespring, W. F.: *Proc. Amer. Math. Soc.* **6** (1955), 211.
75. Størmer, E.: *Acta. Math.* **110** (1963), 233.
76. Størmer, E.: *Lecture Notes in Physics* **29**, Springer-Verlag, Berlin, 1974, pp. 85–106.
77. Stragier, G., Quaegebeur, J., and Verbeure, A.: 'Quantum detailed balance'. Preprint, Leuven, 1983.
78. Takasaki, T. and Tomiyama, J.: 'On the geometry of positive maps in matrix algebras'. Preprint, Niigata.
79. Takasaki, J. and Tomiyama, J.: Work in progress.
80. Takesaki, M.: *Theory of Operator Algebras I.* Springer-Verlag, New York, Heidelberg, Berlin, 1979.
81. Tsui, S. K.: *Trans. Amer. Math. Soc.* **66** (1977), 305.
82. Van Castersen, J. A.: 'Invariant subsets of strongly continuous semigroups'. Preprint, Antwerp, 1983.
83. Vanheuverzwijn, P.: *Ann. Inst. Poincaré* **29** (1978), 123; Errata. **30** (1979), 83.
84. Watanabe, S.: 'Asymptotic behaviour and eigenvalues of dynamical semigroups on operator algebras'.

Acta Applicandae Mathematicae **2**, 353–378. 0167–8019/84/0024–0353$03.90

Stochastic Dilations of Uniformly Continuous Completely Positive Semigroups★

R. L. HUDSON
Mathematics Department, University of Nottingham, University Park, Nottingham NG7 2RD, England

and

K. R. PARTHASARATHY
Indian Statistical Institute, 7 SJS Sansanwal Marg, New Delhi 110016, India

(Received: 20 December 1983)

Abstract. For an arbitrary uniformly continuous completely positive semigroup $(\mathcal{T}_t : t \geqslant 0)$ on the space $B(\mathfrak{h}_0)$ of bounded operators on a Hilbert space $\mathfrak{h}_0$, we construct a family $(U(t) : t \geqslant 0)$ of unitary operators on a Hilbert space $\mathfrak{H}_0 = \mathfrak{h}_0 \otimes \mathfrak{H}$ and a conditional expectation $\mathbb{E}_0$ from $B(\mathfrak{H}_0)$ to $B(\mathfrak{h}_0)$, such that, for arbitrary $t \geqslant 0$, $X \in B(\mathfrak{h}_0)$ $\mathcal{T}_t(X) = \mathbb{E}_0[U(t)X \otimes IU(t)^{\dagger}]$. The unitary operators $U(t)$ satisfy a stochastic differential equation involving a noncommutative generalisation of infinite dimensional Brownian motion. They do not form a semigroup.

AMS (MOS) subject classifications (1980). 46L50, 46L55, 46L60, 47D05, 81H05, 81C20, 81C35, 82A35, 15A66, 60E07, 60G20, 60H05, 34F05, 60H15, 60J65, 60K35, 81D05, 47A20.

Key words. Completely positive semigroup, operators on Hilbert space, conditional expectation, stochastic differential equation, ∞-dimensional Brownian motion, Fock space, Itô product formula, stochastic dilation.

1. Introduction

In [2] we constructed a noncommutative extension of the Itô stochastic calculus for operator-valued processes. Using the duality transformation to identify $L^2(w)$, where w is Wiener measure, with the Boson Fock space $\mathfrak{H} = \Gamma(L^2(0, \infty))$, classical Brownian motion is expressed as the sum $A(t) + A^{\dagger}(t)$ of two mutually noncommuting operator valued processes, which are, respectively, the Fock annihilation and creation operators $A(t) = a(\chi_{[0,t]})$, $A^{\dagger}(t) = a^{\dagger}(\chi_{[0,t]})$. The extended Itô product formula for the calculus based on A and $A^{\dagger}$ is expressed formally by the multiplication table

	$dA^{\dagger}$	dA	dt
$dA^{\dagger}$	0	0	0
dA	dt	0	0
dt	0	0	0

from which the product formula for classical Brownian motion follows as a special case.

★ Part of this work was completed when the first author was visiting research associate at the Center for Relativity, Physics Department, The University of Texas at Austin, Austin, TX 78712, U.S.A., supported in part by NSF PHY 81-01381.

Using this calculus, we showed in [2] that, for given bounded operators L and $\mathscr{H}$ in a Hilbert space $\mathfrak{h}_0$, of which $\mathscr{H}$ is self-adjoint, the stochastic differential equation

$$dU = U(L \otimes dA^\dagger - L^\dagger \otimes dA + (i\mathscr{H} - \tfrac{1}{2}L^\dagger L) \otimes I\, dt), \quad U(0) = I \tag{1.1}$$

has a unique solution which consists of unitary operators in $\mathfrak{H}_0 = \mathfrak{h}_0 \otimes \mathfrak{H}$. Moreover, if $\mathbb{E}_0$ is the vacuum conditional expectation from $B(\mathfrak{H}_0)$ onto $B(\mathfrak{h}_0)$ defined by

$$\langle u, \mathbb{E}_0[T]v\rangle = \langle u \otimes \Psi_0, Tv \otimes \Psi_0\rangle \quad (T \in B(\mathfrak{h}_0), u, v \in \mathfrak{h}_0)$$

where Ψ_0 is the Fock vacuum vector, then the formula

$$\mathscr{T}_t(X) = \mathbb{E}_0[U(t)X \otimes IU(t)^{-1}] \quad (X \in B(\mathfrak{h}_0), t \geqslant 0) \tag{1.2}$$

defines a uniformly continuous semigroup of completely positive maps of which the infinitesimal generator $\mathscr{L}$ is given by

$$\mathscr{L}(X) = i[\mathscr{H}, X] - \tfrac{1}{2}(L^\dagger LX - 2L^\dagger XL + XL^\dagger L) \tag{1.3}$$

Now in [4] it is shown that the general form of the infinitesimal generator of a uniformly continuous semigroup of completely positive maps in $B(\mathfrak{h}_0)$ is

$$\mathscr{L}(X) = i[\mathscr{H}, X] - \tfrac{1}{2}\sum_j (L_j^\dagger LX - 2L_j^\dagger XL_j + XL_j^\dagger L_j) \tag{1.4}$$

where $\mathscr{H} \in B(\mathfrak{h}_0)$ is self-adjoint, and the operators $L_j \in B(\mathfrak{h}_0)$ may be infinite in number, but must be such that $\Sigma_j L_j L_j$ converges strongly. Our purpose in this paper is to construct a stochastic unitary dilation of the semigroup of which (1.4) is the infinitesimal generator, by means of a noncommutative stochastic calculus generalising that of [2].

An intuitive procedure for carrying out this goal would be as follows; introduce independent quantum Brownian motions A_j, corresponding to the terms L_j in (1.4), and satisfying the product rules

	$dA_k^\dagger$	dA_k	dt
$dA_j^\dagger$	0	0	0
dA_j	$\delta_{jk}\, dt$	0	0
dt	0	0	0

and solve the equation

$$dU = U\left(\sum_j L_j \otimes dA_j^\dagger - \sum_j L_j^\dagger \otimes dA_j + \left(i\mathscr{H} - \tfrac{1}{2}\sum_j L_j^\dagger L_j\right) \otimes I\, dt\right),$$

$$U(0) = I. \tag{1.5}$$

However, the operator theoretic difficulties of this approach are formidable when there are infinitely many L_j, and an alternative strategy is called for. This is to introduce the single process $A_L(t) = \Sigma_j L_j^\dagger \otimes A_j$ together with its formal adjoint $A_L(t) = \Sigma_j L_j \otimes A_j^\dagger$, for which the Itô rules are

	$dA_L^\dagger$	dA_L	dt
$dA_L^\dagger$	0	0	0
dA_L	$\Sigma_j L_j^\dagger L_j \otimes I\,dt$	0	0
dt	0	0	0

and (1.5) becomes

$$dU = U\left(dA_L^\dagger - dA_L + \left(i\mathscr{H} - \tfrac{1}{2}\sum_j L_j^\dagger L_j\right)\otimes I\,dt\right),$$

$$U(0) = I.$$

Because adaptedness no longer forces processes to commute with stochastic differentials, the appropriate theory of stochastic integration must now distinguish between the left and the right integral. We turn this complication to advantage by developing the theory of adapted processes and stochastic integrals in such a way that formal adjunction is a symmetry converting the left into the right integral and vice-versa.

In this connection we make constant use of the following extension of the well-known result that an everywhere defined operator in a Hilbert space with a densely defined adjoint is bounded.

THEOREM 1.1. *Let $\mathfrak{h}_0$, $\mathfrak{H}$ be Hilbert spaces and let $\mathfrak{h}_0 \otimes \mathfrak{H}$ be their Hilbert space tensor product. Let $\mathscr{E}$ be a dense subspace of $\mathfrak{H}$ and let $\mathfrak{h}_0 \underline{\otimes} \mathscr{E}$ denote the algebraic tensor product. Let T and $T^\dagger$ be mutually adjoint operators with common domain $\mathfrak{h}_0 \underline{\otimes} \mathscr{E}$, so that for arbitrary $u, v \in \mathfrak{h}_0$, $\Phi, \Psi \in \mathscr{E}$,*

$$\langle u \otimes \Phi, Tv \otimes \Psi\rangle = \langle T^\dagger u \otimes \Phi, v \otimes \Psi\rangle.$$

Then for each $\Phi \in \mathscr{E}$, the operators T_Φ, $T_\Phi^\dagger$ given by

$$T_\Phi u = Tu \otimes \Phi, \qquad T_\Phi^\dagger u = T^\dagger u \otimes \Phi \quad (u \in \mathfrak{h}_0)$$

are bounded.

Proof. Fix $\Psi \in \mathfrak{h}_0 \underline{\otimes} \mathscr{E}$ with $\|\Psi\| \leqslant 1$. The linear map

$$\lambda_\Psi(u) = \langle T^\dagger \Psi, u \otimes \Phi\rangle = \langle \Psi, Tu \otimes \Phi\rangle$$

is bounded on $\mathfrak{h}_0$ since

$$|\langle T^\dagger \Psi, u \otimes \Phi\rangle| \leqslant \|T^\dagger \Psi\|\, \|\Phi\|\, \|u\|.$$

Moreover, the λ_Ψ, $\Psi \in \mathfrak{h}_0 \underline{\otimes} \mathscr{E}$, $\|\Psi\| \leqslant 1$ are pointwise bounded, since

$$|\langle \Psi, Tu \otimes \Phi\rangle| \leqslant \|Tu \otimes \Phi\| \quad \text{for } \|\Psi\| \leqslant 1.$$

Hence, by the uniform boundedness principle, there exists a positive number M such that, for all $\Psi \in \mathfrak{h}_0 \underline{\otimes} \mathscr{E}$ with $\|\Psi\| \leqslant 1$,

$$|\langle \Psi, Tu \otimes \Phi \rangle| \leqslant M \, \|u\|, \quad (u \in \mathfrak{h}_0)$$

and, hence, such that for all $\Psi \in \mathfrak{h}_0 \underline{\otimes} \mathscr{E}$,

$$|\langle \Psi, Tu \otimes \Phi \rangle| \leqslant M \, \|u\| \, \|\Phi\|, \quad (u \in \mathfrak{h}_0).$$

Since $\mathfrak{h}_0 \underline{\otimes} \mathscr{E}$ is dense in $\mathfrak{h}_0 \otimes \mathfrak{H}$, it follows that

$$\|Tu \otimes \Phi\| \leqslant M \, \|u\|, \quad (u \in \mathfrak{h}_0)$$

that is T_Φ is bounded. The argument for $T_\Phi^\dagger$ is similar. □

2. Notation and Preliminaries

Let a separable Hilbert space $\mathfrak{h}_0$ and a finite or countably infinite index set J be given, once and for all. We denote by $\mathfrak{h}$ the direct sum $\mathfrak{h} = \oplus_{j \in J} L^2[0, \infty)$. The *Boson Fock space* over $\mathfrak{h}$ may be conveniently characterised as a pair $(\mathfrak{H}, \Psi)$ comprising a Hilbert space $\mathfrak{H}$ and a map $\Psi: \mathfrak{h} \to \mathfrak{H}$ such that $\{\Psi(f) : f \in \mathfrak{h}\}$ is total in $\mathfrak{H}$ and, for all $f, g \in \mathfrak{h}$,

$$\langle \Psi(f), \Psi(g) \rangle = \exp \langle f, g \rangle.$$

$\Psi(f)$ is called the *exponential vector* or coherent state corresponding to $f \in \mathfrak{h}$. The *vacuum vector* is $\Psi_0 = \Psi(0)$. We denote by $\mathscr{E}$ the dense subspace of $\mathfrak{H}$ spanned algebraically by the exponential vectors.

The operator-valued processes which concern us live in the tensor product $\mathfrak{H}_0 = \mathfrak{h}_0 \otimes \mathfrak{H}$ of $\mathfrak{H}$ with the 'initial space' [2] $\mathfrak{h}_0$. If T and $T^\dagger$ are mutually adjoint operators in $\mathfrak{H}_0$ with domains containing $\mathfrak{h}_0 \underline{\otimes} \mathscr{E}$, then for $f \in \mathfrak{h}$ we denote by $\|T\|_f$ and $\|T^\dagger\|_f$ the bounds of the operators on $\mathfrak{h}_0$

$$u \mapsto Tu \otimes \Psi(f), \qquad u \mapsto T^\dagger u \otimes \Psi(f),$$

which are bounded by Theorem 1.1.

We denote by

$$\mathfrak{h} = \mathfrak{h}_t \oplus \mathfrak{h}^t \tag{2.1}$$

the natural decomposition

$$\mathfrak{h} = \bigoplus_{j \in J} L^2[0, \infty) = \left(\bigoplus_{j \in J} L^2[0, t] \right) \oplus \left(\bigoplus_{j \in J} L^2(t, \infty) \right)$$

and for $f \in \mathfrak{h}$ we write $f = (f_t, f^t)$ for its components in these subspaces. Corresponding to the direct sum decomposition (2.1), there is a tensor product decomposition $\mathfrak{H} = \mathfrak{H}_t \otimes \mathfrak{H}^t$ of $\mathfrak{H}$ into the Fock spaces $\mathfrak{H}_t$ and $\mathfrak{H}^t$ over $\mathfrak{h}_t$ and $\mathfrak{h}^t$ respectively, in which for each $f \in \mathfrak{h}$

$$\Psi(f) = \Psi(f_t) \otimes \Psi(f^t).$$

In this decomposition clearly $\mathscr{E} = \mathscr{E}_t \underline{\otimes} \mathscr{E}^t$, where $\mathscr{E}_t$ and $\mathscr{E}^t$ are the spans of the exponential vectors in $\mathfrak{H}_t$ and $\mathfrak{H}^t$, respectively.

Now let $B(\mathfrak{h}_0; J)$ denote the set of J-tuples of bounded operators in $\mathfrak{h}_0$, $L = (L_j : j \in J)$ for which $\Sigma_j L_j^\dagger L_j$ converges strongly in $B(\mathfrak{h}_0)$. Then $B(\mathfrak{h}_0; J)$ is a complex vector space under component-wise operations. Furthermore, for $L, M \in B(\mathfrak{h}_0; J)$ the sum $\Sigma_j L_j^\dagger M_j$ converges strongly, as is seen from the polarisation identity. For $f = (f_j : j \in J) \in \mathfrak{h}$ and $0 \leqslant s \leqslant t$, since

$$\sum_j \left(\int_s^t f_j I\right)^\dagger \left(\int_s^t f_j I\right) = \sum_j |\langle f, \chi_{(s,t)}\rangle|^2 I \leqslant (t-s)\,\|f\|^2 I \tag{2.2}$$

the J-tuple $(\int_s^t f_j I) \in B(\mathfrak{h}_0; J)$. Hence, the operators $\Sigma_j \int_s^t f_j L_j^\dagger$ and $\Sigma_j \int_s^t \bar{f}_j L_j$ are well defined in $B(\mathfrak{h}_0)$.

3. Processes

DEFINITION 3.1. An *adapted process* is a family of operators $F = (F(t) : t \geqslant 0)$ in $\mathfrak{H}_0$ such that for each $t \geqslant 0$

(a) $D(F(t)) = \mathfrak{h}_0 \underline{\otimes}\, \mathscr{E}_t \underline{\otimes}\, \mathfrak{H}^t$.

(b) There is an operator $F^\dagger(t)$ with the same domain adjoint to $F(t)$.

(c) There are operators $F_1(t)$ and $F_1^\dagger(t)$ on $\mathfrak{h}_0 \underline{\otimes}\, \mathscr{E}_t$ such that $F(t) = F_1(t) \otimes I$, $F^\dagger(t) = F^\dagger(t) = F_1^\dagger(t) \otimes I$.

The *adjoint process* of F is $F^\dagger = (F^\dagger(t) : t > 0)$. A *simple process* is an adapted process of the form

$$F(t) = \sum_{n=0}^{\infty} F_n \chi_{(t_n, t_{n+1})}(t) \quad (t \geqslant 0)$$

for some sequence $0 = t_0 < t_1 < \cdots < t_n \to \infty$. An adapted process is *regular* if there exists a sequence $F^{(n)}$, $n = 1, 2, \ldots$ of simple processes such that, for all $f \in \mathfrak{h}$,

$$\|F(t) - F^{(n)}(t)\|_f, \qquad \|F^\dagger(t) - F^{\dagger(n)}(t)\|_f \xrightarrow[n]{} 0$$

uniformly on compact sets in $(0, \infty)$, and *continuous* if for all $u \in \mathfrak{h}_0$, $f \in \mathfrak{h}$, the maps

$$t \mapsto F(t)u \otimes \Psi(f), \qquad t \mapsto F(t)^\dagger\, u \otimes \Psi(f) \quad \text{are continuous from } [0, \infty) \text{ to } \mathfrak{H}_0.$$

Then every continuous process is regular. We denote by $\mathscr{A}$, $\mathscr{A}_0$, $\mathscr{A}_r$ and $\mathscr{A}_c$, respectively, the sets of adapted, simple, regular and continuous processes.

Now fix $L \in B(\mathfrak{h}_0, J)$, once and for all.

We define operators $A_L(t)$, $t \geqslant 0$, initially with domain $\mathfrak{h}_0 \underline{\otimes}\, \mathscr{E}$, by

$$A_L(t) u \otimes \Psi(f) = \left(\sum_j \int_0^t f_j L_j^\dagger u\right) \otimes \Psi(f).$$

Formally,

$$A_L(t) = \sum_j L_j^\dagger \otimes A_j(t),$$

where we make the identification

$$\mathfrak{H} = \Gamma\left(\left\{\bigoplus_{k=1}^{j} L^2[0,\infty)\right\} \oplus \left\{\bigoplus_{k>j} L^2[0,\infty)\right\}\right)$$

$$= \left\{\bigotimes_{k=1}^{j} \Gamma(L^2[0,\infty))\right\} \otimes \Gamma\left(\bigoplus_{k>j} L^2[0,\infty)\right)$$

and set $A_j(t) = \otimes^{j-1}_{k=1} I \otimes A(t) \otimes I$.

We wish to establish the existence of an operator $A_L^\dagger(t)$ with the same domain adjoint to $A_L(t)$; formally $A_L^\dagger(t) = \Sigma_j L_j \otimes A_j^\dagger(t)$. We introduce the notation

$$\partial_\Delta^j \Psi(f_1, \ldots, f_j, \ldots) = \frac{d}{d\sigma} \Psi(f_1, \ldots, f_j + \sigma\chi_\Delta, \ldots)|_{\sigma=0}$$

where χ_Δ is the indicator function of the finite interval $\Delta \subseteq [0, \infty)$. Then when J is finite, $A_L^\dagger(t)$ is given by the action

$$A_L^\dagger(t) u \otimes \Psi(f) = \sum_j L_j u \otimes \partial^j_{[0,t]} \Psi(f). \tag{3.1}$$

That this sum converges when J is infinite is a corollary of Theorem 3.2. Before stating it we note that, if F is an operator whose domain includes $\mathfrak{h}_0 \underline{\otimes} \mathscr{E}$ such that for each $f \in \mathfrak{h}$, the operator $u \mapsto Fu \otimes \Psi(f)$ is bounded on $\mathfrak{h}_0$, that is $\|F\|_f < \infty$, then for $f, g \in \mathfrak{h}$, $u \in \mathfrak{h}_0$ the sum $\Sigma_j \langle FL_j u \otimes \Psi(f), FL_j u \otimes \Psi(g)\rangle$ converges absolutely. Indeed,

$$\begin{aligned}
&\sum_j |\langle FL_j u \otimes \Psi(f), FL_j u \otimes \Psi(g)\rangle| \\
&\leqslant \left(\sum_j \|FL_j u \otimes \Psi(f)\|^2\right)^{1/2} \left(\sum_j \|FL_j u \otimes \Psi(g)\|^2\right)^{1/2} \\
&\leqslant \|F\|_f \|F\|_g \sum_j \|L_j u\|^2 \\
&= \|F_f\| \; \|F\|_g \left\langle u, \sum_j L_j^\dagger L_j u\right\rangle < \infty.
\end{aligned}$$

THEOREM 3.2. *Let $0 \leqslant s \leqslant t$. Let F and $F^\dagger$ be mutually adjoint operators with domain $\mathfrak{h}_0 \underline{\otimes} \mathscr{E}_s \underline{\otimes} \mathfrak{H}^s$ of form $F_1 \otimes I$ and $F_1^\dagger \otimes I$ where F_1 and $F_1^\dagger$ are operators on $\mathfrak{h}_0 \underline{\otimes} \mathscr{E}_s$. Then in the case when J is infinite, the sum*

$$\sum_j FL_j u \otimes \partial^j_{(s,t]} \Psi(f)$$

converges. Moreover

(a) *For arbitrary $u \in \mathfrak{h}_0$, $f, g \in \mathfrak{h}$.*

$$\left\langle \sum_j FL_j u \otimes \partial^j_{(s,t]} \Psi(f), \sum_j FL_j u \otimes \partial^j_{(s,t]} \Psi(f) \right\rangle$$

$$= \left\langle F \sum_j \int_s^t \bar{f} L_j u \otimes \Psi(f), F \sum_j \int_s^t \bar{g} L_j u \otimes \Psi(g) \right\rangle +$$

$$+ (t-s) \sum_j \langle FL_j u \otimes \Psi(f), FL_j u \otimes \Psi(g) \rangle, \tag{3.2}$$

(b) *for arbitrary* $u, v \in \mathfrak{h}_0$, $f, g \in \mathfrak{h}$

$$\left\langle \sum_j FL_j u \otimes \partial^j_{(s,t]} \Psi(f), v \otimes \Psi(g) \right\rangle = \left\langle u \otimes \Psi(f), \sum_j \int_s^t g_j L_j^\dagger \otimes I F^\dagger v \otimes \Psi(g) \right\rangle.$$

Proof. Assume $J = \mathbb{N}$ and let $\phi_n = \Sigma_{j=1}^n FL_j u \otimes \partial^j_{(s,t]} \Psi(f)$. Then, for $m \geq n$,

$$\| \phi_m - \phi_n \|^2$$

$$= \sum_{j,k=n+1}^m \langle FL_j u \otimes \partial^j_{(s,t]} \Psi(f), FL_k u \otimes \partial^k_{(s,t]} \Psi(f) \rangle$$

$$= \sum_{j,k=n+1}^m \langle F_1 L_j u \otimes \Psi(f_s), F_1 L_k u \otimes \Psi(f_s) \rangle \times$$

$$\times \frac{\partial^2}{\partial\sigma\,\partial\tau} \langle \Psi(f_1^s, \ldots, f_j^s + \sigma\chi_{(s,t]}, \ldots), \Psi(g_1^s, \ldots, g_k^s + \tau\chi_{(s,t]}, \ldots) \rangle \Big|_{\substack{\sigma=0\\ \tau=0}}$$

$$= \sum_{j,k=n+1}^m \langle F_1 L_j u \otimes \Psi(f_s), F_1 L_k u \otimes \Psi(f_s) \rangle \frac{\partial^2}{\partial\sigma\,\partial\tau} \times$$

$$\times \exp \langle (f_1^s, \ldots, f_j^s + \sigma\chi_{(s,t]}, \ldots), (g_1^s, \ldots, g_k^s + \tau\chi_{(s,t]}, \ldots) \rangle \big|_{\substack{\sigma=0\\ \tau=0}}$$

$$= \sum_{j,k=n+1}^n \langle F_1 L_j u \otimes \Psi(f_s), F_1 L_k u \otimes \Psi(f_s) \rangle \times$$

$$\times \left\{ \int_s^t f_j \int_s^t \overline{f_k} + (t-s)\delta_{jk} \right\} \| \Psi(f^s) \|^2$$

$$= \left\{ \left\| F_1 \sum_{j=n+1}^m \int_s^t \overline{f_j} L_j u \otimes \Psi(f_s) \right\|^2 + \right.$$

$$\left. + (t-s) \sum_{j=n+1}^m \| F_1 L_j u \otimes \Psi(f_s) \|^2 \right\} \| \Psi(f^s) \|^2$$

$$= \left\| F \sum_{j=n+1}^m \int_s^t \overline{f_j} L_j u \otimes \Psi(f) \right\|^2 + (t-s) \sum_{j=n+1}^m \| FL_j u \otimes \Psi(f) \|^2$$

$$\leq \| F \|_f^2 \left\{ \left\| \sum_{j=n+1}^m \int_s^t \bar{f} L_j u \right\|^2 + (t-s) \sum_{j=n+1}^m \langle u, L_j^\dagger L_j u \rangle \right\} \xrightarrow[m,n]{} 0. \tag{3.3}$$

Hence, (Φ_n) converges as asserted. A similar calculation to that leading to (3.3) establishes (a). To prove (b) we have (assuming $J = N$ is infinite)

$$\left\langle \sum_j FL_j u \otimes \partial^j_{(s,t]} \Psi(f), v \otimes \Psi(g) \right\rangle$$

$$= \lim_n \sum_{j=1}^n \langle F_1 L_j u \otimes \Psi(f_s), v \otimes \Psi(g_s)\rangle \times$$

$$\times \frac{\mathrm{d}}{\mathrm{d}\sigma} \langle \Psi(f_1^s, \ldots, f_j^s + \sigma\chi_{(s,t]}, \ldots), \Psi(g^s)\rangle \big|_{\sigma=0}$$

$$= \lim_n \sum_{j=1}^n \langle u \otimes \Psi(f_s), L_j^\dagger \otimes IF_1^\dagger v \otimes \Psi(g_s)\rangle \int_s^t g_j \langle \Psi(f^s), \Psi(g^s)\rangle$$

$$= \lim_n \left\langle u \otimes \Psi(f), \sum_{j=1}^n \int_s^t g_j L_j^\dagger \otimes IF^\dagger v \otimes \Psi(g) \right\rangle$$

$$= \left\langle u \otimes \Psi(f), \sum_j \int_s^t g_j L_j^\dagger \otimes IF^\dagger v \otimes \Psi(g) \right\rangle. \qquad \square$$

Taking $s = 0$, $F = I$ in the theorem, we see that the sum (3.1) converges, and defines an operator $A_L^\dagger(t)$ adjoint to $A_L(t)$ on the domain $\mathfrak{h}_0 \underline{\otimes}\, \mathscr{E}$.

The operators $A_L(t)$ and $A_L^\dagger(t)$ are clearly of form $A_1 \otimes I$ and $A_1^\dagger \otimes I$, respectively, on $(\mathfrak{h}_0 \underline{\otimes}\, \mathscr{E}_t) \underline{\otimes}\, \mathscr{E}^t$. As such they extend naturally to mutually adjoint operators on $(\mathfrak{h}_0 \underline{\otimes}\, \mathscr{E}_t) \underline{\otimes}\, \mathfrak{H}^t$, which constitute mutually adjoint adapted processes. Furthermore, these processes are additive, in the sense that for $0 \leqslant s \leqslant t$, $A_L(t) - A_L(s)$ and $A_L^\dagger(t) - A_L^\dagger(s)$ are of form $I \otimes A_2 \otimes I$, $I \otimes A_2^\dagger \otimes I$ on $(\mathfrak{h}_0 \underline{\otimes}\, \mathscr{E}_s) \underline{\otimes}\, \mathscr{E}_t^s \underline{\otimes}\, \mathfrak{H}^t$, where $\mathscr{E}_t^s$ is the span of the exponential vectors in $\mathfrak{H}_t^s = \Gamma(\oplus_{j \in J} L^2(s, t])$. Thus $A_L(t) - A_L(s)$ and $A_L^\dagger(t) - A_L^\dagger(s)$ extend naturally to operators, for which we use the same symbols, on $(\mathfrak{h}_0 \otimes \mathfrak{H}_s) \underline{\otimes}\, \mathscr{E}_t^s \underline{\otimes}\, \mathfrak{H}^t$. Then if F and $F^\dagger$ satisfy the hypotheses of Theorem 3.2, the operators $F(A_L(t) - A_L(s))$, $(A_L(t) - A_L(s))F$, and $(A_L^\dagger(t) - A_L^\dagger(s))F$ are well defined on $\mathfrak{h}_0 \underline{\otimes}\, \mathscr{E}_t \underline{\otimes}\, \mathfrak{H}^t$. We define the operator $F(A_L^\dagger(t) - A_L^\dagger(s))$ on the same domain using Theorem 3.2 by

$$F(A_L^\dagger(t) - A_L^\dagger(s))u \otimes \Psi(f) = \sum_j FL_j u \otimes \partial^j_{(s,t]} \Psi(F).$$

Theorem 3.2 (b) shows that the operators $F(A_L^\dagger(t) - A_L^\dagger(s))$ and $(A_L(t) - A_L(s))F^\dagger$ are mutually adjoint, and straightforward calculation shows that the same is true of $F(A_L(t) - A_L(s))$ and $(A^\dagger(t) - A^\dagger(s))F^\dagger$.

We note that the identity (3.2) can be restated as

$$\langle F(A_L^\dagger(t) - A_L^\dagger(s))u \otimes \Psi(f), F(A_L^\dagger(t) - A_L^\dagger(s))u \otimes \Psi(g)\rangle$$
$$= \langle F(A_L(t) - A_L(s))u \otimes \Psi(f), F(A_L(t) - A_L(s))u \otimes \Psi(g)\rangle +$$
$$+ (t - s) \sum_j \langle FL_j u \otimes \Psi(f), FL_j u \otimes \Psi(g)\rangle. \tag{3.3}$$

4. Stochastic Integrals of Simple Processes

DEFINITION 4.1. Let $F, G, H \in \mathscr{A}_0$ and write

$$F = \sum_{n=0}^{\infty} F_n \chi_{[t_n, t_{n+1})}, \qquad G = \sum_{n=0}^{\infty} G_n \chi_{[t_n, t_{n+1})},$$

$$H = \sum_{n=0}^{\infty} H_n \chi_{[t_n, t_{n+1})} \tag{4.1}$$

where $0 = t_0 < t_1 < \cdots < t_n \xrightarrow[n]{} \infty$. The families of operators $M = (M(t) : t \geqslant 0)$, $N = (N(t) : t \geqslant 0)$ with domains $D(M(t)) = D(N(t)) = \mathfrak{h}_0 \underline{\otimes}\, \mathscr{E}_t \underline{\otimes}\, \mathfrak{H}^t$ defined by $M(0) = 0$, $N(0) = 0$,

$$M(t) = M(t_n) + F_n(A_L^\dagger(t) - A_L^\dagger(t_n)) + G_n(A_L(t) - \\ - A_L(t_n)) + (t - t_n)H_n$$

$$N(t) = N(t_n) + (A_L^\dagger(t) - A_L^\dagger(t_n))F_n + (A_L(t) - \\ - A_L(t_n))G_n + (t - t_n)H_n$$

for $t_n < t \leqslant t_{n+1}$, are called the *right* and *left stochastic integrals* of (F, G, H), and denoted by

$$M(t) = \int_0^t (F\,\mathrm{d}A_L^\dagger + G\,\mathrm{d}A_L + H\,\mathrm{d}\tau),$$

$$N(t) = \int_0^t (\mathrm{d}A_L^\dagger F + \mathrm{d}A_L G + H\,\mathrm{d}\tau).$$

Clearly M and N are adapted processes and

$$\left[\int_0^t (F\,\mathrm{d}A_L^\dagger + G\,\mathrm{d}A_L + H\,\mathrm{d}\tau)\right]^\dagger = \int_0^t (\mathrm{d}A_L^\dagger G^\dagger + \mathrm{d}A_L F^\dagger + H^\dagger\,\mathrm{d}\tau). \tag{4.2}$$

We describe by the differential relations

$$\mathrm{d}M = F\,\mathrm{d}A_L^\dagger + G\,\mathrm{d}A_L + H\,\mathrm{d}t, \qquad \mathrm{d}N = \mathrm{d}A_L^\dagger F + \mathrm{d}A_L G + H\,\mathrm{d}t$$

the situation that, for $t \geqslant 0$,

$$M(t) = M_0 \otimes I + \int_0^t (F\,\mathrm{d}A_L^\dagger + G\,\mathrm{d}A_L + H\,\mathrm{d}t),$$

$$N(t) = N_0 \otimes I + \int_0^t (\mathrm{d}A_L^\dagger F + \mathrm{d}A_L G + H\,\mathrm{d}t)$$

where $M_0, N_0 \in B(\mathfrak{h}_0)$.

THEOREM 4.2. *Let $F, G, H \in \mathscr{A}_0$ and*

$$\mathrm{d}M = F\,\mathrm{d}A_L^\dagger + G\,\mathrm{d}A_L + H\,\mathrm{d}t, \qquad \mathrm{d}N = \mathrm{d}A_L^\dagger F + \mathrm{d}A_L G + H\,\mathrm{d}t.$$

Then for arbitrary $u, v \in \mathfrak{h}_0$, $f, g \in \mathfrak{h}$ the functions on $(0, \infty)$

$$t \mapsto \langle u \otimes \Psi(f), M(t) v \otimes \Psi(g)\rangle, \qquad t \mapsto \langle u \otimes \Psi(f), N(t) v \otimes \Psi(g)\rangle$$

are absolutely continuous, with generalised derivatives

$$\frac{\mathrm{d}}{\mathrm{d}t}\langle u \otimes \Psi(f), M(t) v \otimes \Psi(g)\rangle \tag{4.3}$$

$$= \left\langle u \otimes \Psi(f), \left[F(t) \sum_j \overline{f_j(t)} L_j \otimes I + G(t) \sum_j g_j(t) L_j^\dagger \otimes I + H(t)\right] v \otimes \Psi(g)\right\rangle,$$

$$\frac{\mathrm{d}}{\mathrm{d}t}\langle u \otimes \Psi(f), N(t) v \otimes \Psi(g)\rangle \tag{4.4}$$

$$= \left\langle u \otimes \Psi(f), \left[\sum_j \overline{f_j(t)} L_j \otimes I F(t) + \sum_j g_j(t) L_j^\dagger \otimes I G(t) + H(t)\right] v \otimes \Psi(g)\right\rangle.$$

Proof. We give the proof only for the case of the right integral. Assume F, G and H are given by (4.1) and that $t \in (t_n, t_{n+1})$. Then

$$\begin{aligned}
&\langle u \otimes \Psi(f), M(t) v \otimes \Psi(g)\rangle \\
&= \langle u \otimes \Psi(f), M(t_n) v \otimes \Psi(g)\rangle + \langle u \otimes \Psi(f), F_n(A_L^\dagger(t) - A_L^\dagger(t_n)) v \otimes \Psi(g)\rangle + \\
&\quad + \langle u \otimes \Psi(f), G_n(A_L(t) - A_L(t_n)) v \otimes \Psi(g)\rangle + \\
&\quad + \langle u \otimes \Psi(f), (t - t_n) H_n v \otimes \Psi(g)\rangle .
\end{aligned} \tag{4.5}$$

The second term can be written as

$$\begin{aligned}
&\langle u \otimes \Psi(f), F_n(A_L^\dagger(t) - A_L^\dagger(t_n)) v \otimes \Psi(g)\rangle \\
&= \langle (A_L(t) - A_L(t_n)) F_n^\dagger u \otimes \Psi(f), v \otimes \Psi(g)\rangle \\
&= \left\langle \sum_j \int_{t_n}^t f_j L_j^\dagger \otimes I F_n^\dagger u \otimes \Psi(f), v \otimes \Psi(g)\right\rangle \\
&= \left\langle F_n^\dagger u \otimes \Psi(f), \sum_j \int_{t_n}^t \bar{f}_j L_j \otimes I v \otimes \Psi(g)\right\rangle \\
&= \sum_j \int_{t_n}^t \langle F_n^\dagger u \otimes \Psi(f), \overline{f(\tau)} L_j \otimes I v \otimes \Psi(g)\rangle\,\mathrm{d}\tau .
\end{aligned} \tag{4.6}$$

Now for arbitrary $\phi \in \mathfrak{H}_0$, by Schwarz's inequality,

$$\begin{aligned}
&\sum_j \int_{t_n}^t |\langle \phi, \overline{f_j(\tau)} L_j v \otimes \Psi(g)\rangle|\,\mathrm{d}\tau \\
&= \sum_j |\langle \chi_{(t_n, t]}, f_j\rangle|\,|\langle \phi, L_j v \otimes \Psi(g)\rangle|
\end{aligned}$$

$$\leqslant (t - t_n)^{1/2} \sum_j \|f_j\| \, \|L_j v\| \, \|\phi\| \, \|\Psi(g)\|$$

$$\leqslant (t - t_n)^{1/2} \left(\sum_j \|f_j\|^2\right)^{1/2} \left(\sum_j \|L_j v\|^2\right)^{1/2} \|\phi\| \, \|\Psi(g)\| < \infty.$$

Hence, by the dominated convergence theorem, we may reverse the order of summation and integration in (4.6) and write the second term in (4.5) as

$$\int_{t_n}^{t} \sum_j \langle F_n^\dagger u \otimes \Psi(f), \overline{f_j(\tau)} L_j \otimes I v \otimes \Psi(g) \rangle \, \mathrm{d}\tau$$

which is manifestly absolutely continuous as a function of t, with generalised derivative

$$\sum_j \langle F_n^\dagger u \otimes \Psi(f), \overline{f(t)} L_j \otimes I v \otimes \Psi(g) \rangle$$

$$= \langle u \otimes \Psi(f), F(t) \sum \overline{f_j(t)} L_j \otimes I v \otimes \Psi(g) \rangle$$

since $F(t) = F_n$ for $t \in (t_n, t_{n+1})$. A similar argument shows that the third term in (4.5) is absolutely continuous as function of t with generalised derivative $\langle u \otimes \Psi(f), G(t) \Sigma_j g_j(t) L_j^\dagger \otimes I v \otimes \Psi(g) \rangle$. Since the first term is constant and the fourth differentiable with derivative $\langle u \otimes \Psi(f), H(t) v \otimes \Psi(g) \rangle$, the proof is complete. □

THEOREM 4.3. *Under the hypotheses of Theorem* 4.2, *if* $0 \leqslant s \leqslant t$, $\phi \in \mathfrak{h}_0 \otimes \mathfrak{H}_s$, $f, g \in \mathfrak{h}$, $v \in \mathfrak{h}_0$,

$$\langle \phi \otimes \Psi(f^s), (M(t) - M(s)) v \otimes \Psi(g) \rangle$$

$$= \int_s^t \Big\langle \phi \otimes \Psi(f^s), \Big[F(\tau) \sum_j \overline{f_j(\tau)} L_j \otimes I + G(\tau) \sum_j g_j(\tau) L_j^\dagger \otimes I + H(\tau) \Big] v \otimes \Psi(g) \Big\rangle \, \mathrm{d}\tau,$$

$$\langle \phi \otimes \Psi(f^s), (N(t) - N(s)) v \otimes \Psi(g) \rangle$$

$$= \int_s^t \Big\langle \phi \otimes \Psi(f^s), \Big[\sum_j \overline{f_j(\tau)} L_j \otimes I F(\tau) + \sum_j g_j(\tau) L_j^\dagger \otimes I G(\tau) + H(\tau) \Big] v \otimes \Psi(g) \Big\rangle \, \mathrm{d}\tau.$$

Proof. We give the proof for the right integral. Assume first that

$$\phi = u \otimes \Psi(f_s^{(1)}) \tag{4.7}$$

for $u \in \mathfrak{h}_0$, $f^{(1)} \in \mathfrak{h}$. We obtain the theorem in this case by replacing f in (4.3) by $f^{(1)} \chi_{[0,s]} + f \chi_{(s,\infty)}$ and integrating from s to t. Since vectors of the form (4.7) are total we obtain the general case by passing to limits of finite linear combinations. □

THEOREM 4.4. *Let* $F, G, H \in \mathscr{A}_0$ *and*

$$M(t) = \int_0^t (F \, \mathrm{d}A_L^\dagger + G \, \mathrm{d}A_L + H \, \mathrm{d}\tau),$$

$$N(t) = \int_0^t (\mathrm{d}A_L^\dagger F + \mathrm{d}A_L G + H\,\mathrm{d}\tau).$$

Then for arbitrary $u \in \mathfrak{h}_0$ and $f, g \in \mathfrak{h}$ the functions

$$t \mapsto \langle M(t)u \otimes \Psi(f), M(t)u \otimes \Psi(g)\rangle, \qquad t \mapsto \langle N(t)u \otimes \Psi(f), N(t)u \otimes \Psi(g)\rangle$$

are absolutely continuous, with generalised derivatives

$$\begin{aligned}
&\frac{\mathrm{d}}{\mathrm{d}t}\langle M(t)u \otimes \Psi(f), M(t)u \otimes \Psi(g)\rangle \\
&= \Big\langle M(t)u \otimes \Psi(f), \Big[F(t)\sum_j \overline{f_j(t)}L_j \otimes I + \\
&\quad + G(t)\sum_j g_j(t)L_j^\dagger \otimes I + H(t)\Big]u \otimes \Psi(g)\Big\rangle + \\
&\quad + \Big\langle\Big[F(t)\sum_j \overline{g_j(t)}L_j \otimes I + G(t)\sum_j f_j(t)L_j^\dagger \otimes I + H(t)\Big]u \otimes \Psi(f), \\
&\qquad M(t)u \otimes \Psi(g)\Big\rangle + \sum_j \Big\langle F(t)L_j u \otimes \Psi(f), F(t)L_j u \otimes \Psi(g\Big\rangle,
\end{aligned} \tag{4.8}$$

$$\begin{aligned}
&\frac{\mathrm{d}}{\mathrm{d}t}\langle N(t)u \otimes \Psi(f), N(t)u \otimes \Psi(g)\rangle \\
&= \Big\langle N(t)u \otimes \Psi(f), \Big[\sum_j \overline{f_j(t)}L_j \otimes IF(t) + \\
&\quad + \sum_j g_j(t)L_j^\dagger \otimes IG(t) + H(t)\Big]u \otimes \Psi(g)\Big\rangle + \\
&\quad + \Big\langle\Big[\sum_j \overline{g_j(t)}L_j \otimes IF(t) + \sum_j f_j(t)L_j^\dagger \otimes IG(t) + J(t)\Big]u \otimes \Psi(f), \\
&\quad N(t)v \otimes \Psi(g)\Big\rangle + \sum_j \langle L_j \otimes IF(t)u \otimes \Psi(f), L_j \otimes IF(t)u \otimes \Psi(g)\rangle.
\end{aligned} \tag{4.9}$$

Proof. We prove the case (4.8) of the right integral; (4.9) is similar. We assume, F, G and H given by (4.1) and $t \in (t_n, t_{n+1})$. Then

$$\begin{aligned}
&\langle M(t)u \otimes \Psi(f), M(t)u \otimes \Psi(g)\rangle \\
&= \langle [M(t_n) + F_n(A_L^\dagger(t) - A_L^\dagger(t_n)) + G_n(A_L(t) - A_L(t_n)) + \\
&\quad + (t - t_n)H_n]u \otimes \Psi(f), \\
&\quad [M(t_n) + F_n(A_L^\dagger(t) - A_L^\dagger(t_n)) + G_n(A_L(t) - A_L(t_n)) + \\
&\quad + (t - t_n)H_n u \otimes \Psi(g)\rangle.
\end{aligned}$$

We replace $A_L(t) - A_L(t_n)$ by its actions $\Sigma_j \int_{t_n}^t f_j L_j^\dagger \otimes I$ on the left and $\Sigma_j \int_{t_n}^t g_j L_j^\dagger \otimes I$ on the right. Similarly, using the commutativity of $A_L(t) - A_L(t_n)$ with $F_n^\dagger, M(t_n), F_n, G_n$ and H_n, as in proof of Theorem 4.2 we replace $A_L^\dagger(t) - A_L^\dagger(t_n)$ by its adjoint actions $\Sigma_j \int_{t_n}^t \bar{g}_j L_j \otimes I$ on the left and $\Sigma_j \int_{t_n}^t \bar{f}_j L_j \otimes I$ on the right. Since $A_L(t) - A_L(t_n)$ fails to commute with $A_L^\dagger(t) - A_L^\dagger(t_n)$, we use (3.3) to deal with the crossterm

$$\langle F_n(A_L^\dagger(t) - A_L^\dagger(t_n))u \otimes \Psi(f), F_n(A_L^\dagger(t) - A_L^\dagger(t_n))u \otimes \Psi(g)\rangle.$$

We obtain in this way

$$\begin{aligned}
&\langle M(t)u \otimes \Psi(f), M(t)u \otimes \Psi(g)\rangle \\
&= \Big\langle \Big[M(t_n) + F_n \sum_j \int_{t_n}^t \bar{g}_j L_j \otimes I + \\
&\quad + G_n \sum_j \int_{t_n}^t f_j L_j^\dagger \otimes I + (t - t_n) H_n\Big] u \otimes \Psi(f), \\
&\quad \Big[M(t_n) + F_n \sum_j \int_{t_n}^t \bar{f}_j L_j \otimes I + \\
&\quad + G_n \sum_j \int_{t_n}^t g_j L_j^\dagger \otimes I + (t - t_n) H_n\Big] u \otimes \Psi(g)\rangle + \\
&\quad + (t - t_n) \sum_j \langle F_n L_j u \otimes \Psi(f), F_n L_j u \otimes \Psi(g)\Big\rangle.
\end{aligned}$$

Taking the summations and integrations out of the inner product and reversing their order, as is possible by the argument used in the proof of Theorem 4.2, we see that $\langle M(t)u \otimes \Psi(f), M(t)u \otimes \Psi(g)\rangle$ is absolutely continuous, with generalised derivative

$$\begin{aligned}
&\Big\langle \Big[M(t_n) + F_n \sum_j \int_{t_n}^t \bar{g}_j L_j \otimes I + G_n \sum_j \int_{t_n}^t f_j L_j^\dagger \otimes I + (t - t_n) H_n\Big] u \otimes \Psi(f), \\
&\Big[F_n \sum_j \overline{f_j(t)}\, L_j \otimes I + G_n \sum_j g_j(t)\, L_j^\dagger \otimes I + H_n\Big] u \otimes \Psi(g)\Big\rangle + \\
&\quad + \Big\langle \Big[F_n \sum_j \overline{g_j(t)}\, L_j \otimes I + G_n \sum_j f_j(t)\, L_j^\dagger \otimes I + H_n\Big] u \otimes \Psi(f), \\
&\Big[M(t_n) + F_n \sum_j \int_{t_n}^t \bar{f}_j L_j \otimes I + \\
&\quad + G_n \sum_j \int_{t_n}^t g_j L_j^\dagger \otimes I + (t - t_n) H_n\Big] u \otimes \Psi(g\Big\rangle + \\
&\quad + \sum_j \langle F_n L_j u \otimes \Psi(f), F_n L_j u \otimes \Psi(f)\rangle.
\end{aligned}$$

Once more, extracting summations and integrations from the inner product, reversing their order and applying Theorem 3.4, having observed first that

$$\left[F_n \sum_j \overline{f_j(t)} L_j \otimes I + G_n \sum_j g_j(t) L_j^\dagger \otimes I + H_n\right] u \otimes \Psi(g)$$
$$= \phi_1 \otimes \Psi(g^{t_n})$$
$$\left[F_n \sum_j \overline{g_j(t)} L_j \otimes I + G_n \sum_j f_j t) L_j^\dagger \otimes I + H_n\right] u \otimes \Psi(f)$$
$$= \phi_2 \otimes \Psi(f^{t_n})$$

for $\phi_1, \phi_2 \in \mathfrak{h}_0 \otimes \mathfrak{H}_{t_n}$ and that for $\tau \in (t_n, t)$

$$F_n = F(\tau), \qquad G_n = G(\tau), \qquad H_n = H(\tau),$$

we obtain the result. □

5. Stochastic Integrals of Regular Processes

In the following we use the identity

$$\left\|\sum_j z_j L_j^\dagger\right\| = \left\|\sum_j \bar{z}_j L_j\right\| \leqslant \left(\sum_j |z_j|^2\right)^{1/2} \left\|\sum_j L_j^\dagger L_j\right\|^{1/2} \tag{5.1}$$

for scalars $z_j, j \in J$ with $\Sigma_j |z_j|^2 < \infty$. This holds because for arbitrary $u \in \mathfrak{h}_0$

$$\left\|\sum_j z_j L_j u\right\|^2$$
$$= \sum_{j,k} z_j \bar{z}_k \langle L_j u, L_k u\rangle$$
$$\leqslant \sum_{j,k} |z_j|\,|z_k|\,\|L_j u\|\,\|L_k u\|$$
$$= \left(\sum_j |z_j|\,\|L_j u\|\right)^2$$
$$\leqslant \sum_j |z_j|^2 \sum_j \|L_j u\|^2$$
$$\leqslant \sum_j |z_j|^2 \left\|\sum_j L_j^\dagger L_j\right\| \|u\|^2.$$

Now let $F, G, H \in \mathscr{A}_0$ and $M(t) = \int_0^t (F\,\mathrm{d}A_L^\dagger + G\,\mathrm{d}A_L + H\,\mathrm{d}\tau)$ so that, according to (4.2),

$$M(t)^\dagger = \int_0^t (\mathrm{d}A_L^\dagger G^\dagger + \mathrm{d}A_L F^\dagger + H^\dagger\,\mathrm{d}\tau)$$

From Theorem 4.4 we have, for arbitrary $u \in \mathfrak{h}_0$, $f \in \mathfrak{h}$,

$$\frac{\mathrm{d}}{\mathrm{d}t} \| M(t)u \otimes \Psi(f) \|^2$$

$$= 2\,\mathrm{Re} \left\langle M(t)u \otimes \Psi(f), \left[F(t) \sum_j \overline{f_j(t)} L_j \otimes I + \right.\right.$$

$$\left.\left. + G(t) \sum_j f_j(t) L_j^\dagger \otimes I + H(t) \right] u \otimes \Psi(f) \right\rangle +$$

$$+ \left\| \sum_j F(t) L_j u \otimes \Psi(f) \right\|^2$$

$$\leqslant 2 \| M(t)u \otimes \Psi(f) \| \left[\| F(t) \|_f \left\| \sum_j \overline{f_j(t)} L_j \right\| + \| G(t) \|_f \left\| \sum_j f_j(t) L_j^\dagger \right\| + \| H(t) \|_f \right] \times$$

$$\times \| u \| + \| F(t) \|_f^2 \left\| \sum_j L_j^\dagger L_j \right\| \| u \|^2$$

$$\leqslant 2 \| M(t)u \otimes \Psi(f) \| \left[\left(\sum_j | f_j(t) |^2 \right)^{1/2} \left\| \sum_j L_j^\dagger L_j \right\|^{1/2} (\| F(t) \|_f + \| G(t) \|_f) + \| H(t$$

$$\times \| u \| + \| F(t) \|_f^2 \left\| \sum_j L_j^\dagger L_j \right\| \| u \|^2,$$

where we use (5.1). Using the arithmetic-geometric mean inequality gives

$$\frac{\mathrm{d}}{\mathrm{d}t} \| M(t)u \otimes \Psi(f) \|^2$$

$$\leqslant 2 \| M(t)u \otimes \Psi(f) \|^2 \sum_j | f_j(t) |^2 + \left\| \sum_j L_j^\dagger L_j \right\| (\| F(t) \|_f^2 + \| G(t) \|_f^2) \| u \|^2 +$$

$$+ \| M(t)u \otimes \Psi(f) \|^2 + \| H(t) \|_f^2 \| u \|^2 + \| F(t) \|_f^2 \left\| \sum_j L_j^\dagger L_j \right\| \| u \|^2$$

$$= \| M(t)u \otimes \Psi(f) \|^2 \left\{ 2 \sum_j | f_j(t) |^2 + 1 \right\} +$$

$$+ \left[\left\| \sum_j L_j^\dagger L_j \right\| (2 \| F(t) \|_f^2 + \| G(t) \|_f^2) + \| H(t) \|_f^2 \right] \| u \|^2.$$

Multiplying by the integrating factor $\exp(-2 \| f_t \|^2 - t)$ and integrating the differential inequality, we obtain

$$\| M(t)u \otimes \Psi(f) \|^2$$

$$\leqslant \int_0^t \exp\{ 2 \| f_t \|^2 - 2 \| f_\tau \|^2 + t - \tau \} \times$$

$$\times \left[\left\| \sum_j L_j^\dagger L_j \right\| (2 \| F(\tau) \|_f^2 + G(\tau) \|_f^2) + \| H(\tau) \|_f^2 \right] \| u \|^2 \, \mathrm{d}\tau. \tag{5.2}$$

The corresponding estimate for the left integral,

$$\| M^\dagger(t)u\otimes\Psi(f)\|^2$$

$$\leqslant \int_0^t \exp\{2\|f_t\|^2 - 2\|f_\tau\|^2 + t - \tau\}\times$$

$$\times\left[\left\|\sum_j L_j^\dagger L_j\right\|(\|F^\dagger(\tau)\|_f^2 + 2\|G^\dagger(\tau)\|_f^2) + \|H^\dagger(\tau)\|_f^2\right]\|u\|^2\,\mathrm{d}\tau \tag{5.3}$$

is proved similarly.

Now let F, G, H be regular processes, so that these exist simple processes $F^{(n)}$, $G^{(n)}$, $H^{(n)}$, $n = 1, 2, \ldots$ such that for each $f\in\mathfrak{h}$

$$\|F(\tau) - F^{(n)}(\tau)\|_f,\ \|G(\tau) - G^{(n)}(\tau)\|_f,\ \|H(\tau) - H^{(n)}(\tau)\|_f \xrightarrow[n]{} 0$$

and

$$\|F(\tau)^\dagger - F^{(n)}(\tau)^\dagger\|_f,\ \|G(\tau)^\dagger - G^{(n)}(\tau)^\dagger\|_f\ \|H(\tau)^\dagger - H^{(n)}(\tau)^\dagger\|_f \xrightarrow[n]{} 0$$

uniformly on each finite interval. Then if

$$M^{(n)}(t) = \int_0^t (F^{(n)}\,\mathrm{d}A_L^\dagger + G^{(n)}\,\mathrm{d}A_L + H^{(n)}\,\mathrm{d}\tau),$$

the estimates (5.2) and (5.3) show that, for each $u\in\mathfrak{h}_0$, $f\in\mathfrak{h}$ and $t>0$, the sequences $M^{(n)}(t)u\otimes\Psi(f)$, $M^{(n)}(t)^\dagger u\otimes\Psi(f)$, $n = 1, 2, \ldots$ converge to limits independent of the choice of approximating simple processes. We define the stochastic integrals

$$M(t) = \int_0^t (F\,\mathrm{d}A_L + G\,\mathrm{d}A_L + H\,\mathrm{d}\tau),$$

$$M^\dagger(t) = \int_0^t (\mathrm{d}A_L^\dagger G^\dagger + \mathrm{d}A_L F^\dagger + H^\dagger\,\mathrm{d}\tau)$$

in the first instance on the domain $\mathfrak{h}_0\underline{\otimes}\mathscr{E}$ by

$$M(t)u\otimes\Psi(f) = \lim_n M^{(n)}(t)\otimes\Psi(f),$$

$$M^\dagger(t)u\otimes\Psi(f) = \lim_n M^{(n)}(t)^\dagger u\otimes\Psi(f).$$

The operators $M(t)$ and $M^\dagger(t)$ inherit from $M^{(n)}(t)$ and $M^{(n)}(t)^\dagger$ the property of being of form $M_1(t)\otimes I$, $M_1(t)^\dagger\otimes I$ on $(\mathfrak{h}_0\underline{\otimes}\mathscr{E}_t)\underline{\otimes}\mathscr{E}^t$ and, hence, extend naturally to the domain $(\mathfrak{h}_0\underline{\otimes}\mathscr{E}_t)\underline{\otimes}\mathfrak{H}^t$. $M = (M(t): t\geqslant 0)$ is then an adapted process of which $M^\dagger = (M^\dagger(t): t\geqslant 0)$ is the adjoint process.

The estimates (5.2) and (5.3), together with their generalisations obtained by replacing F, G, H by $F\chi_{(s,\infty)}$, $G\chi_{(s,\infty)}$, $H\chi_{(s,\infty)}$ respectively, namely

$$\|(M(t) - M(s))u\otimes\Psi(f)\|^2$$

$$\leqslant \int_s^t \exp\{2\|f_t\|^2 - 2\|f_\tau\|^2 + t - \tau\} \times$$

$$\times \left[\left\|\sum_j L_j^\dagger L_j\right\| (2\|F(\tau)\|_f^2 + \|G(\tau)\|_f^2) + \|H(\tau)\|_f^2\right] \|u\|^2 \, d\tau. \tag{5.4}$$

$$\|(M(t)^\dagger - M(s)^\dagger)u \otimes \Psi(f)\|^2$$

$$\leqslant \int_s^t \exp\{2\|f_t\|^2 - 2\|f_\tau\|^2 + t - \tau\} \times$$

$$\times \left[\left\|\sum_j L_j^\dagger L_j\right\| (\|F^\dagger(\tau)\|_f^2 + 2\|G^\dagger(\tau)\|_f^2) + \|H^\dagger(\tau)\|_f^2\right] \|u\|^2 \, d\tau, \tag{5.5}$$

persist in the transition to the limit, showing that the processes M and $M^\dagger$ are continuous, hence, regular and, in particular, that the maps $\tau \mapsto M(\tau)u \otimes \Psi(f)$, $\tau \mapsto M(\tau)u \otimes \Psi(f)$ are continuous and, hence, bounded on each finite interval $[0, t]$. From this it follows that we may pass to the limit in the integrated forms of (4.3), (4.4), (4.8) and (4.9) and deduce that Theorems 4.2, 4.3 and 4.4 hold for arbitrary F, G, $H \in \mathscr{A}_r$. We summarise.

THEOREM 5.1. *Theorems* 4.2, 4.3 *and* 4.4 *hold for arbitrary* $F, G, H \in \mathscr{A}_r$. *Furthermore, if*

$$M(t) = \int_0^t (F \, dA_L^\dagger + G \, dA_L + H \, d\tau)$$

then

$$M^\dagger(t) = \int_0^t (dA_L^\dagger G^\dagger + dA_L F^\dagger + H^\dagger \, d\tau)$$

and the estimates (5.2) *and* (5.3) *are satisfied.*

6. The Unitary Process

Let $\mathscr{H}$ be a bounded self-adjoint operator in $\mathfrak{h}_0$, fixed once and for all.

THEOREM 6.1. *There exists a unique adapted process* $(U(t) : t \geqslant 0)$ *satisfying*

$$dU = U\left(dA_L^\dagger - dA_L + \left(i\mathscr{H} - \tfrac{1}{2}\sum_j L_j^\dagger L_j\right) \otimes I \, dt\right), \quad U(0) = I. \tag{6.1}$$

Proof. We establish existence by iteration. Thus, define $U_0(t) \equiv I$ and, assuming that the regular process $(U_n(t) : t \geqslant 0)$ has been defined, define

$$U_{n+1}(t) = I + \int_0^t U_n(\tau)\left(dA_L^\dagger - dA_L + \left(i\mathscr{H} - \frac{1}{2}\sum_j L_j^\dagger L_j\right) \otimes I \, d\tau\right). \tag{6.2}$$

The process U_{n+1} is then continuous, hence, regular. We write

$$U_{n+1}(t) - U_n(t) = \int_0^t (U_r(\tau) - U_{n-1}(\tau)) \times \times \left(dA_L^\dagger - dA_L + \left(i\mathcal{H} - \tfrac{1}{2}\sum_j L_j^\dagger L_j \right) \otimes I \, d\tau \right)$$

and use the estimate (5.2) to write, for $u \in \mathfrak{h}_0$, $f \in \mathfrak{h}$, $t > 0$

$$\| U_{n+1}(t) - U_n(t))u \otimes \Psi(f) \|^2$$
$$\leqslant \int_0^t \exp\{2\|f_t\|^2 - 2\|f_\tau\|^2 + t - \tau\}$$
$$\left[3 \left\| \sum_j L_j^\dagger L_j \right\| \| U_n(\tau) - U_{n-1}(\tau) \|_f^2 + + \left\| (U_n(\tau) - U_{n-1}(\tau)) \left(i\mathcal{H} - \tfrac{1}{2}\sum_j L_j^\dagger L_j \right) \otimes I \right\|_f \right] \| u \|^2 \, d\tau$$
$$\leqslant C \int_0^t \exp\{2\|F_t\|^2 - 2\|F_\tau\|^2 + t - \tau\} \| U_n(\tau) - U_{n-1}(\tau) \|_f^2 \, \| u \|^2 \, d\tau$$

where

$$C = 3 \left\| \sum_j L_j^\dagger L_j \right\| + \left\| i\mathcal{H} - \tfrac{1}{2}\sum_j L_j^\dagger L_j \right\|.$$

Hence

$$\| U_{n+1}(t) - U_n(t) \|_f^2$$
$$\leqslant C \int_0^t \exp\{2\|f_t\|^2 - 2\|f_\tau\|^2 + t - \tau\} \| U_n(\tau) - U_{n-1}(\tau) \|_f^2 \, d\tau. \tag{6.3}$$

By induction on n we obtain that

$$\| \mathrm{U}_n(\mathrm{t}) - \mathrm{U}_{n-1}(\mathrm{t}) \|_{\mathrm{f}}^2 \leqslant (\mathrm{n}!)^{-1} \mathrm{C}^n \mathrm{t}^n \exp(2\|f_t\|^2 + t).$$

From this and from the corresponding argument for the adjoint processes based on (5.3) it is clear that, for $u \in \mathfrak{h}_0$ and $f \in \mathfrak{h}$, the limits

$$\begin{aligned} U(t)u \otimes \Psi(f) &= \lim_n U_n(t)u \otimes \Psi(f), \\ U^\dagger(t)u \otimes \Psi(f) &= \lim_n U_n^\dagger(t)u \otimes \Psi(f), \end{aligned} \tag{6.4}$$

exist and define mutually adjoint adapted processes. Moreover the convergence in (6.4) is uniform for t in bounded intervals, enabling us to take strong limits in (6.2) and conclude that $(U(t) : t \geqslant 0)$ satisfies (6.1).

If $(V(t) : t \geqslant 0)$ is a second adapted process satisfying (6.1), then, from the estimate (5.2) we obtain as above

$$\| U(t) - V(t) \|_f^2 \leqslant C \int_0^t \exp\{2\|f_t\|^2 - 2\|f_\tau\|^2 + t - \tau\} \| U(\tau) - V(\tau) \|_f^2 \, d\tau. \tag{6.5}$$

Since $U - V$ is a stochastic integral, the map $\tau \mapsto (U(\tau) - V(\tau))u \otimes \Psi(f)$ is continuous for each $u \in \mathfrak{h}_0$, $f \in \mathfrak{h}$, and hence bounded on $[0, t]$. Hence, by the uniform boundedness principle there exists $M > 0$ such that, for all $\tau \in [0, t]$,

$$\| U(\tau) - V(\tau) \|_f^2 \leqslant M.$$

But then by iterating (6.5) we find that $\| U(t) - V(t) \|_f = 0$. This being so for all $f \in \mathfrak{h}$ shows that $U = V$. □

The adjoint process $U^\dagger$ to U satisfies

$$\mathrm{d}U^\dagger = \left(- \mathrm{d}A_L^\dagger + \mathrm{d}A_L - \left(i\mathscr{H} + \tfrac{1}{2} \sum_j L_j^\dagger L_j \right) \otimes I \, \mathrm{d}t \right) U^\dagger$$

in view of Theorem 5.1. We apply (4.9) to write, for arbitrary $u \in \mathfrak{h}_0$, $f, g \in \mathfrak{h}$,

$$\begin{aligned} &\frac{\mathrm{d}}{\mathrm{d}t} \langle U^\dagger(t)u \otimes \Psi(f), U^\dagger(t)u \otimes \Psi(g) \rangle \\ &= \Big\langle U^\dagger(t)u \otimes \Psi(f), \Big[- \sum_j \overline{f_j(t)} L_j \otimes I + \\ &\quad + \sum_j g_j(t) L_j^\dagger \otimes I - \Big(i\mathscr{H} + \tfrac{1}{2} \sum_j L_j^\dagger L_j \Big) \otimes I \Big] U^\dagger(t)u \otimes \Psi(g) \Big\rangle + \\ &\quad + \Big\langle \Big[- \sum_j \overline{g_j(t)} L_j \otimes I + \sum_j f_j(t) L_j^\dagger \otimes I - \\ &\quad - \Big(i\mathscr{H} + \tfrac{1}{2} \sum_j L_j^\dagger L_j \Big) \otimes I \Big] U^\dagger(t)u \otimes \Psi(f), U^\dagger(t)u \otimes \Psi(g) \Big\rangle + \\ &\quad + \sum_j \langle L_j \otimes I U^\dagger(t)u \otimes \Psi(f), L_j \otimes I U^\dagger(t)u \otimes \Psi(g) \rangle \\ &= 0. \end{aligned}$$

Since $U^\dagger(0) = I$ we conclude that, for all $t \geqslant 0$,

$$\langle U^\dagger(t)u \otimes \Psi(f), U^\dagger(t)u \otimes \Psi(g) \rangle = \langle u \otimes \Psi(f), u \otimes \Psi(g) \rangle.$$

By polarisation we obtain that

$$\langle U^\dagger(t)u \otimes \Psi(f), U^\dagger(t)v \otimes \Psi(g) \rangle = \langle u \otimes \Psi(f), v \otimes \Psi(g) \rangle$$

for arbitrary $u, v \in \mathfrak{h}_0$, $f, g \in \mathfrak{h}$. Thus $U^\dagger(t)$ is isometric.

THEOREM 6.2. *Each $U(t)$, $t \geqslant 0$, is unitary.*

Proof. Since $U(t)$ is the adjoint of an isometry it is bounded. To prove it is unitary we need only prove that its action on a total family of vectors is isometric. For these we choose the vectors $u \otimes \Psi(f)$ where $u \in \mathfrak{h}_0$ is arbitrary and $f = (f_j)$ has only finitely many nonzero components, each of which is piecewise constant. For such vectors $u \otimes \Psi(f)$ and $v \otimes \Psi(g)$ by (4.8)

$$\frac{\mathrm{d}}{\mathrm{d}t} \langle U(t)u \otimes \Psi(f), U(t)v \otimes \Psi(g)\rangle$$

$$= \Big\langle U(t)u \otimes \Psi(f), U(t)\Big[\sum_j \overline{f_j(t)} L_j \otimes I -$$

$$- \sum_j g_j(t) L_j^\dagger \otimes I + \Big(i\mathscr{H} - \tfrac{1}{2} \sum_j L_j^\dagger L_j \Big) \otimes I \Big] v \otimes \Psi(g) \Big\rangle +$$

$$+ \Big\langle U(t) \Big[\sum_j \overline{g_j(t)} L_j \otimes I - \sum_j f_j(t) L_j^\dagger \otimes I +$$

$$+ \Big(i\mathscr{H} - \tfrac{1}{2} \sum_j L_j^\dagger L_j \Big) \otimes I \Big] u \otimes \Psi(f), U(t)v \otimes \Psi(g) \Big\rangle +$$

$$+ \sum_j \langle U(t) L_j u \otimes \Psi(f), U(t) L_j v \otimes \Psi(g)\rangle.$$

It follows that the bounded operator $K_{f,g}(t)$ on $\mathfrak{h}_0$ defined by

$$\langle u, K_{f,g}(t)v\rangle = \langle U(t)u \otimes \Psi(f), U(t)v \otimes \Psi(g)\rangle \; (u, v \in \mathfrak{h}_0)$$

satisfies the weak sense differential equation

$$\frac{\mathrm{d}}{\mathrm{d}t} K_{f,g}(t) = \Big[K_{f,g}, \sum_j \overline{f_j(t)} L_j - \sum_j g_j(t) L_j^\dagger + i\mathscr{H} \Big] -$$

$$- \tfrac{1}{2} \sum_j (L_j^\dagger L_j K_{f,g} - 2 L_j^\dagger K_{f,g} L + K_{f,g} L_j^\dagger L_j)$$

with initial condition

$$K_{f,g}(0) = \langle \Psi(f), \Psi(g)\rangle \, I.$$

Since $K_{f,g} \equiv \langle \Psi(f), \Psi(g)\rangle \, I$ satisfies this equation we may appeal to the uniqueness theorem for the differential equation

$$\frac{\mathrm{d}K}{\mathrm{d}t} = \mathscr{L} K$$

in the Banach space $B(\mathfrak{h}_0)$, $\mathscr{L}$ being a bounded operator in $B(\mathfrak{h}_0)$, within each interval of constancy of the functions f_j and g_j, to conclude that $K_{f,g}(t)$ is indeed equal to $\langle \Psi(f), \Psi(g)\rangle \, I$ for all t. Hence, $U(t)$ is isometric as required. □

THEOREM 6.3. *Let $s \geqslant 0$ and let T be a bounded operator of form $I \otimes T_1 \otimes I$ where $T_1 \in B(\mathfrak{h}_s)$. Then for $t \geqslant s$, $U^\dagger(s)U(t)$ commutes with T.*

Proof. Define processes J and K by

$$J(t) = \begin{cases} 0 & \text{if } t \leqslant s \\ U^\dagger(s)U(t) & \text{if } t > s. \end{cases} \qquad K(t) = [T, J(t)]$$

These clearly inherit adaptedness and regularity from U. Subtracting the corresponding equation with t replaced by s from

$$U(t) = I + \int_0^t U(\tau)\left(\mathrm{d}A_L^\dagger - \mathrm{d}A_L + \left(i\mathscr{H} - \tfrac{1}{2}\sum_j L_j^\dagger L_j\right) \otimes I\,\mathrm{d}\tau\right)$$

and multiplying by $U^\dagger(s)$ gives

$$J(t) = I\chi_{[s,\infty)}(t) + \int_0^t J(\tau)\left(\mathrm{d}A_L^\dagger - \mathrm{d}A_L + \left(i\mathscr{H} - \tfrac{1}{2}\sum_j L_j^\dagger L_j\right) \otimes I\,\mathrm{d}\tau\right).$$

Using Theorem 4.3 we deduce from this that, for $t > s$, $u, v \in \mathfrak{h}_0$, $f, g \in \mathfrak{h}$,

$$\begin{aligned}
&\langle u \otimes \Psi(f), K(t) v \otimes \Psi(g)\rangle \\
&= \langle T^\dagger u \otimes \Psi(f), J(t) v \otimes \Psi(g)\rangle - \langle J^\dagger(t) u \otimes \Psi(f), Tu \otimes \Psi(f)\rangle \\
&= \int_s^t \Big\langle T^\dagger u \otimes \Psi(f), J(\tau)\Big(\sum_j \overline{f_j(\tau)} L_j \otimes I - \\
&\quad - \sum_j g_j(\tau) L_j^\dagger \otimes I + \Big(i\mathscr{H} - \tfrac{1}{2}\sum_j L_j^\dagger L_j\Big) \otimes I\Big) v \otimes \Psi(g)\Big\rangle\,\mathrm{d}\tau - \\
&\quad - \int_s^t \Big\langle u \otimes \Psi(f), J(\tau)\Big(\sum_j \overline{f_j(\tau)} L_j \otimes I - \\
&\quad - \sum_j g_j(\tau) L_j^\dagger \otimes I + \Big(i\mathscr{H} - \tfrac{1}{2}\sum_j L_j^\dagger L_j\Big) \otimes I\Big) Tv \otimes \Psi(g)\Big\rangle\,\mathrm{d}\tau \\
&= \int_0^t \Big\langle u \otimes \Psi(f), K(\tau)\Big(\sum_j \overline{f_j(\tau)} L_j \otimes I - \\
&\quad - \sum_j g_j(\tau) L_j^\dagger \otimes I + \Big(i\mathscr{H} - \tfrac{1}{2}\sum_j L_j^\dagger L_j\Big) \otimes I\Big) v \otimes \Psi(g)\Big\rangle\,\mathrm{d}\tau.
\end{aligned}$$

Since this holds trivially for $t < s$, we conclude that K satisfies

$$\mathrm{d}K = K\left(\mathrm{d}A_L^\dagger - \mathrm{d}A_L + \left(i\mathscr{H} - \tfrac{1}{2}\sum_j L_j^\dagger L_j\right) \otimes I\,\mathrm{d}t\right), \quad K(0) = 0.$$

But then adding K to the solution U of (6.1) would yield a different solution $U + K$, contradicting uniqueness unless $K \equiv 0$. □

7. The Reduced Semigroup

Let S be a contraction on $\mathfrak{h}$. There is a contraction $\Gamma(S)$ on $\mathfrak{H}$ called the second quantisation [3] of S whose action on exponential vectors is $\Gamma(S)\Psi(f) = \Psi(Sf)$.

We denote by $\gamma(S)$ the operator $I \otimes \Gamma(S)$ in $\mathfrak{H}_0$. From corresponding properties of second quantisations [3] we have

$$\gamma(S_1 S_2) = \gamma(S_1)\gamma(S_2), \qquad \gamma(S^\dagger) = \gamma(S)^\dagger, \quad \gamma(I) = I \tag{7.1}$$

for arbitrary contractions S_1, S_2, S on $\mathfrak{h}$. Also if $S = S_1 \otimes S_2$ is the direct sum of contractions S_1 and S_2, $\Gamma(S) = \Gamma(S_1) \otimes \Gamma(S_2)$.

We denote by S_t, $t \geqslant 0$, the shift in $\mathfrak{h}$

$$S_t f_j(\tau) = \begin{cases} 0 & \text{if } \tau < t \\ f_j(\tau - t) & \text{if } \tau \geqslant t. \end{cases}$$

S_t is isometric and $S_t S_t^\dagger$ is the projector E^t onto $\mathfrak{h}^t$.

THEOREM 7.1. *For arbitrary* s, $t \geqslant 0$

$$U(t) = \gamma(S_s)^\dagger\, U(s)^\dagger\, U(s + t)\gamma(S_s).$$

Proof. Fix $s \geqslant 0$ and consider the family of bounded operators

$$V(t) = \gamma(S_s)^\dagger\, U(s)^\dagger\, U(s + t)\gamma(S_s), \quad t \geqslant 0.$$

We prove that this is an adapted process, that is each $V(t)$ is of the form $V(t) = V_1(t) \otimes I$ on $(\mathfrak{h}_0 \otimes \mathfrak{H}_t) \otimes \mathfrak{H}^t$. To do this write S_s as the direct sum $S_s = S_1 \oplus S_2$ of its restrictions $S_1 \colon \mathfrak{h}_t \to \mathfrak{h}_{s+t}$, $S_2 \colon \mathfrak{h}^t \to \mathfrak{h}^{s+t}$. Correspondingly

$$\Gamma(S_s) = \Gamma(S_1) \otimes \Gamma(S_2), \qquad \Gamma(S_s)^\dagger = \Gamma(S_1)^\dagger \otimes \Gamma(S_2)^\dagger,$$

where $\Gamma(S_1)$ maps $\mathfrak{H}_t$ to $\mathfrak{H}_{s+t}$ and $\Gamma(S_2)$ $\mathfrak{H}^t$ to $\mathfrak{H}^{s+t}$. Because U is adapted we can write $U^\dagger(s)U(s + t) = U_1 \otimes I$ for some operator U_1 on $\mathfrak{h}_0 \otimes \mathfrak{H}_{s+t}$. Thus,

$$\begin{aligned} V(t) &= \gamma(S_s)^\dagger\, U(s)^\dagger\, U(s + t)\gamma(S_s) \\ &= ((I \otimes \Gamma(S_1)^\dagger) \otimes \Gamma(S_2)^\dagger) U_1 \otimes I((I \otimes \Gamma(S_1)) \otimes \Gamma(S_2)) \\ &= (I \otimes \Gamma(S_1)^\dagger) U_1 (I \otimes \Gamma(S_1)) \otimes \Gamma(S_2)^\dagger\, \Gamma(S_2) \\ &= V_1 \otimes I, \end{aligned}$$

where $V_1 = I \otimes \Gamma(S_1)^\dagger\, U_1 I \otimes \Gamma(S_1)$ is a bounded operator on $\mathfrak{h}_0 \otimes \mathfrak{H}_t$, as required.

The adapted process V inherits regularity from U. Let us show that it satisfies the stochastic differential equation (6.1); by the uniqueness of the solution we shall then be able to conclude that $U = V$. Applying Theorem 4.3, in which we take $s = 0$, $\phi = u$ and

$$\mathrm{d}M = V\left(\mathrm{d}A_L^\dagger - \mathrm{d}A_L + \left(i\mathscr{H} - \tfrac{1}{2}\sum_j L_j L_j\right) \otimes I\, \mathrm{d}t\right),$$

for arbitrary $t > 0$, u, $v \in \mathfrak{h}_0$, f, $g \in \mathfrak{h}$,

$$\left\langle u \otimes \Psi(f), \int_0^t V(\tau)\left(\mathrm{d}A_L^\dagger - \mathrm{d}A_L + \left(i\mathscr{H} - \tfrac{1}{2}\sum_j L_j^\dagger L_j\right) \otimes I\, \mathrm{d}\tau\right) v \otimes \Psi(g)\right\rangle$$

$$= \int_0^t \Big\langle u \otimes \Psi(f), \gamma(S_s^\dagger) U(s)^\dagger U(\tau + s) \gamma(S_s) \times$$

$$\times \Big[\sum_j \overline{f_j(\tau)} L_j - \sum_j g_j(\tau) L_j^\dagger + i\mathscr{H} - \tfrac{1}{2} \sum_j L_j^\dagger L_j \Big] \otimes Iv \otimes \Psi(g) \Big\rangle \, d\tau$$

$$= \int_0^t \Big\langle U(s) u \otimes \Psi(S_s f), U(\tau + s) \times$$

$$\times \Big[\sum_j \overline{f_j(\tau)} L_j - \sum_j g_j(\tau) L_j^\dagger + i\mathscr{H} - \tfrac{1}{2} \sum_j L_j^\dagger L_j \Big] \otimes Iv \otimes \Psi(S_s g) \Big\rangle \, d\tau$$

$$= \int_0^t \Big\langle U(s) u \otimes \Psi(S_s f), U(\tau + s) \times$$

$$\times \Big[\sum_j \overline{S_s f_j(\tau + s)} L_j - \sum_j S_s g_j(\tau + s) L_j^\dagger + i\mathscr{H}_0 - \tfrac{1}{2} \sum_j L_j^\dagger L_j \Big] \otimes Iv \otimes \Psi(S_s g) \Big\rangle \, d\cdot$$

$$= \int_s^{s+t} \Big\langle U(s) u \otimes \Psi(S_s f), U(\tau) \times$$

$$\times \Big[\sum_j \overline{S_s f_j(\tau)} L_j - \sum_j S_s g_j(\tau) L_j^\dagger + i\mathscr{H}_0 - \tfrac{1}{2} \sum_j L_j^\dagger L_j \Big] \otimes Iv \otimes \Psi(S_s g) \Big\rangle \, d\tau$$

$$= \langle U(s) u \otimes \Psi(S_s f), (U(s + t) - U(s)) v \otimes \Psi(S_s g) \rangle,$$

where we use the adaptedness of U to write $U(s)u \otimes \Psi(S_s f)$ in the form $\phi \otimes \Psi((S_s f)_s)$ for $\phi \in \mathfrak{h}_0 \otimes \mathfrak{H}_s$, so that Theorem 4.3 is applicable again, and (6.1),

$$= \langle u \otimes \Psi(S_s f), (U^\dagger(s) U(s + t) - I) v \otimes \Psi(S_s g) \rangle$$

$$= \langle u \otimes \Psi(f), (\gamma(S_s)^\dagger U^\dagger(s) U(s + t) \gamma(S_s) - I) v \otimes \Psi(g) \rangle$$

$$= \langle u \otimes \Psi(f), (V(t) - I) v \otimes \Psi(g) \rangle,$$

where we use the isometry of S_s and (7.1) to write

$$\gamma(S_s)^\dagger I \gamma(S_s) = \gamma(S_s^\dagger S_s) = \gamma(I) = I.$$

It follows that V satisfies (6.1) and the proof is complete. □

For each $t \geqslant 0$ we define a conditional expectation map $\mathbb{E}_t$ from $B(\mathfrak{H}_0)$ onto $B(\mathfrak{h}_0 \otimes \mathfrak{H}_t)$ ($= B(\mathfrak{h}_0)$ when $t = 0$) as follows. For $T \in B(\mathfrak{H}_0)$, $\mathbb{E}_t(T)$ is the unique bounded operator on $\mathfrak{h}_0 \otimes \mathfrak{H}_t$ such that, for arbitrary $\phi_1, \phi_2 \in \mathfrak{h}_0 \otimes \mathfrak{H}_t$

$$\langle \phi_1, \mathbb{E}_t(T) \phi_2 \rangle = \langle \phi_1 \otimes \Psi_0^t, T \phi_2 \otimes \Psi_0^t \rangle$$

where Ψ_0^t is the vacuum in $\mathfrak{H}^t$. We write $\mathbb{E}^t(T) = \mathbb{E}_t(T) \otimes I$, where I is the identity in $\mathfrak{H}^t$. The maps $\mathbb{E}^t$ have the easily verified properties of conditional expectations

$$\text{(a)} \quad \mathbb{E}^s \mathbb{E}^t = \mathbb{E}^s \quad \text{for } 0 \leqslant s \leqslant t, \tag{7.2}$$

$$\text{(b)} \quad \mathbb{E}^t(S_1 T S_2) = S_1 \mathbb{E}^t(T) S_2$$

if S_1 and S_2 are both of form $S \otimes I$ for $S \in B(\mathfrak{h}_0 \otimes \mathfrak{H}_s)$

(c) $\mathbb{E}^t(I) = I$.

We also note

(d) If for $s \geqslant 0$, T commutes with all operators of form $I \otimes S \otimes I$ on $\mathfrak{h}_0 \otimes \mathfrak{H}_s \otimes \mathfrak{H}^s$, then $\mathbb{E}^s(T) = \mathbb{E}^0(T)$.

To prove (d) observe that under the given hypothesis, for $S \in B(\mathfrak{H}_s)$,

$$\begin{aligned}(\mathbb{E}_s(T) I \otimes S) \otimes I &= \mathbb{E}^s(T)\,(I \otimes S \otimes I)\\ &= \mathbb{E}^s(TI \otimes S \otimes I) \quad \text{by (b)}\\ &= \mathbb{E}^s(I \otimes S \otimes IT)\\ &= (I \otimes S\mathbb{E}_s(T)) \otimes I\end{aligned}$$

reversing the previous steps, hence, $\mathbb{E}_s(T)$ commutes with $I \otimes S$ for all $S \in B(\mathfrak{H}_s)$. But then $\mathbb{E}_s(T)$ is necessarily of form $S_1 \otimes I$ with $S_1 \in B(\mathfrak{h}_0)$, and so by (a), (b) and (c)

$$\mathbb{E}^0(T) = \mathbb{E}^0\mathbb{E}^s(T) = \mathbb{E}^0(S_1 \otimes I) = S_1 \otimes I\mathbb{E}^0(I) = S_1 \otimes I = \mathbb{E}^s(T).$$

Finally we note that, since the second quantisations $\Gamma(T_j)$ map the vacuum to itself,

(e) $\mathbb{E}^0(\gamma(T_1) T \gamma(T_2)) = \mathbb{E}^0(T)$ for arbitrary contractions T_1, T_2 on $\mathfrak{h}$.

We are ready for our main Theorem.

THEOREM 7.3. *For* $t \geqslant 0$ *define* $\mathcal{T}_t$: $B(\mathfrak{h}_0) \to B(\mathfrak{h}_0)$ *by*

$$\mathcal{T}_t(X) = \mathbb{E}_0[U(t) X \otimes IU(t)^\dagger], \quad X \in B(\mathfrak{h}_0).$$

Then $(\mathcal{T}_t : t \geqslant 0)$ *is a uniformly continuous one-parameter semigroup of completely positive maps, whose infinitesimal generator*

$$\mathcal{L} = \left.\frac{\mathrm{d}\mathcal{T}_t}{\mathrm{d}t}\right|_{t=0}$$

is given by

$$\mathcal{L}(X) = i[\mathcal{H}, X] - \tfrac{1}{2} \sum_j (L_j^\dagger L_j X - 2L_j^\dagger X L_j + X L_j^\dagger L_j). \tag{7.3}$$

Proof. Being the product of a conditional expectation with a unitary conjugation, both of which are necessarily completely positive, $\mathcal{T}_t$ is also completely positive.

To prove the semigroup property, for s, $t \geqslant 0$ and $X \in B(\mathfrak{h}_0)$ write

$$\begin{aligned}\mathcal{T}_{s+t}(X) \otimes I &= \mathbb{E}^0[U(s+t) X \otimes IU(s+t)^\dagger]\\ &= \mathbb{E}^0\mathbb{E}^s[U(s)U(s)^\dagger\, U(s+t) X \otimes IU(s+t)^\dagger\, U(s)U(s)^\dagger]\\ &= \mathbb{E}^0[U(s)\mathbb{E}^s\{U(s)^\dagger\, U(s+t) X \otimes IU(s+t)^\dagger\, U(s)\}\, U(s)^\dagger]\\ &= \mathbb{E}^0[U(s)\mathbb{E}^0\{U(s)^\dagger\, U(s+t) X \otimes IU(s+t)^\dagger\, U(s)\}\, U(s)^\dagger]\end{aligned} \tag{7.4}$$

using properties (a), (b) and (d) of conditional expectations, respectively, the use of (d) being justified by Theorem 6.3. On the other hand, by Theorem 7.1 and (e),

$$\begin{aligned}\mathcal{T}_t(X)\otimes I &= \mathbb{E}^0[U(t)X\otimes IU(t)^\dagger]\\ &= \mathbb{E}^0[\gamma(S_s)^\dagger U(s)^\dagger U(s+t)\gamma(S_s)X\otimes I\gamma(S_s)^\dagger U^\dagger(s+t)U(s)\gamma(S_s)]\\ &= \mathbb{E}^0[U(s)^\dagger U(s+t)\gamma(S_s)\gamma(S_s^\dagger)X\otimes IU^\dagger(s+t)U(s)]\\ &= \mathbb{E}^0[U(s)^\dagger U(s+t)\gamma(E^s)X\otimes IU^\dagger(s+t)U(s)].\end{aligned}$$

Now by Theorem 6.3, since $\gamma(E^t) = I\otimes\Gamma(0)\otimes I$ in $\mathfrak{h}_0\otimes\mathfrak{H}_s\otimes\mathfrak{H}^s$, $\gamma(E^t)$ commutes with $U(s)^\dagger U(s+t)$.

Hence, using (e) again,

$$\begin{aligned}\mathcal{T}_t(X)\otimes I &= \mathbb{E}^0[\gamma(E^s)U(s)^\dagger U(s+t)X\otimes IU^\dagger(s+t)U(s)]\\ &= \mathbb{E}^0[U(s)^\dagger U(s+t)X\otimes IU^\dagger(s+t)U(s)].\end{aligned}$$

Substituting in (7.4) we obtain

$$\begin{aligned}\mathcal{T}_{s+t}(X)\otimes I &= \mathbb{E}^0[U(s)\,(\mathcal{T}_t(X)\otimes I)U(s)^\dagger]\\ &= \mathcal{T}_s[\mathcal{T}_t(X)]\otimes I\end{aligned}$$

and so $\mathcal{T}_{s+t} = \mathcal{T}_s\mathcal{T}_t$.

To complete the proof use the polarised form of (4.9), in which we set $f = g = 0$, to write

$$\begin{aligned}&\frac{\mathrm{d}}{\mathrm{d}t}\langle u, \mathcal{T}_t(X)v\rangle\\ &= \frac{\mathrm{d}}{\mathrm{d}t}\langle U^\dagger(t)u\otimes\Psi_0, (X\otimes I)U^\dagger(t)v\otimes\Psi_0\rangle\\ &= \left\langle \mathrm{U}^\dagger(\mathrm{t})\mathrm{u}\otimes\Psi_0, -\mathrm{X}\left(\mathrm{i}\mathscr{H} + \tfrac{1}{2}\sum_j L_j^\dagger L_j\right)\otimes IU^\dagger(t)v\otimes\Psi_0\right\rangle +\\ &\quad + \left\langle -\left(i\mathscr{H} + \tfrac{1}{2}\sum_j L_j^\dagger L_j\right)\otimes IU^\dagger(t)u\otimes\Psi_0, U^\dagger(t)v\otimes\Psi_0\right\rangle +\\ &\quad + \sum_j\langle L_j\otimes IU^\dagger(t)u\otimes\Psi_0, XL_j\otimes IU^\dagger(t)v\otimes\Psi_0\rangle\\ &= \langle u, \mathcal{T}_t\mathscr{L}(X)v\rangle,\end{aligned}$$

where $\mathscr{L}$ is given by (7.3). From this it is clear that $\mathcal{T}_t$ is uniformly continuous and has infinitesimal generator $\mathscr{L}$. □

References

1. Bratteli, O. and Robinson, D.: *Operator Algebras and Statistical Mechanics*, Volume II, Springer-Verlag, Berlin, 1981.
2. Hudson, R. L. and Parthasarathy, K. R.: 'Quantumn Itô's Formula and Stochastic Evolutions', *Comm. Math. Phys.* **93** (1984), 301–323.
3. Evans, D. and Lewis, J. T.: *Dilations of Irreversible Evolutions in Algebraic Quantum Theory*, Comm. Dublin Inst. for Adv. Studies, Series A **24** (1974).
4. Lindblad, G.: 'On the Generators of Quantum Dynamical Semigroups', *Comm. Math. Phys.* **48** (1946) 119–130.

Acta Applicandae Mathematicae **2**, 379–390. 0167–8019/84/0024–0379$01.80

Order Properties of Attractive Spin Systems

L. L. HELMS
Department of Mathematics, University of Illinois, 1409 W. Green St., Urbana, IL 61801, U.S.A.

(Received: 15 February 1984)

Abstract. Methods of constructing semigroups of operators describing interacting particle systems are reviewed. A simple proof is given showing the existence of semigroups of operators on the space of bounded Borel measurable functions for nonnegative continuous attractive spin rates along with a proof of the existence of invariant measures for the semigroups so constructed.

AMS (MOS) subject classifications (1980). Primary 47D05; Secondary 60J25, 82A05.

Key words. Markov semigroups, invariant measures, attractive spin systems.

1. Introduction

Statistical equilibrium distributions have been around for some time in the statistical mechanics literature. In 1963, Glauber [4] introduced a model for the time evolution of a physical system having a given distribution as its limiting behavior for large time. Mathematical research in this area can be marked by a paper of Dobrushin [1] in which existence of time-evolving models and ergodic theorems were obtained in a rather general setting and a paper of Spitzer [25] in which a variety of time-evolving models were proposed to explain specific asymptotic distributions. The latter models were described by perfectly reasonable operators acting on functions.

It did not take long for investigators to find out that the problem of showing that the operators generate a semigroup of operators is sometimes not well-posed. We will review some of the methods used to prove that certain operators generate semigroups and what can be done in case the problem is not well-posed with particular emphasis on results involving an order structure. The reader wanting to learn more about the subject would best start with the papers of Liggett [21], Spitzer [25], Stroock [27], and Sullivan [29].

2. Finite Spin Models

A classical procedure for constructing time-dependent probability models is embodied in 'Q-matrix' theory (c.f. [3]). Let Q be a real $n \times n$ matrix with entries q_{ij}, $1 \leqslant i, j \leqslant n$, satisfying $q_{ij} \geqslant 0$ whenever $i \neq j$ and $\Sigma_j q_{ij} = 0$ for each i. Then for each $t \geqslant 0$ a stochastic matrix $P(t) = \{p_{ij}(t)\}$ can be defined by putting $P(t) = \exp(tQ)$. The family of transition functions $\{p_{ij}(t) : t \geqslant 0\}$ can then be used to construct a continuous parameter Markov chain with state space $\{1, 2,\ldots, n\}$. The chain so constructed has the following interpretation in terms of the q_{ij}. Suppose $i \neq j$. If at some given time the chain is in the state i, then the probability that there will be a jump within a time interval of length Δt to j is $q_{ij}\Delta t + \mathcal{O}(\Delta t)$; the probability of two or more jumps in the interval of length Δt is $\mathcal{O}(\Delta t)$.

The limiting behavior of the transition functions $p_{ij}(t)$ as t becomes large is well understood and is easily described in the particular case of irreducible Q; that is, given $i \neq j$, either $q_{ij} > 0$ or there is a finite chain $i = i_0, i_1, \ldots, i_p = j$ such that $q_{i_0 i_1} \cdot q_{i_1 i_2} \cdot \ldots \cdot q_{i_{p-1} i_p} \neq 0$. In this case, there are probabilities $\mu_1, \mu_2, \ldots, \mu_n$ with $\Sigma_j \mu_j = 1$ such that $\lim_{t \to \infty} p_{ij}(t) = \mu_j$, independently of i and, in fact, the convergence is exponentially fast. The measure $\mu = (\mu_1, \mu_2, \ldots, \mu_n)$, represented as a row vector, is the unique probability density satisfying the equation $\mu Q = 0$ or, equivalently, $\mu P(t) = \mu$ for all $t \geqslant 0$. Since the integers $1, 2, \ldots, n$ can serve as labels, the above results can be applied to any finite set of states.

Finite spin systems are constructed using Q-matrix theory in the following way. If Γ is a nonempty finite set, the points of Γ will be thought of as particle sites and each particle can have one of two spins -1 and $+1$. An element $\eta \in S_\Gamma = \{-1, +1\}^\Gamma$ is called a configuration with $\eta(x)$, $x \in \Gamma$, representing the spin of a particle at $x \in \Gamma$. In utilizing the above construction, the configurations in S_Γ will play the role of the states of which there are $2^{|\Gamma|}$ in number where $|\Gamma|$ is the number of points in Γ. To carry out the construction we need only specify a Q-matrix on $S_\Gamma \times S_\Gamma$. Suppose the spin system is in the configuration $\eta \in S_\Gamma$ at some given time. We will permit only jumps from η to configurations which differ from η at just one site in small time intervals. For each $x \in \Gamma$, let η_x be the configuration obtained from η by changing the spin $\eta(x)$ at x to $-\eta(x)$. Transitions of the type $\eta \to \eta_x$, $x \in \Gamma$, will be permitted but transitions $\eta \to \zeta$, where ζ differs from η at two or more sites, will not be permitted in small time intervals. In defning a Q-matrix Ω_Γ on $S_\Gamma \times S_\Gamma$, we will thus put $\Omega_\Gamma(\eta, \zeta) = 0$ whenever ζ differs from η at two or more sites.

In order to specify $\Omega_\Gamma(\eta, \eta_x)$ for $x \in \Gamma$, we will let $c(x, \eta)$ be a nonnegative real-valued function on $\Gamma \times S_\Gamma$ and put $\Omega_\Gamma(\eta, \eta_x) = c(x, \eta)$ for $x \in \Gamma$. Since the row sums of a Q-matrix must be equal to zero, we complete the definition of Ω_Γ by putting $\Omega_\Gamma(\eta, \eta) = -\Sigma_{x \in \Gamma} c(x, \eta)$ for each $\eta \in S_\Gamma$. The $c(x, \eta)$ are called rates. Since $c(x, \eta)$ can depend upon spins at sites other than x, the $c(x, \eta)$ can produce an interaction between spins at different sites. Having defined the Q-matrix Ω_Γ, we now let $P_\Gamma(t) = \exp(t\Omega_\Gamma)$ for $t \geqslant 0$ and obtain a family of transition functions $\{P_\Gamma(t, \eta, \zeta) : t \geqslant 0\}$ on $S_\Gamma \times S_\Gamma$ which in turn determine a continuous parameter Markov chain with state space S_Γ. If the $c(x, \cdot)$ are strictly positive, then Ω_Γ is an irreducible Q-matrix and the above remarks regarding the limiting behavior of the $P_\Gamma(t, \eta, \zeta)$ for large t are applicable.

As a practical matter, the determination of the limiting measure μ as a solution of the matrix equation $\mu\Omega_\Gamma = 0$ is not always easy. In some instances, depending upon the form of the rates $c(x, \cdot)$, there is a general procedure in [25] for determining μ. We will illustrate such a case by means of a finite stochastic Ising model.

Let $\Lambda = Z^d$, the d-dimensional integer lattice. For $x = (x_1, \ldots, x_d) \in \Lambda$ and $r > 0$, let $\|x\| = \max_{1 \leqslant i \leqslant d} |x_i|$ and $B_{x,r} = \{y \in \Lambda : \|y - x\| \leqslant r\}$. As before, we consider the set of configurations $S = \{-1, +1\}^\Lambda$ with the discrete topology on $\{-1, +1\}$ and the product topology on S. Define functions $c(x, \eta)$ on $\Lambda \times S$ by

$$c(x, \eta) = \exp\left[-\frac{\beta}{2} \sum_{\|y-x\|=1} \eta(y)\eta(x) - h\eta(x)\right] \tag{2.1}$$

where $\beta > 0$ plays the role of inverse temperature and h is a real parameter specifying the strength of an external field. If $\Gamma = B_{x,r}$ for some positive integer r and $\phi \in S$, let $[\eta, \phi]_\Gamma$, $\eta \in S_\Gamma$, denote the element of S which is equal to η on Γ and ϕ on $\Lambda \sim \Gamma$. The configuration ϕ plays the role of a boundary condition outside Γ. The family of rates $\{c(x, [\eta, \phi]_\Gamma) : x \in \Gamma\}$ then determines a Q-matrix Ω_Γ^ϕ on $S_\Gamma \times S_\Gamma$ which when applied to a real-valued function f on S_Γ gives

$$\Omega_\Gamma^\phi f(\eta) = \sum_{x \in \Gamma} c(x, [\eta, \phi]_\Gamma)\Delta_x f(\eta), \tag{2.2}$$

where $\Delta_x f(\eta) = f(\eta_x) - f(\eta)$. The operator Ω_Γ^ϕ determines a semigroup of operators $\{T_\Gamma^\phi(t) : t \geqslant 0\}$ by putting $T_\Gamma^\phi(t) = \exp(t\Omega_\Gamma^\phi)$. Choosing $\alpha > 0$ such that $\Omega_\Gamma^\phi + \alpha I$ is a positive operator and using the fact that $T_\Gamma^\phi(t) = \exp(-\alpha t)\exp(t(\Omega_\Gamma^\phi + \alpha I))$, each $T_\Gamma^\phi(t)$ is a positive operator. Also, since $\Omega_\Gamma^\phi 1 = 0$, $T_\Gamma^\phi(t)1 = 1$ and $\{T_\Gamma^\phi(t) : t \geqslant 0\}$ is a uniformly continuous positive contraction semigroup of operators. The limiting distribution of the semigroup is easily determined from the form of the rates as in [25] and is given by

$$\mu_\Gamma^\phi(\eta) = Z(\Gamma, \phi)^{-1} \exp\left[\frac{\beta}{2} \sum_{x \in \Gamma} \sum_{\|y-x\|=1} [\eta, \phi]_\Gamma(y)[\eta, \phi]_\Gamma(x) + h \sum_{x \in \Gamma} \eta(x)\right] \tag{2.3}$$

where $Z(\Gamma, \phi)^{-1}$ is chosen so that μ_Γ^ϕ is a probability measure on S_Γ. Two boundary configurations $\phi = +1$ on Λ and $\phi = -1$ on Λ are of particular interest. The associated Q-matrices, semigroups, and invariant measures will be denoted by $\Omega_\Gamma^\pm$, $T_\Gamma^\pm(t)$, and $\mu_\Gamma^\pm$.

3. Infinite Spin Systems

In the construction of infinite spin systems, we will use $C(S)$, the Banach space of continuous real-valued functions on the compact space S with the sup norm, and the space $B(S)$ of bounded real-valued Borel measurable functions on S with the sup norm. A strongly continuous, positive, contraction semigroup of operators on $C(S)$ will be called a Feller semigroup. A positive contraction semigroup on $B(S)$ will be called a Markov semigroup. If μ is a measure on the Borel subsets of S and $f \in B(S)$, μf will denote the integral of f with respect to μ. In proving that a sequence of probability measures $\{\mu_n\}$ on S converges in the weak* topology to a probability measure μ on S, we will use the fact that it suffices to show that $\lim_{n\to\infty} \mu_n f = \mu f$, where f is the indicator function of a set E of the form $E = \{\eta \in S : \eta(x) = +1, x \in A\}$ where A is a finite subset of Λ. This follows from the fact that the indicator function of any finite-dimensional cylinder set in S can be expressed as a linear combination of such indicator functions and the set of finite linear combinations of indicator functions of the cylinder sets is dense in $C(S)$. If J is a finite subset of Λ, $\mathscr{F}(J)$ will denote the set of functions f on S such that $f(\eta)$ depends only upon $\eta|_J$, the restriction of η to J, and $\mathscr{F}$ will denote the union of all such $\mathscr{F}(J)$.

If $\{c(x,\cdot): x \in \Lambda\}$ is a family of nonnegative continuous rates on S, we define an operator Ω on $\mathscr{F}$ by putting

$$\Omega f(\eta) = \sum_{x \in \Lambda} c(x, \eta)\Delta_x f(\eta), \quad \eta \in S, f \in \mathscr{F}. \tag{3.1}$$

If $f \in \mathscr{F}(J)$, J finite, then $\Delta_x f(\eta) = 0$ for $x \notin J$ and the sum in (3.1) reduces to a finite sum and $\Omega f \in C(S)$. Since $\mathscr{F}$ is dense in $C(S)$, Ω is a densely defined operator on $C(S)$ and is the analog of a Q-matrix. Several questions can be posed:

(i) Is Ω the generator of a Feller semigroup of operators $\{T(t): t \geqslant 0\}$?
(ii) If so, is the semigroup unique?
(iii) If unique, does the semigroup have an invariant measure?
(iv) Is the invariant measure unique?

The answer to (i) is affirmative for a large class of rates $c(x,\cdot)$ but, generally speaking, the answers to the uniqueness questions are negative.

The above questions were first addressed by Dobrushin in [1]. In this paper the Cauchy initial value problem $\partial u/\partial t = \Omega u$, $u(0) = f$, is solved for particular f, namely indicator functions of finite-dimensional cylinder sets in S, by constructing approximate solutions $u^{(n)}$ corresponding to finite spin systems on S_{Γ_n} with the Γ_n increasing to exhaust Λ. This necessitates showing that the $u^{(n)}$ converge uniformly on $S \times [0, t]$, $0 \leqslant t < +\infty$. Although Dobrushin's paper is the source of several important ideas, the methods involve some hard analysis and are not appropriate here. It should also be mentioned that Dobrushin's paper was followed by [11] in which Holley uses the approximation theorem in [19] to prove the existence of Feller semigroups generated by operators describing the time evolution of infinitely many particles moving on the integers.

For the time being we will assume that the $c(x,\cdot)$ are uniformly bounded. Then the operator Ω defined by (3.1) has a natural domain, containing $\mathscr{F}$, which consists of those real-valued functions f on S for which $|||f||| = \Sigma_{x\in\Lambda} \|\Delta_x f\| < +\infty$. The set of such functions will be denoted by $C^1(S)$. For $f \in C^1(S)$, the series defining Ωf converges uniformly and $\Omega f \in C(S)$. Since $\mathscr{F} \subset C^1(S) \subset C(S)$ and $\mathscr{F}$ is dense in $C(S)$, Ω with domain $C^1(S)$ is densely defined. It is easy to show that Ω is dissipative; that is, $\|I - \lambda\Omega f\| \geqslant \|f\|$ for $\lambda \geqslant 0$ and $f \in C^1(S)$ (see [21]). This implies that Ω has a closed extension which is also dissipative. The following theorem is due to Liggett [20].

THEOREM 1. *If* $\sup_x[\|c(x,\cdot)\| + |||c(x,\cdot)|||] < +\infty$, *then the closure of* Ω *generates a unique Feller semigroup* $\{T(t): t \geqslant 0\}$ *on* $C(S)$ *which maps* $C^1(S)$ *into itself.*

The hypothesis of this theorem is known as Liggett's condition. The proof of the theorem involves showing that the range of $I - \lambda\Omega$ is dense in $C(S)$ for sufficiently small λ. This is accomplished by approximating the $c(x,\cdot)$ by $c_n^\phi(x,\cdot)$, $n \geqslant 1$, defined for $\eta \in S$ by

$$c_n^\phi(x,\eta)=\begin{cases}c(x,[\eta,\phi]_{\Gamma_n}) & \text{if } x\in\Gamma_n=\{x:\|x\|\leqslant n\}\\ 0 & \text{if } x\notin\Gamma_n,\end{cases}\tag{3.2}$$

where ϕ is a fixed boundary condition, and using the operator Ω_n^ϕ determined by the $c_n^\phi(x,\cdot)$. Each Ω_n^ϕ is a bounded operator on $C(S)$ which generates a Feller semigroup $\{T_n^\phi(t): t\geqslant 0\}$. If $g\in C^1(S)$, then the functions f_n^ϕ defined by

$$f_n^\phi = R_n^\phi(\lambda)g=\int_0^\infty e^{-t}T_n^\phi(\lambda t)g\,dt,\quad n\geqslant 1,$$

satisfy the equation $f_n^\phi-\lambda\Omega_n^\phi f_n^\phi=g$ and it can be shown that the sequence $\{f_n^\phi\}$ is equicontinuous and uniformly bounded so that the Arzelà–Ascoli Theorem can be applied to obtain a convergent subsequence with limit $f\in C^1(S)$ which satisfies the equation $f-\lambda\Omega f=g$ for sufficiently small λ. A lucid proof of this fact can be found in [27]. Note that the solution f of this equation is unique for if $h-\lambda\Omega h=0$ then $0=\|h-\lambda\Omega h\|\geqslant\|h\|$ by dissipativity. In particular, the solution f is independent of the boundary condition ϕ. An application of the Hille–Yosida Theorem then shows that Ω has a closed extension which is the infinitesimal generator of a Feller semigroup. Uniqueness of the semigroup is a by-product of the Hille–Yosida Theorem. Another by-product of the proof is that

$$\lim_{n\to\infty}\sup_{0\leqslant s\leqslant t}\|T_n^\phi(s)f-T(s)f\|=0,\quad f\in C(S), t\geqslant 0.\tag{3.3}$$

This result follows from a version of the Trotter–Kato Theorem due to Kurtz [19].

Liggett also proves in [20] a version of Theorem 1 for operators describing dynamic gas models of particles moving about in a countable set.

Sullivan [28] has a unified treatment of the results of Dobrushin and Liggett which also produces ergodic theorems. The ergodic results will be discussed in Section 5. An elementary proof that Ω generates a semigroup of operators in the case for which there is an $L>0$ such that $c(x,\cdot)\in\mathscr{F}(B_{x,L})$, $x\in\Lambda$, can be found in [14]. This proof is based on defining the semigroup directly as an exponential series. An alternative approach to spin systems based on nonstandard hyperfinite models can be found in [9, 10]. An L_2 treatment can be found in [16, 17, 27].

Holley and Stroock produced the first example of rates $c(x,\cdot)$ for which Ω does not generate a unique Feller semigroup [16]. Gray and Griffeath show in [7] that there are continuous $c(x,\cdot)$ for which there is a continuum of distinct Feller semigroups having generators which are extensions of the operator Ω. This might seem to contradict the fact that Ω, as a dissipative operator, has a unique closure $\tilde{\Omega}$ but $\tilde{\Omega}$ cannot be the infinitesimal generator of a Feller semigroup. Gray has probabilistic necessary and sufficient conditions for Ω, constructed from continuous $c(x,\cdot)$, to generate a unique Feller semigroup [5].

Holley and Stroock abandoned the semigroup approach in [16] in favor of a martingale approach to spin systems, thereby constructing probability models of spin systems assuming only that the $c(x,\cdot)$ are continuous. As a consequence, they are able to use a result of Krylov [18] to show that there is a Markov semigroup of operators $\{T(t) : t \geqslant 0\}$ for which

$$T(t)f = f + \int_0^t T(s)\Omega f \mathrm{d}s, \quad f \in \mathscr{F}, t \geqslant 0 .$$

Using monotone convergence methods, we will construct Markov semigroups directly for a special class of $c(x, \cdot)$ in the next section.

4. Monotone Convergence

A useful tool for comparing two semigroups, called coupling, has its origin in the papers of Dobrushin [1] and Vasershtein [30] and makes use of the natural partial ordering of S whereby $\eta \leqslant \zeta$ means $\eta(x) \leqslant \zeta(x)$ for all $x \in \Lambda$.

Consider a finite $\Gamma \subset \Lambda$ and two families of nonnegative functions $\{c_i(x, \cdot) : x \in \Gamma\}, i = 1, 2$, in $\mathscr{F}(\Gamma)$ satisfying

$$[c_1(x, \eta) - c_2(x, \zeta)][\eta(x) + \zeta(x)] \geqslant 0, \quad x \in \Gamma \tag{4.1}$$

whenever $\eta(x) \leqslant \zeta(x)$ for all $x \in \Gamma$. Letting $c_i(x, \cdot) = 0$ for $x \in \Lambda \sim \Gamma$, bounded operators Ω_i on $\mathscr{F}$ can be defined as in (2.2) which generate Feller semigroups $T_i(t), i = 1, 2$. To compare the two semigroups, an operator Ω is defined for $f \in C(S \times S)$ as in [21] by putting

$$\begin{aligned} \Omega f(\eta, \zeta) = & \sum_{x \in \Lambda} c_1(x, \eta)[f(\eta_x, \zeta) - f(\eta, \zeta)] + \\ & + \sum_{x \in \Lambda} c_2(x, \zeta)[f(\eta, \zeta_x) - f(\eta, \zeta)] + \\ & + \sum_{x : \eta(x) = \zeta(x)} \min[c_1(x, \eta), c_2(x, \zeta)] \times \\ & \times [f(\eta_x, \zeta_x) - f(\eta_x, \zeta) - f(\eta, \zeta_x) + f(\eta, \zeta)] . \end{aligned} \tag{4.2}$$

It is clear that Ω is a bounded operator on $C(S \times S)$ and generates a Feller semigroup $T(t)$ on $C(S \times S)$. The operator Ω accomplishes a coupling of the two operators so that

(i) if $f(\eta, \zeta)$ depends only on η, then $\Omega f(\eta, \zeta) = \Omega_1 f(\eta, \zeta)$ and $T(t)f(\eta, \zeta) = T_1(t)f(\eta, \zeta)$

(ii) if $f(\eta, \zeta)$ depends only on ζ, then $\Omega f(\eta, \zeta) = \Omega_2 f(\eta, \zeta)$ and $T(t)f(\eta, \zeta) = T_2(t)f(\eta, \zeta)$.

Let $C_i(S)$ and $C_d(S)$ denote those $f \in C(S)$ which are increasing and decreasing, respectively, on S with $\mathscr{F}_i$, $\mathscr{F}_d$, $\mathscr{F}_i(A)$, and $\mathscr{F}_d(A)$ defined similarly. A consequence of the coupling is that $T_1(t)f(\eta) \leqslant T_2(t)f(\zeta)$ whenever $f \in C_i(S)$ and $\eta \leqslant \zeta$. In particular,

$$T_1(t)f(\eta) \leqslant T_2(t)f(\eta) \tag{4.3}$$

for $t \geqslant 0, f \in C_i(S)$. Proofs of the above statements can be found in [21].

Consider now a family of nonnegative continuous rates $\{c(x, \cdot) : x \in \Lambda\}$ on S. We will assume that the $c(x, \cdot)$ are *attractive*, meaning that

$$[c(x, \eta) - c(x, \zeta)][\eta(x) + \zeta(x)] \geqslant 0, \quad x \in \Lambda \tag{4.4}$$

whenever $\eta \leqslant \zeta$. Now let Γ be a nonempty finite subset of Λ and let $\phi \in S$ be a boundary condition. Define $c_\Gamma^\phi(x, \cdot)$, $x \in \Lambda$, as in (3.2) and corresponding bounded operators Ω_Γ^ϕ on $C(S)$ which determine Feller semigroups $T_\Gamma^\phi(t)$. If we take $c_1(x, \eta) = c_\Gamma^\phi(x, \eta)$ and $c_2(x, \zeta) = c_\Gamma^\phi(x, \zeta)$ in (4.1) so that the $T_1(t)$ and $T_2(t)$ in (4.3) are both $T_\Gamma^\phi(t)$, it then follows from (4.3) that

$$T_\Gamma^\phi(t) f(\eta) \leqslant T_\Gamma^\phi(t) f(\zeta) \quad \text{whenever } \eta \leqslant \zeta, f \in C_i(S). \tag{4.5}$$

If $\psi \in S$ is a second boundary condition with $\phi \leqslant \psi$ and we take $c_1(x, \eta) = c_\Gamma^\phi(x, \eta)$ and $c_2(x, \zeta) = c_\Gamma^\psi(x, \zeta)$, then (4.1) is satisfied and

$$T_\Gamma^\phi(t) f(\eta) \leqslant T_\Gamma^\psi(t) f(\zeta) \quad \text{whenever } \eta \leqslant \zeta, f \in C_i(S). \tag{4.6}$$

When using $\phi = +1$ as a boundary condition, we will replace ϕ by '+' so that $T_\Gamma^\phi(t) = T_\Gamma^+(t)$ with a similar change for $\phi = -1$. Suppose now that Γ and Σ are nonempty finite subsets of Λ with $\Gamma \subset \Sigma$. Let $M(x) = \sup_\eta c(x, \eta)$,

$$d(x, \eta) = \begin{cases} c_\Gamma^+(x, \eta) & \text{if } x \in \Gamma \\ M(x) & \text{if } x \in \Sigma \sim \Gamma,\ \eta(x) = -1 \\ 0 & \text{if } x \in \Sigma \sim \Gamma,\ \eta(x) = +1 \quad \text{or} \quad x \in \Lambda \sim \Sigma, \end{cases}$$

and $S(t)$ the Feller semigroup determined by the $d(x, \cdot)$. Taking $c_1(x, \eta) = c_\Sigma^+(x, \eta)$ and $c_2(x, \zeta) = d(x, \zeta)$, condition (4.1) is satisfied and $T_\Sigma^+(t) f(\eta) = S(t) f(\eta)$ whenever $\eta \in S$, $f \in C_i(S)$. Since the spins at sites in Γ do not interact with those in $\Sigma \sim \Gamma$ when controlled by the $d(x, \cdot)$, $S(t) f = T_\Gamma^+(t) f$ whenever $f \in \mathscr{F}_i(\Gamma)$. Therefore, $T_\Sigma^+(t) f \leqslant T_\Gamma^+(t) f$ whenever $f \in \mathscr{F}_i(\Gamma)$; that is, $T_\Gamma^+(t) f$ eventually decreases as a function of increasing Γ. It can be shown in the same way that $T_\Gamma^-(t) f$ eventually increases as a function of increasing Γ whenever $f \in \mathscr{F}_i$. Also $T_\Gamma^-(t) f \leqslant T_\Gamma^+(t) f$ whenever $f \in \mathscr{F}_i(\Gamma)$ by (4.6). Suppose now that $\{\Gamma_n\}$ is an increasing sequence of finite subsets of Λ which exhausts Λ. We will change $T_{\Gamma_n}^\pm(t)$ and $c_{\Gamma_n}^\pm(x, \cdot)$ to $T_n^\pm(t)$ and $c_n^\pm$, respectively. For $f \in \mathscr{F}_i$, $t \geqslant 0$, $\eta \in S$, we can therefore define

$$T^\pm(t) f(\eta) = \lim_{n \to \infty} T_n^\pm(t) f(\eta). \tag{4.7}$$

For such f, $T^+(t) f$ is u.s.c. on S and $T^-(t) f$ is l.s.c. on S with $T^-(t) f \leqslant T^+(t) f$. We will now show that $T^\pm(t)$ can be extended to $B(S)$. First note that there are probability measures $P_n^\pm(t, \eta, \cdot)$ on the Borel subsets of S such that $T_n^\pm(t) f(\eta) = \int f(\xi) P_n^\pm(t, \eta, d\xi), f \in C(S)$. Let $P^\pm(t, \eta, \cdot)$ be weak* cluster points of the sequences $\{P_n^\pm(t, \eta, \cdot)\}$. If A is a finite subset of Λ, then the indicator function χ_E of the cylinder set

$$E = \{\eta \in S : \eta(x) = +1, x \in A\}$$

belongs to $\mathscr{F}_i(A)$ and

$$\lim_{n \to \infty} P_n^\pm(t, \eta, E) = \lim_{n \to \infty} T_n^\pm(t) \chi_E(\eta) = T^\pm(t) \chi_E(\eta) P^\pm(t, \eta, E).$$

By the remarks at the beginning of Section 3, $P^{\pm}(t, \eta, \cdot)$ are actually weak* limits. It follows that $T^{\pm}(t)f = \lim_{n\to\infty} T_n^{\pm}(t)f$ is defined for all $f \in C(S)$ and is equal to $\int f(\xi)P^{\pm}(t, \eta, d\xi)$. Thus, $T^{\pm}(t)$ maps $C(S)$ into $B(S)$. Each $T^{\pm}(t)$ can be extended to $B(S)$ by putting $T^{+}(t)f(\eta) = \int f(\xi)P^{\pm}(t, \eta, d\xi)$ for $f \in B(S)$ and is clearly a positive contraction operator on $B(S)$. The assertion about invariant measures in the following theorem will be proved in the next section.

THEOREM 2. *There are Markov semigroups* $\{T^{\pm}(t) : t \geqslant 0\}$ *on* $B(S)$ *such that*

$$T^{-}(t)f \leqslant T^{+}(t)f \quad \text{for } f \in \mathscr{F}_i$$

and

$$T^{\pm}(t)f = f + \int_0^t T^{\pm}(s)\Omega f\, ds \quad \text{for } f \in \mathscr{F}.$$

Moreover, there are probability measures $\mu^{\pm}$ *such that* $\mu^{\pm}T^{\pm}(t) = \mu^{\pm}$ *for all* $t \geqslant 0$.

Proof. We first prove the semigroup property for the $T^{+}(t)$, the proof for $T^{-}(t)$ being similar. Fix $t, s \geqslant 0$ and $\eta \in S$. We will make use of the fact that $T_\Gamma^{+}(t)f \in \mathscr{F}_i(\Gamma)$ whenever $f \in \mathscr{F}_i(\Gamma)$. Suppose $f \in \mathscr{F}_i$. Then $f \in \mathscr{F}_i(\Gamma_n)$ for all n larger than some n_0; for $m \geqslant n \geqslant n_0$,

$$T_n^{+}(t+s)f = T_n^{+}(s)T_n^{+}(t)f \geqslant T_m^{+}(s)T_n^{+}(t)f$$

and

$$T^{+}(t+s)f \leqslant T_m^{+}(t+s)f = T_m^{+}(t)T_m^{+}(s)f \leqslant T_m^{+}(t)T_n^{+}(s)f.$$

Letting $m \to \infty$ first and then $n \to \infty$, we obtain $T^{+}(t+s)f = T^{+}(t)T^{+}(s)f$ for $f \in \mathscr{F}_i$; this property can then be extended to $f \in B(S)$. To prove the second assertion note that

$$T_n^{+}(t)f = f + \int_0^t T_n^{+}(s)\Omega_n^{+} f\, ds, \quad f \in \mathscr{F}. \tag{4.8}$$

If $f \in \mathscr{F}(\Gamma_n)$ for $n \geqslant n_0$, then $\Omega_n^{+} f(\eta) = \Sigma_{x\in\Gamma_n} c(x, [\eta, +]_{\Gamma_n})\Delta_x f(\eta)$.
Since the $c(x, \cdot)$, $x \in \Gamma_{n_0}$, are uniformly continuous on S, the sequence $\{\Omega_n^{+} f\}$ converges uniformly to Ωf. We can therefore take the limit in (4.8) as $n \to \infty$ since the measures $P_n^{+}(s, \eta, \cdot)$ converge to $P^{+}(s, \eta, \cdot)$ in the weak* topology.

5. Invariant Measures

If $\{T(t) : t \geqslant 0\}$ is a Feller semigroup, it is a simple matter to show that it has an invariant measure by defining for each $t \geqslant 0$ a probability measure μ_t by means of the equation

$$\frac{1}{t}\int_0^t T(s)f\, ds = \int f\, d\mu_t, \quad f \in C(S),$$

and selecting a sequence of real numbers $t_n \uparrow +\infty$ such that $\{\mu_{t_n}\}$ converges in the weak* topology to a probability measure μ which is easily seen to be invariant for the semigroup. It is also easy to see that the set $\mathscr{I}$ of invariant probability measures is nonempty, convex, and compact in the weak* topology.

Holley has made several important contributions to the study of invariant measures in [11, 12, 13]. Assuming that the continuous rates $c(x, \cdot)$ are attractive and the associated operator Ω generates a unique Feller semigroup $\{T(t) : t \geqslant 0\}$ which can be approximated as in Section 3 by a sequence of Feller semigroups $T_n^{\phi_n}(t)$ with associated invariant measures $\mu_n^{\phi_n}$, Holley shows in [13] that any weak* cluster point μ of the $\mu_n^{\phi_n}$ sequence is invariant for the $T(t)$ semigroup and that if μ is the same for all choices of the $\Gamma_n \uparrow \Lambda$ and ϕ_n, then the semigroup is ergodic in the sense that $\lim_{t\to\infty} T(t) f(\eta) = \mu f$ for all $\eta \in S$, $f \in C(S)$.

The stochastic Ising model corresponding to the attractive rates $c(x, \cdot)$ defined by (2.1) provides an example of a spin system having more than one invariant measure. It is easily checked that Liggett's condition is satisfied for these rates since they are finite range and so (3.3) holds for the approximating semigroups $T_n^{\pm}(t)$ with unique invariant measures $\mu_n^{\pm}$, considered as measures on S, given as in (2.3). Letting $\mu^{\pm}$ be weak* cluster points of the $\mu_n^{\pm}$ sequences, $\mu^{\pm}$ are invariant measures for the unique semigroup $\{T(t) : t \geqslant 0\}$ corresponding to the operator Ω determined by the $c(x, \cdot)$ rates. If $h = 0$ in (2.1) and (2.3) and $d = 2$ or 3, then there is a critical $\beta_c > 0$ such that $\mu^+ \neq \mu^-$ whenever $\beta > \beta_c$. This result is due to Dobrushin [2] and Griffiths [8]. The stochastic Ising model has been extensively studied in the physics and mathematics literature. See [15, 23, 24] for related results.

It is desirable to have theorems with hypotheses which can be verified by examining the rates $c(x, \cdot)$ directly as, for example, in Theorem 1. The following theorem is of this nature and is due to Dobrushin [1] and Sullivan [28]. We will need the constant

$$\gamma = \inf_{x\in\Lambda} \inf_{\eta\in S} [c(x, \eta) + c(x, \eta_x)] - \sup_{x\in\Lambda} \sum_{y\neq x} \|\Delta_y c(x, \cdot)\|.$$

THEOREM 3. *Suppose the* $c(x, \cdot)$, $x \in \Lambda$, *satisfy Liggett's condition and* $\gamma > 0$. *Then* Ω *generates a unique Feller semigroup* $\{T(t) : t \geqslant 0\}$ *having a unique invariant measure* μ *for which*

$$\left\| T_f f - \int f \, d\mu \right\| \leqslant \left(\frac{M}{\gamma}\right) \exp(-\gamma t) |\!|\!| f |\!|\!|, \quad f \in C^1(S), t \geqslant 0,$$

where $M = \sup_x \|c(x, \cdot)\|$.

In the non-ergodic case we would like to determine the set $\mathscr{I}_e$ of extreme points of the set $\mathscr{I}$ of invariant probability measures for specific spin models. In order to state a result along these lines we will look at an ordering of measures introduced by Holley.

If μ_1 and μ_2 are probability measures on the Borel subsets of S, we define $\mu_1 \leqslant \mu_2$ to mean that there is a probability measure ν on the Borel subsets of $S \times S$ which has marginals μ_1 and μ_2 for which $\nu\{(\eta, \xi) : \eta \leqslant \zeta\} = 1$. If $\mu_1 \leqslant \mu_2$ and f is a bounded, increasing, Borel measurable function on S, then $\mu_1 f \leqslant \mu_2 f$. Note that if ν^+ assigns unit mass to the configuration $\eta = +1$ and ν^- assigns unit mass to $\eta = -1$ and ν is any probability measure on S, then $\nu^- \leqslant \nu \leqslant \nu^+$ in the sense of this ordering.

Suppose that $\{c_i(x,\cdot): x \in \Lambda\}$, $i = 1, 2$, are two families of nonnegative continuous rates which determine unique Feller semigroups $\{T_i(t): t \geqslant 0\}$, $i = 1, 2$, and satisfy (4.1). If μ_1 and μ_2 are probability measures on S with $\mu_1 \leqslant \mu_2$, then $\mu_1 T_1(t) \leqslant \mu_2 T_2(t)$ where the $\mu_i T_i(t)$ are interpreted as measures (c.f. [21]). In particular, if $\{c(x,\cdot): x \in \Lambda\}$ is a family of nonnegative attractive continuous rates which determine a unique Feller semigroup $\{T(t): t \geqslant 0\}$ and μ is any probability measure on S, then $\nu^- T(t) \leqslant \mu T(t) \leqslant \nu^+ T(t)$. Since $\nu^- \leqslant \nu^- T(t)$ and $\nu^+ T(t) \leqslant \nu^+$, it follows from the semigroup property that the $\nu^- T(t)$ increase with t and the $\nu^+ T(t)$ decrease as t increases and both have weak* limits μ^- and μ^+, respectively, which are invariant measures for the $T(t)$ semigroup. The μ^- and μ^+ of the next theorem are the measures constructed in this way. The theorem is due to Liggett [22].

THEOREM 4. *Let $\{c(x,\cdot): x \in Z\}$ be nonnegative attractive translation invariant rates on $\{-1, +1\}^Z$ such that $c(x,\cdot) \in \mathscr{F}(\{x-1, x\, x+1\})$ and $c(x,\eta) + c(x,\eta_x) > 0$ for all $x \in Z$, $\eta \in S$. Then $\mathscr{I}_e = \{\mu^-, \mu^+\}$.*

We will use the above ordering of measures to complete the proof of Theorem 2 in which it was assumed that the $c(x,\cdot)$ are nonnegative attractive continuous rates and the $T^{\pm}(t)$ were defined by (4.7). We will carry out the proof for the $T^+(t)$ semigroup. Recall that the $T_n^+(t)f$ eventually decrease for $f \in \mathscr{F}_i$. Since each $\{T_n^+(t): t \geqslant 0\}$ is a Feller semigroup, $\nu^+ T_n^+(t)$ has a weak* limit μ_n^+ as $t \to \infty$ which is an invariant measure for $T_n^+(t)$ according to the remarks preceding Theorem 4. If $f \in \mathscr{F}_i$ and n is sufficiently large, then $\nu^+ T_n^+(t)f \geqslant \nu^+ T_{n+1}^+(t)f$ and letting $t \to \infty$, $\mu_n^+ f \geqslant \mu_{n+1}^+ f$. It follows that the sequence $\{\mu_n^+\}$ has a weak* limit μ^+ which we will show is an invariant measure for the $T^+(t)$ semigroup. If $f \in \mathscr{F}_i$, then $T_n^+(t)f$ eventually decreases. If $f \in \mathscr{F}_i(\Gamma_{n_0})$ and $m \geqslant n \geqslant n_0$, then $\mu^+ f \leqslant \mu_m^+ f = \mu_m^+ T_m^+(t)f \leqslant \mu_m^+ T_n^+(t)f$. Letting $m \to \infty$ first and then $n \to \infty$, $\mu^+ f \leqslant \mu^+ T^+(t)f$. Since $T_n^+(t)f \in \mathscr{F}_i$, $\mu_m^+ T_n^+(t)f \leqslant \mu_n^+ T_n^+(t)f = \mu_n^+ f$, letting $m \to \infty$ first and then $n \to \infty$, $\mu^+ T^+(t)f \leqslant \mu^+ f$ and the two are equal.

6. Notes and Remarks

No claim is made that Theorem 2 is new since Markov semigroups on $B(S)$ have been constructed in [16] and the invariant measures have been constructed in [13] assuming that Ω generates a unique Feller semigroup of operators. The monotone convergence methods used in the proof provide a means of constructing Markov semigroups when the Hille–Yosida Theorem is not applicable. The essential ingredient in the method is a sequence of approximating Markov semigroups of operators which converges monotonically on a generating positive cone with each semigroup leaving invariant the cone and having an invariant measure.

There are many open problems associated with spin systems. As pointed out in [22], the μ^+ and μ^- of Theorem 4 are probably equal under the stronger condition that the $c(x,\cdot)$ are strictly positive on S. Gray and Griffeath ask in [6] if the continuous rates $c(x,\cdot)$ always determine a Feller semigroup or a strong Markov process. They also ask if it is possible that the closure of the operator Ω, although not the generator of a semigroup, has an extension which is a generator. Except for some results in [16], virtually nothing has been done with Borel measurable rates.

For the sale of simplicity, only two spins -1 and $+1$ were allowed in the above discussion. More general spins are allowed in [6, 13, 28] and, in some cases, the $c(x, \eta)$ are measures on the set of possible spins.

References

1. Dobrushin, R. L.: 'Markov Processes with a Large Number of Locally Interacting Components', *Problems of Information Transmission* **7** (1971), 149–164.
2. Dobrushin, R. L.: 'The Existence of a Phase Transition in the Two- and Three-dimensional Ising Models', *Teorija Verojatin i ee Prim.* **10** (1965), 209–230.
3. Doob, J. L.: *Stochastic Processes*, John Wiley, New York, 1953.
4. Glauber, R.: 'The Statistics of the Stochastic Ising Model', *J. Math. Physics* **4** (1963), 294–307.
5. Gray, L.: 'Controlled Spin-flip Systems', *Ann. Prob.* **6** (1978), 953–974.
6. Gray, L. and Griffeath, D.: 'On the Uniqueness of Certain Interacting Particle Systems', *Z. Wahr. und Verw. Geb.* **35** (1976), 75–86.
7. Gray, L. and Griffeath, D.: 'On the Uniqueness and Nonuniqueness of Proximity Processes', *Ann. Prob.* **5** (1977), 678–692.
8. Griffiths, R.: 'Peierl's Proof of Spontaneous Magnetization in a Two-dimensional Ising Ferromagnet', *Phys. Rev.* **136A** (1964), 437–439.
9. Helms, L. L.: *Hyperfinite Spin Models*, Lecture Notes in Mathematics, No. 983 Springer-Verlag, New York, 1983, pp. 15–26.
10. Helms, L. L. and Loeb, P. A.: 'Applications of Nonstandard Analysis to Spin Models', *J. Math. Anal. Appl.* **69** (1979), 341–352.
11. Holley, R.: 'A Class of Interactions in an Infinite Particle System', *Adv. Math.* **5** (1970), 291–309.
12. Holley, R.: 'Free Energy in a Markovian Model of a Lattice Spin System', *Comm. Math. Phys.* **23** (1971), 87–99.
13. Holley, R.: 'An Ergodic Theorem for Interacting Systems with Attractive Interactions', *Z. Wahr. und Verw. Geb.* **24** (1972), 325–334.
14. Holley, R.: 'Markovian Interaction Processes with Finite Range Interactions', *Ann. Math. Stat.* **43** (1972), 1961–1967.
15. Holley, R.: 'Recent Results on the Stochastic Ising Model', *Rocky Mountain J. Math.* **4** (1974), 479–496.
16. Holley, R. and Stroock, D.: 'A Martingale Approach to Infinite Systems of Interacting Processes', *Ann. Prob.* **4**, (1976), 195–228.
17. Holley, R. and Stroock, D.: 'L_2 Theory for the Stochastic Ising Model', *Z. Wahr. und Verw. Geb.* **35** (1976), 87–101.
18. Krylov, N.V.: 'On the Selection of a Markov Process from a System of Processes and the Construction of Quasi-diffusion Processes', *Math. USSR Izvestija.* **7** (1973), 691–709.
19. Kurtz, T.: 'Extensions of Trotter's Operator Semigroup Approximation Theorems', *J. Funct. Anal.* **3** (1969), 354–375.
20. Liggett, T. M.: 'Existence Theorems for Infinite Particle Systems', *Trans. Am. Math. Soc.* **165** (1972), 471–481.
21. Liggett, T. M. *The Stochastic Evolution of Infinite Systems of Interacting Particles, École d'Eté de Probabilités de Saint-Flour VI-1976*, Lecture Notes in Mathematics, No. 598, Springer-Verlag, New York, 1976, pp. 187–248.
22. Liggett, T. M.: 'Attractive Nearest Neighbor Spin Systems on the Integers', *Ann. Prob.* **6** (1978), 629–636.
23. Preston, C.J.: *Gibbs States on Countable Sets*, Cambridge Univ. Press, London, 1974.
24. Ruelle, D.: *Statistical Mechanics*, W. A. Benjamin, New York, 1969.
25. Spitzer, F.: 'Interaction of Markov Processes', *Adv. Math.* **5** (1970), 246–290.
26. Spitzer, F.: 'Random Fields and Interacting Particle Systems', MAA Summer Seminar, Williamtown, Mass., 1971.

27. Stroock, D. W.: *Lectures on Infinite Interacting Systems*, Kyoto University, Kinokuniya Book Store, Tokyo, 1978.
28. Sullivan, W. G.: 'A Unified Existence and Ergodic Theorem for Markov Evolution of Random Fields', *Z. Wahr. und Verw. Geb.* **31** (1974), 47–56.
29. Sullivan, W. G.: 'Markov Processes for Random Fields', *Comm. Dublin Inst. Adv. Studies*, series A, No. 23, 1975.
30. Vasershtein, L. N.: 'Processes over Denumerable Products of Spaces Describing Large Systems, *Problems of Information Transmission* **3** (1969), 47–52.

Book Reviews

E. B. Davies, *One-Parameter Semigroups*, Academic Press, New York, 1980.

The book under review is about one-parameter semigroups of linear operators. This might seem to be a rather special topic, but in fact the theory is closely related to fundamental ideas in science. This review will first present an overview of the general theory and then a description of some features of the book.

A one-parameter semigroup acting on a Banach space is a continuous action

$$T : [0, \infty) \times B \to B \tag{1}$$

of the semigroup $[0, \infty)$ on the Banach space B. More explicitly, the map

$$(t, f) \mapsto T_t f \tag{2}$$

is a semigroup homomorphism,

$$T_0 f = f, \qquad T_{s+t} f = T_s(T_t f), \tag{3}$$

acts linearly,

$$T_t(f + g) = T_t f + T_t g, \qquad T_t(af) = aT_t f, \tag{4}$$

and is jointly continuous in t and f.

A one-parameter semigroup is a natural mathematical object. The true importance of the notion, however, comes from the fact that we live in a world where time evolution is a central fact of experience. In almost all applications the parameter t represents time and the operator T_t maps the present into the future. The continuity of the evolution is an abstraction of the concept of well-posed problem formulated by Hadamard. The requirement of linearity deserves additional comment.

Many time evolutions are linear, at least to a good approximation. Obviously there should also be a theory of non-linear semigroups, and indeed such a theory exists. The restriction to linear evolutions is justified by the existence of several important subjects where linearity is fundamental. The first of these is Markov processes. The semigroup is defined by taking the expectation of a function of the process; the expectation is linear by definition. The second is quantum mechanics. Here the origin of linearity is more subtle.

Investigations into the foundations of quantum mechanics have shown that under reasonably natural assumptions the automorphisms of the Hilbert space of state vectors must be linear or conjugate-linear. When the automorphism is connected to the identity in a continuous group of automorphisms the conjugate-linear case is excluded. Automorphisms in other components may in fact be conjugate linear;

the most striking example is time reversal. A third example with linear time evolution is electromagnetism. From the current point of view the linearity is due to the fact that the gauge group is abelian. The equations of more encompassing gauge theories have non-linear terms that arise from commutators.

The general theory of one-parameter semigroups is rather complete. The central theme is the representation of the semigroup as

$$T_t = \exp(tZ) \tag{5}$$

where

$$Z\colon \mathrm{Dom}(Z) \to B \tag{6}$$

is a linear operator from a dense linear subspace Dom(Z) of B to B. The difficulty is that the spectrum of Z may extend to ∞, where the exponential function has an essential singularity. The most obvious definitions of the exponential function do not work in general. The series expansion need not converge. Spectral theory may be useless. Both difficulties are illustrated by the following example.

Let B be the Hilbert space $L^2(0, a)$. Define a semigroup by

$$\begin{aligned} T_t f(x) &= f(x+t), \quad \text{if } x+t \leqslant a, \\ &= 0, \qquad\quad\ \ \text{otherwise.} \end{aligned} \tag{7}$$

This moves the function to the left at speed one, making it zero to fill in the gap generated from the right end point a. The function simply disappears as it reaches the left end point 0. After time a it is identically zero. This shows that the semigroup can not be analytic. The generator is

$$Z = \mathrm{d}/\mathrm{d}x \tag{8}$$

with the boundary condition

$$f(a) = 0, \quad \text{for } f \text{ in } \mathrm{Dom}(Z). \tag{9}$$

This operator has no spectrum in the complex plane. The only singularity of the resolvent $(z - Z)^{-1}$ is at $z = \infty$, so spectral analysis is useless.

Another approach to the construction of a semigroup from its generator might be to use the forward difference operator $F(t/n)$ given by

$$F(t/n) - I = (t/n)Z. \tag{10}$$

The semigroup would be defined by

$$T_t f = \lim(n \to \infty) F(t/n)^n. \tag{11}$$

This is extremely unstable and will not work in general. However it suggests using a more sophisticated difference scheme, such as the backward difference operator $F(t/n)$ defined by

$$F(t/n) - I = (t/n)Z\, F(t/n). \tag{12}$$

This has a greater chance of being stable. In fact,

$$F(t/n) = (I - (t/n)Z)^{-1}. \tag{13}$$

The inverse that occurs in this formula complicates calculations but improves stability. The condition on Z that is necessary and sufficient for it to generate a semigroup $T_t = \exp(tZ)$ with

$$\| T_t \| \leqslant M \exp(ta) \tag{14}$$

is

$$\| (\lambda - Z)^{-n} \| \leqslant M(\lambda - a)^{-n} \tag{15}$$

for all $\lambda > 0$ and all integers $n \geqslant 1$. This is precisely what is needed to give a uniform bound on the norm of $F(t/n)^n$. In fact, the choice $\lambda = n/t$ gives

$$\| F(t/n)^n \| \leqslant M(1 - (t/n)a)^{-n}. \tag{16}$$

In the limit as n approaches infinity this gives the bound (14) on the semigroup defined by the backward difference scheme (12) or (13) and the limit (11).

An irritating feature of the condition (15) is that the estimate must hold for all powers $n \geqslant 1$ of the resolvent operator $(\lambda - Z)^{-1}$. In the special case $M = 1$ the estimate for $n = 1$ implies the estimate for all $n \geqslant 1$. Everything reduces to a single estimate on the resolvent itself. The theory is simpler and cleaner in this case. The notion of dissipative operator acting in a Banach space is particularly helpful in obtaining an estimate on the resolvent $(\lambda - Z)^{-1}$ from a condition on Z itself.

The definition of dissipative operator involves the dual space B^* of the Banach space B. If f is in B, then there exists ϕ in B^* with $\| \phi \| = 1$ and $\langle f, \phi \rangle = \| f \|$. Such a ϕ gives f its greatest possible value as a function on the unit ball of B^*. An operator Z is dissipative if for all f in $\mathrm{Dom}(Z)$ there exists ϕ giving f its greatest possible value and with

$$\mathrm{Re}\langle Zf, \varnothing \rangle \leqslant 0. \tag{17}$$

The necessary and sufficient condition for a densely defined operator Z to be the generator of a contraction semigroup (one with $M = 1$ and $a = 0$) is that Z be dissipative and that the range of $\lambda - Z$ be equal to the entire Banach space B for all $\lambda > 0$.

The condition of being dissipative is a strong constraint on the operator Z and the Banach space B. An example will make this clear. Take $B = C(X)$, the real Banach space of continuous functions on the compact space X. If f is in $C(X)$, then there is a point a in X with $\pm f(a) = \| f \|$. The point measure δ_a gives f its greatest possible value. The operator Z is dissipative if

$$\pm \langle Zf, \delta_a \rangle = \pm Zf(a) \leqslant 0. \tag{18}$$

This is obviously a version of the maximum principle.

The Davies book gets quickly to the essentials of the theory. The main results that hold in the general setting are presented in the first three chapters. Many of the theorems give necessary and sufficient conditions for a property of the semigroup in terms of the generator. The next chapter is on the Hilbert space case. The generator of a one-parameter unitary group is i times a self-adjoint operator. In quantum mechanics self-adjoint operators are fundamental objects and much of the Hilbert space theory has been developed in the context of quantum mechanics.

Chapter 5 deals with limits as λ approaches zero of semigroups with generators of the form $Z/\lambda + A$. In this limit the long-time behavior of the semigroup generated by Z dominates, and A plays the role of a very weak perturbation. This theory and its extensions have applications in various areas of statistical mechanics.

The last three chapters also treat special topics. These are contraction semigroups on Hilbert space, positivity-preserving semigroups on function spaces (with application to the large time limiting behavior of Markov semigroups), and spectral properties of groups of isometries.

Throughout the book the examples are chosen more to illustrate the theory than to give a realistic picture of possible applications. However, it is clear that the author has the applications in mind. The key points are there in a compact, neat package.

It should be mentioned that linear semigroup theory is a subject on which there are now a relatively large number of books. In addition to the numerous titles in the bibliography of the book under review, there are works by Faris [1], Fattorini [2]. Nelson [3], Pazy [4], Zaidman [5], and others. The number of journal articles is overwhelming. The book by Fattorini on the abstract Cauchy problem lists 117 pages of references and does not even begin to cover the specialized literature of quantum mechanics and probability.

The Davies book has the advantage that it does not attempt to cover all topics. It is a tasteful selection from the wealth of linear semigroup theory. The reader leaves with the impression that semigroups are useful but not formidable.

References

1. Faris, W. G., *Self-Adjoint Operators*, Springer-Verlag, Berlin, 1975.
2. Fattorini, H. O., *The Cauchy Problem*, Addison-Wesley, Reading, Mass., 1983.
3. Nelson, E., *Topics in Dynamics I: Flows*, Princeton Univ. Press, Princeton, NJ, 1969.
4. Pazy, A., *Semigroups of Linear Operators and Applications to Partial Differential Equations*, Springer-Verlag, New York, 1983.
5. Zaidman, S. D., *Abstract Differential Equations*, Pitman, London, 1979.

University of Arizona WILLIAM G. FARIS

L. Asimow and A. J. Ellis: *Convexity Theory and its Applications in Functional Analysis*, London Math. Society Monographs, Vol. 16, Academic Press, London, 1980.

One of the great attractions of functional analysis to its devotees has always been the interplay between technicalities involving convergence, compactness and such notions on the one hand, and geometrical intuition, generalized from three dimensions to infinite dimensions, on the other. In historical terms this interplay was first taken to its ultimate limits for Hilbert spaces, and the wealth of results available for Hilbert spaces and operators on them is still unmatched. Other areas of functional analysis were subsequently treated in a similar spirit, and rather deep structural theories were gradually built up for the classical Banach spaces, for locally convex topological linear spaces, and for various types of Banach algebras, particularly C^*-algebras and sup-norm algebras. Each of these subjects has now developed so far that it is possible to spend one's entire working career on it alone.

In all of these areas the spatial properties of compact convex sets in locally convex topological linear spaces turn out to have great importance, and in the late 1950s the study of such sets began to emerge as a separate discipline within functional analysis. One of the first issues to be clarified was the representability of a point α in a compact convex set X as the barycentre of a boundary measure. By this last term is meant a probability measure μ on X which is concentrated in a suitable sense on the set δX of all extreme points of X; the sense is a little complicated to specify in general but if X is metrisable then δX is a G_δ set and one requires

$$\mu(\delta X) = 1.$$

One says that α is the barycentre of μ if

$$f(\alpha) = \int_X f(x)\,\mathrm{d}\mu(x)$$

for all continuous affine functions $f : X \to \mathbb{R}$. If X is finite-dimensional,the existence of a boundary measure μ with barycentre α is easy to prove for each $\alpha \in X$, and indeed μ can have finite support. The existence of such measures for general compact convex sets was proved in gradually increasing generality in the 1950s by several people, among whom Choquet stands out. It turned out that the techniques involved in these proofs were central to a deep analysis of the general structure of compact convex sets, and an enormous amount of effort was devoted to investigating these developments throughout the 1960s, culminating in the publication in 1971 of the definitive text of E. M. Alfsen (1971).

While this book could hardly be bettered as a systematic account of the geometric and functional analytic structure of compact convex sets, and in particular of simplexes (those compact convex sets for which the boundary measure of every point is unique), it gave almost no information about the application of these ideas to the various fields from which the subject had originally sprung. This was rather a pity, but it was clear that in 1970 no connected account of applications which would be of permanent value could be given.

Since that time there have been three important areas (at least) of functional analysis which involve either applications or developments of the theory as laid out by Alfsen. Two of these, the applications to sup-norm algebras and to C^*-algebras, have reached a certain level of maturity and are discussed at length in the book of Asimow and Ellis under review. The third, namely the spectral theory of positive operators on ordered Banach spaces, is in such a rapid state of development that a coherent account seems impossible at present. It is to be hoped that in due course someone will have the energy and sense of service to the mathematical community to write a monograph on this subject, extending the material in Schaefer (1974) to one-parameter semigroups of positive operators and to ordered Banach spaces other than Banach lattices.

We turn now to a more detailed inspection of the contents of the present book. The first chapter gives an account of the basic separation properties for convex sets, the existence of many extreme points for convex sets (the Krein–Milman theorem), and the existence of maximal representing measures (boundary measures) for points within compact convex sets. This material is classical in the sense that it was all known by 1961. The text is fairly condensed, and is directed towards the mature mathematician, in that one is faced with a series of technical definitions, theorems and proofs, with little attempt at making explicit the geometric motivation. The failure of Asimow and Ellis, as of Alfsen before them, to include more than a single diagram in their book, when they emphasise their desire to promote the geometric perspective, is caused by the obvious difficulty of representing infinite-dimensional objects on paper. The reviewer has deluded himself that he can imagine these objects, a little cloudily it is true, but there seems to be no satisfactory way of communicating this imagination.

Chapter 2 studies the duality of ordered Banach spaces. The relevance of this topic is that by representing a given compact convex set as the base of a convex cone in a Banach space, one can set up an equivalence between certain properties of compact convex sets and other properties of ordered Banach spaces. Moreover, many of the convex sets important in applications arise in precisely this manner. In particular the simplexes, mentioned above, are characterised by the property that the ordered Banach space constructed as outlined above is a Banach lattice and an L-space. The abstract study of Banach lattices (and a slightly more general property known as the Riesz interpolation property) thus has direct consequences for the structure of simplexes. These are investigated in detail in Chapter 3, where the deep facial structure theory of simplexes is developed. Most of this material may also be found in Alfsen (1971) or Wong and Ng (1973), but Section 7 on the uniqueness and universality of the Poulsen simplex is of very recent origin.

The remaining two chapters of the book of Asimow and Ellis break new ground, and are to the reviewer's knowledge the only connected accounts of these subjects from the point of view of convexity theory. Chapter 4 studies the properties of complex function spaces and in particular sup-norm algebras. In contrast to the theory discussed so far these are *complex* Banach spaces. Thus the study of this application requires the extension of all the ideas so far discussed to the complex case, so that new concepts such as complex state spaces and complex representing measures occur. The authors have made very important contributions to this topic, most of which has been developed since 1970, and their account is an authoritative one. By means of this theory, they are able, among other things, to extend Bishop's classical results on peak points for function algebras to a much more general context. They also describe the characterisation by Effros and others of the complex Lindenstrauss spaces.

The final Chapter of the book is devoted to convexity theory of C^*-algebras. The authors assume that the reader is familiar with the basic theory of C^*-algebras and in particular with the definition of the state space of a C^*-algebra, with the GNS construction and with the Jacobsen topology in the primitive ideal space. Much of the chapter is devoted to the relationships between ideals defined in the algebraic sense and ideals defined in the order-theoretic sense, and was developed between 1967 and 1975. The reviewer had originally assumed that a single chapter such as this could not hope to compete with other massive works devoted entirely to C^*-algebras, such as Pedersen (1979). However, the perspective is substantially different, and the authors present a number of results on the facial structure and geometry of the state space which are not to be found in Pedersen's monograph.

This last chapter may in fact be regarded as providing the prerequisites for reading the important work of Alfsen, Schultz and co-workers on the spectral theory of compact convex sets. This work was directed towards giving a geometrical characterisation of those compact convex sets which are the state spaces of JB algebras and more specially of C^*-algebras. The authors write down without proof the theorem in which this line of development culminated on the last page of the book, but their reference is not quite correct, the actual reference being Alfsen *et al.* (1980). Araki (1981) has also given an alternative treatment of a part of this theory.

It will be seen from the above comments that the reviewer considers the book to be a very valuable addition to the mathematical literature. It presents connected accounts of developments in the subject between 1960 and 1980 in a well-argued, if somewhat severe style. The historical notes are of great value and the authors have obviously taken trouble to provide a comprehensive account of developments in the two areas of application which are described.

One must admit that by the nature of the subject the book is not an easy one to pick up and browse through. The authors have constructed a quite complicated machine and have taken it on two substantial journeys. They have described the view seen during each journey and have discovered certain modifications needed for it to function effectively. It is to be hoped that the machine will make many more interesting trips in the coming years.

References

Alfsen, E. M.: *Compact Convex Sets and Boundary Integrals*, Springer-Verlag, 1971.

Alfsen, E. M., Hancke-Olsen, H. and Schultz, F. W.: 'State Space of C^*-Algebras', *Acta Math.* **144** (1980), 267–305.

Araki, H.: 'On a Characterization of the State Space of Quantum Mechanics', *Comm. Math. Phys.* **75** (1980), 1–24.
Pedersen, G. K.: *C*-Algebras and their Automorphism Groups*, Academic Press, 1979.
Schaeffer, H. H.: *Banach Lattices and Positive Operators*, Springer-Verlag, 1974.
Wong, Y. C. and Ng, K. F.: *Partially ordered Topological Vector Spaces*, Clarendon Press, Oxford, 1973.

King's College, London E. B. DAVIES

Publications Received

Asimov, L. and A. J. Ellis: *Convexity Theory and its Applications in Functional Analysis*, Academic Press, London, 1980.

Conway, John B.: *Subnormal Operators*, Pitman Advanced Publishing Program, Boston, London, and Melbourne, 1981.

Doob, J. L.: *Classical Potential Theory and Its Probabilistic Counterpart*, Springer-Verlag, New York, Heidelberg, Berlin, and Tokyo, 1984.

Krishnan, V.: *Nonlinear Filtering and Smoothing: an Introduction to Martingales, Stochastic Integrals and Estimation*, John Wiley, New York, Chichester, Brisbane, Toronto, and Singapore, 1984.

Power, S. C.: *Hankel Operators on Hilbert Space*, Pitman Advanced Publishing Program, Boston, London, and Melbourne, 1982.

ANNOUNCEMENT

Positive Semigroups of Operators, and Applications
Editors: O. Bratteli and P. E. T. Jørgensen

Please note that a hardback edition of this special double issue of *Acta Applicandae Mathematicae,* Vol. 2 Nos. 3 and 4 (September/December 1984), is available from the publishers.
ISBN 90–277–1839–3.
Price: Dfl. 65,– / $ 24.50 / £ 16.50.

TABLE OF CONTENTS